Foundations of Engineering Mechanics

M. Z. Kolovsky, Nonlinear Dynamics of Active and Passive Systems of Vibration Protection

Springer-Verlag Berlin Heidelberg GmbH

M. Z. Kolovsky

Nonlinear Dynamics of Active and Passive Systems of Vibration Protection

Translated by A.K. Belyaev

With 156 Figures

Springer

Series Editors:
V. I. Babitsky
Loughborough University
Department of Mechanical Engineering
LE11 3TU Loughborough, Leicestershire
Great Britain

J. Wittenburg
Universität Karlsruhe (TH)
Institut für Mechanik
Kaiserstrasse 12
D-76128 Karlsruhe / Germany

Author:
Mikhail Z. Kolovsky
Kondratievsky 56 - 24
195197 St. Petersburg / Russia

Translator:
A.K. Belyaev
Johannes-Kepler-Universität Linz
Institut für Mechanik und Maschinenlehre
A-4040 Linz / Austria

Cataloging-in-Publication Data applied for

Die Deutsche Bibliothek - CIP-Einheitsaufnahme
Kolovskij; Mikhail Z.: Nonlinear dynamics of active and passive systems of vibration protection / M.Z. Kolovsky. Transl. by A.K. Belyeav.
(Foundations of engineering mechanics)
ISBN 978-3-662-22236-2 ISBN 978-3-540-49143-9 (eBook)
DOI 10.1007/978-3-540-49143-9

Originally published by Springer-Verlag Berlin Heidelberg New York in 1999
Softcover reprint of the hardcover 1st edition 1999

Typesetting: Camera-ready copy from translator
Cover-Design: de'blik, Berlin
SPIN 10705979 62/3020 5 4 3 2 1 0 Printed on acid-free paper

Preface

With progress in technology, the problem of protecting human-beings, machines and technological processes from sources of vibration and impact has become of utmost importance. Traditional "classical" methods of protection, based upon utilising elastic passive and dissipative elements, turn out to be inefficient in many situations and can not completely satisfy the complex and often contradictory claims imposed on modern vibration protection systems which must provide high performance at minimum dimensions. For these reasons, active vibration protection systems, which are actually systems of automatic control with independent power sources, are widely used nowadays.

Appearing and developing active systems require that traditional approaches to the analysis and synthesis of vibration protection systems must be revised. Firstly, there exists the necessity to re-state the problem of vibration protection from mechanical actions as an equivalent problem in closed-loop control systems design, which is to be solved by the methods of control theory. Furthermore, it turns out that certain inherent properties of active systems must be taken into account for a proper design. In the majority of cases, the dynamic models of the objects to be protected and the bases to which these objects are to be attached must be revised. They are no longer considered as rigid bodies but elastic bodies with weak dissipation.

The dynamic modelling of vibration protection systems is also complicated by the necessity to include nonlinearity. The classical linear theory of elastic suspension, implying that each isolator possesses linear elastic and dissipative properties, is limited due to dimensional constraints. Any actual isolator can have a linear characteristic only for a certain range of deflections, the latter being referred to as the *region of linearity*. Any actual vibration protection system has *restrainers* limiting the maximum admissible geometric size of the linearity region. It is clear that the linear theory can be applied only for those excitations for which the isolators'defl ections remain within the linearity region. However, it is not always feasible to meet this condition.

Dimensional constraints lead to a series of phenomena of apparent nonlinear character which do not manifest themselves in the framework of a linear dynamic model. In particular, the assumption that the steady-state forced vibration of a system under harmonic or polyharmonic excitation is independent of initial conditions is no longer valid. In turn this means that the uniqueness of the steady-state forced vibration is no longer ensured.

Any vibration protection system does not function properly when its vibration exceeds the linearity region. The very possibility of the occurrence of such vibrations must be excluded for a properly designed system. Analysis shows that this can be achieved by increasing the dissipative forces. However, the performance of vibration protection systems deteriorates with the growth of damping. This gives rise to the problem of seeking the optimal damping, and one of the most important tasks of the theory of vibration protection is to find the solution.

The above problems came into being in the early sixties because of the need to design facilities aimed at protecting items of equipment mounted aboard aircraft, spacecraft, vessels and other moving objects from vibration and impact excitations. The results of these theoretical and experimental investigations were covered in two monographs by the present author, namely M.Z. Kolovsky "Nonlinear theory of vibration protection systems", Nauka, Moscow, 1966 and M.Z. Kolovsky "Automatic control of vibration protection systems", Nauka, Moscow, 1976. The contents of these books is the basis for the present book which is offered to the reader in English. The author anticipates that the statements contained in the present book this will be useful for a wide circle of readers whose scientific interests are related to the problems of vibration theory and control theory and his practical recommendations will be used by engineers developing modern vibration protection systems.

The present book consists of six Chapters. The first two Chapters are concerned with the theory of linear active and passive systems. Some approximate methods of analysis for nonlinear systems are outlined in Section 1 of Chapter 3. Here, along with the widely used methods of harmonic balance, Galerkin's method, the method of linearisation with respect to the distribution function is developed. Application of this method is especially efficient for analysis of polyharmonic processes. The method of statistical linearisation is also explained. The rest of Chapter 3, as well as Chapters 4 and 5, deal with the analysis of nonlinear active and passive systems of vibration protection. Chapter 6 is devoted to the optimisation of vibration protection systems.

At the time when the monographs in Russian were first prepared and published, the author was supported by his teachers A.I. Lurie and I.B. Barger, the bright images of whom have been held in remembrance throughout the author's life. The author is also deeply grateful to A.A. Pervozvansky for his valuable advice and to V.I. Babitsky who edited both of the original monographs and contributed much to their improvement. Composition of the present book involved the tasks of combining the above two monographs and translating them from Russian into English. Both of these turned out to be challenging problems. A great deal of help in these tasks was given by A.K. Belyaev, to whom the author is very thankful.

Translator's Preface

Firstly, I would like to thank the author for giving me the opportunity to translate into English the two monographs of his which result in the present book. From a professional perspective, this was a very interesting and cognitive experience.

Secondly, I am thankful to Professor Hans Irschik, from the Johannes Kepler University of Linz, Austria. The translation was carried out during my stay at his Institute, throughout which he provided invaluable advice and contributed greatly to the improvement of the manuscript.

Thirdly, I appreciate the assistance of my son, Nikita Belyaev, from the Technical University of St. Petersburg, Russia for the considerable technical support he gave during the translation.

And finally, I would like to express my sincere gratitude to Dr. Stewart McWilliam, from the University of Nottingham, UK who took the trouble of editing the manuscript which I translated into English. I am greatly obliged to him not only for his thorough correction of the galley-proofs but also for his many useful and profound suggestions on the manuscript.

Contents

1

Dynamic characteristics and efficiency of vibration protection systems

1.1 Statement of the problem of protection from dynamic excitations

We consider an arbitrary mechanical system which is subjected to dynamic forces and kinematic excitations, see Fig. 1.1. Time-varying prescribed forces are applied at points $B_1, ..., B_k$ and are termed the *force excitations.* Their projections onto the coordinates axes, $F_1(t), ..., F_{3k}(t)$, form a $3k$–dimensional vector $\mathbf{F}(t)$. *Kinematic excitations* imply the given time-dependent displacements of some points $A_1, ..., A_s$. Projections of these displacements onto the coordinate axes are the components of $3s$–dimensional

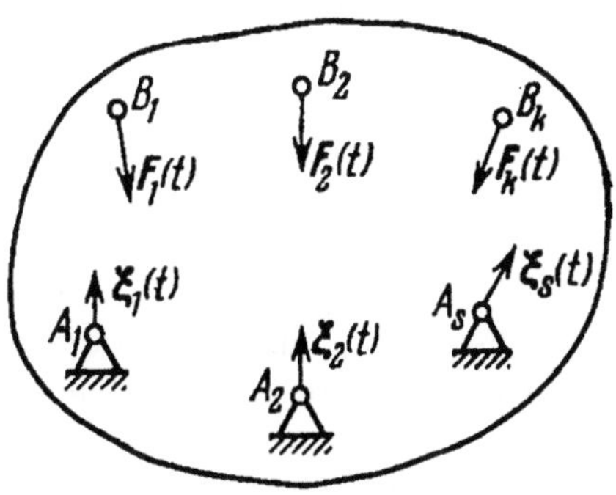

FIGURE 1.1.

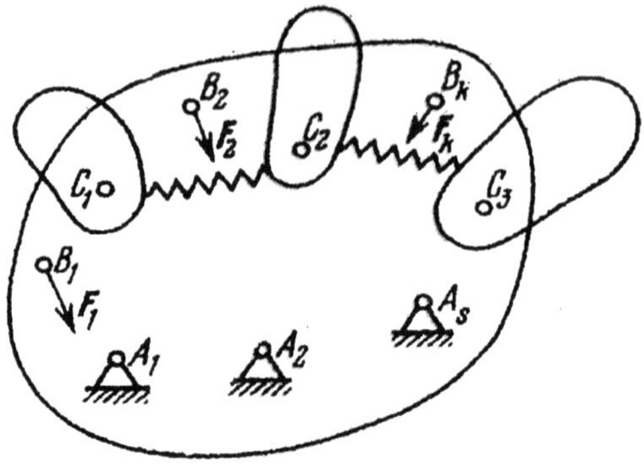

FIGURE 1.2.

vector $\boldsymbol{\xi}(t)$. A vibration field of displacement is said to be observed in the system when the displacement of the system, due to dynamic excitations, is of an oscillatory nature. Clearly, vibration fields of velocity, acceleration, forces, stresses etc. are also observed in the system.

Various branches of technology require a reduction in the intensity of the vibration fields, i.e. reduction of the amplitude or mean values of certain typical parameters of the field. The facilities aimed at decreasing the intensity of the vibration fields are referred to as *systems of vibration protection.*

There exist two fundamentally different principal methods of vibration protection. The first method implies an attachment of another mechanical system to the object to be protected, see Fig. 1.2. The parameters of the attached system are chosen in such a way that a decrease in the vibration of the object is ensured. This means of vibration protection is called *vibration absorption*, and the attached systems are called *dynamic vibration absorbers.*

The second method consists of dividing the original system into two parts, see Fig. 1.3a, with further connection of these parts by means of additional mechanical systems, Fig 1.3b. This method is called *vibration isolation*, where the devices connecting these parts are often called *vibration isolators.* Usually, one of the separated parts turns out to be a "free" body disconnected from the other mechanical systems (e.g. Part I in Fig. 1.3a). This part is termed the *object to be protected* whereas the second part is referred to as the *base* or the *support structure.* When the dynamic excitations are applied to the base, the objective of vibration isolation is to protect the object and if the excitations are exerted directly on the object, the vibration isolators must ensure vibration protection of the base. In the past, it was conventional to refer to the first case as "passive" vibration isolation and to the second case as "active" vibration isolation. As these conceptions nowadays have another meaning, in what follows we will speak about vibration protection of the base or the object from dynamic exci-

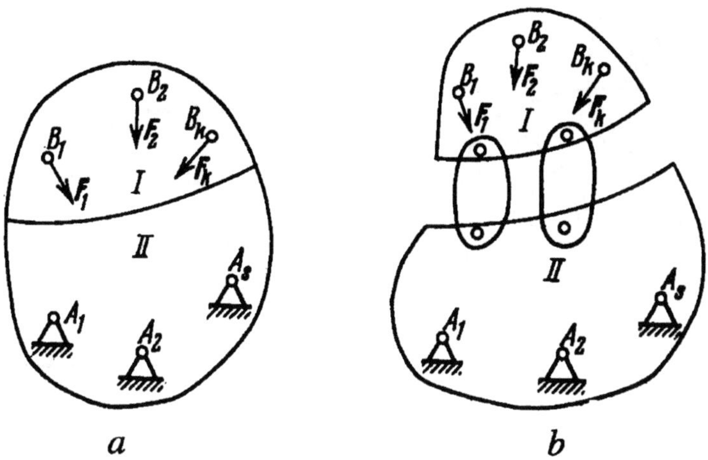

FIGURE 1.3.

tations. In the first case, the object is isolated from the vibrating support while in the second case the base is isolated from the vibrating object.

In addition to the above methods, vibration protection can be realised at the expense of reducing the dynamic excitations and modifying the object or the base. These two methods are not directly related to development of special systems of vibration protection, for this reason, they will be not considered in what follows.

It is easy to see that the action of any vibration protection facility may be considered as creating additional dynamic excitations which provide a required change in the vibration field. From this perspective, the problem of vibration protection can be deemed as the problem of control of the motion of the protected mechanical system and the dynamic excitations resulting in the parameters' change can be viewed as "control forces". Such interpretation of the vibration protection facility proves to be convenient for the statement and solving of various problems of vibration protection. This allows those approaches used in the theory of automatic control to be utilised for analysis and synthesis of the vibration protection systems.

In principle, the vibration protection facility can produce both force and kinematic excitations. However, the latter are not encountered often and thus will be not considered in this book.

We consider now a few systems of vibration protection which allows us to acquaint with features of the approach to vibration protection as a problem of control in detail.

1.1.1 One-dimensional dynamic vibration absorber

Figure 1.4a shows a system of vibration protection which produces force $U(t)$. This force is applied to point C of the object and the direction of the

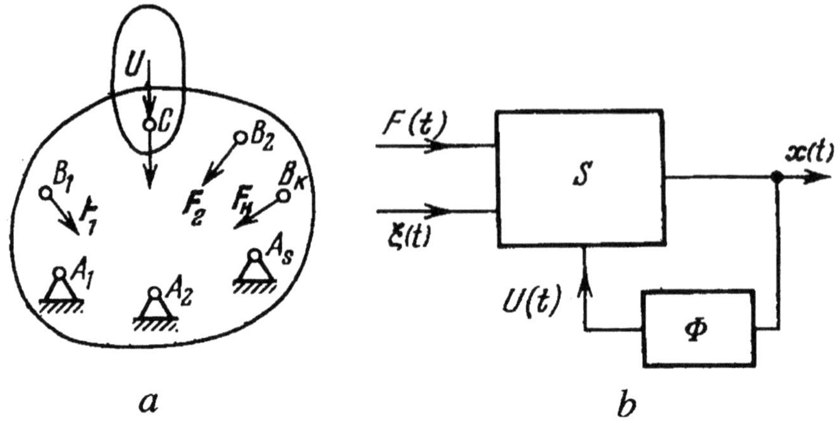

FIGURE 1.4.

force does not change. This facility is a vibration absorber and is ordinarily intended for reducing displacement $x\,(t)$ of point C along the direction of the force U. The block diagram of the system is depicted in 1.4b, where $x\,(t)$ denotes the "output" of the system, and $U\,(t)$ is the force produced by the absorber.

In other words, a functional dependence

$$U\,(t) = \Phi\{x(t)\} \tag{1.1}$$

takes place in which Φ is an operator. The system depicted in 1.4a is a system of the closed-loop type. As an example, we consider a simple dynamic vibration absorber which consists of a spring with spring stiffness c and a mass m. In this case

$$U\,(t) = cy(t), \tag{1.2}$$

where y is the displacement relative to the attachment point C. On the other hand, the governing equation for the absorber's mass m may be written as follows

$$mp^2(x+y) = -cy, \tag{1.3}$$

where p is an operator which differentiates with respect to time. Solving the latter equation for y, we obtain

$$y = -\frac{mp^2}{mp^2+c}x. \tag{1.4}$$

Substituting eq. (1.4) into eq. (1.2) yields

$$U(t) = -\frac{cmp^2}{mp^2+c}x\,(t) = \Phi(p)x(t). \tag{1.5}$$

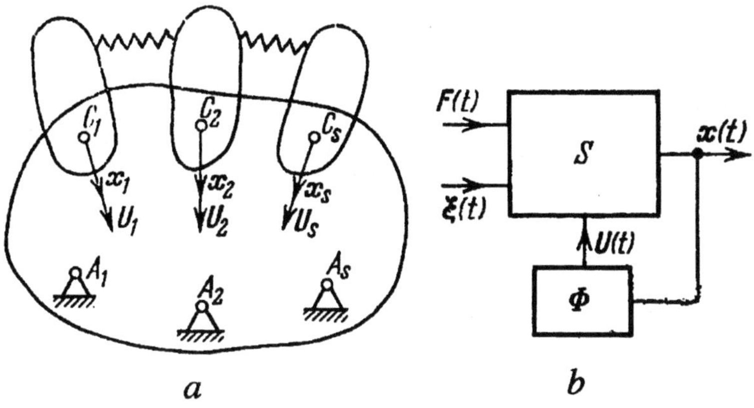

FIGURE 1.5.

The operator Φ is linear in this particular case and is a rational function (i.e. quotient of polynomials) of the operator of differentiation with respect to time p. In the system considered, a single control force was formed. Consistent with this, such a system is referred to as a one-dimensional system in the commonly approved terminology.

1.1.2 Multidimensional dynamic vibration absorber

Figure 1.5a shows a dynamic vibration absorber of a more complex structure which is attached to the object at the points C_1, ..., C_s. Such a multidimensional absorber produces s control forces $U_1(t), ..., U_s(t)$ which are represented in vector form as $\mathbf{U}(t)$, which is referred to as the vector of control forces. The block diagram of the system is depicted in Fig. 1.5b. The system output $\mathbf{x}(t)$ which is a vector with components $x_1(t), ..., x_s(t)$ forms the vector of control forces as follows

$$\mathbf{U}(t) = \Phi\{\mathbf{x}(t)\}. \tag{1.6}$$

Hence, the dynamic vibration absorber of a complex structure may be viewed as a controller of a multidimensional closed-loop control system.

1.1.3 Uniaxial vibration isolator

Let us consider a vibration isolator attached to a system to be isolated from the vibration source at point A and to the vibrating support at point C (see Fig 1.6a). Provided that the forces R and Q at the attachment points act along the same direction, this vibration isolator is termed to be uniaxial and the above direction is said to be its axis. Any uniaxial vibration isolator may be deemed as a mechanical two-port whose outputs are the forces R

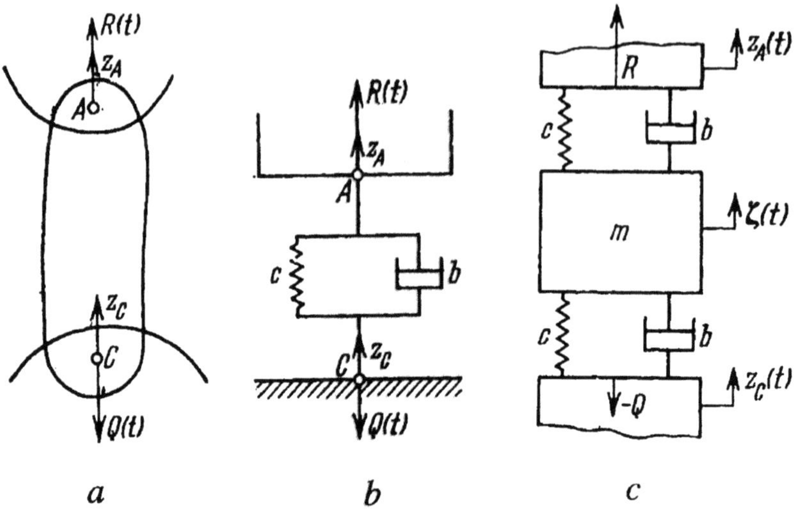

FIGURE 1.6.

and Q and the inputs are the displacements' components along the above axes z_A and z_C (see Fig 1.6a). In other words,

$$\left.\begin{aligned} R &= \Phi_{11}\{z_A(t)\} + \Phi_{12}\{z_C(t)\}, \\ Q &= \Phi_{21}\{z_A(t)\} + \Phi_{22}\{z_C(t)\}, \end{aligned}\right\} \tag{1.7}$$

with Φ_{ik} being some operators. Let us determine these operators for some simple particular cases.

Figure 1.6b shows a simple vibration isolator consisting of a spring of stiffness c and a linear dashpot with the coefficient of viscous damping b. In this case,

$$R(t) = -Q(t) = -(c + bp)(z_A - z_C) \tag{1.8}$$

and hence Φ_{ik} are the following linear operators

$$\Phi_{11} = \Phi_{22} = -\Phi_{12} = -\Phi_{21} = -(c + bp). \tag{1.9}$$

Figure 1.6b shows a two-stage vibration isolator, whose reactions are given by

$$\left.\begin{aligned} R &= -(c + bp)(z_A - \zeta), \\ Q &= (c + bp)(\zeta - z_C), \end{aligned}\right\} \tag{1.10}$$

where ζ denotes the displacement of the intermediate mass m. The governing equation for the latter is as follows

$$mp^2\zeta = -(Q + R). \tag{1.11}$$

Eliminating ζ from eqs. (1.10) and (1.11) yields

$$\left.\begin{aligned} R &= -\frac{c+bp}{mp^2+2bp+2c}[(mp^2+bp+c)z_A-(bp+c)z_C], \\ Q &= -\frac{c+bp}{mp^2+2bp+2c}[-(bp+c)z_A+(mp^2+bp+c)z_C]. \end{aligned}\right\} \qquad (1.12)$$

For any linear vibration isolator, one obtains in analogous fashion that

$$\left.\begin{aligned} R &= \gamma_{11}(p)z_A+\gamma_{12}(p)z_C, \\ Q &= \gamma_{21}(p)z_A+\gamma_{22}(p)z_C. \end{aligned}\right\} \qquad (1.13)$$

Let $R_0(t)$ denote the force which acts on the object at the attachment point in the case of the rigid attachment, thus $-R_0(t)$ is the force which acts on the base. The following force differences

$$U(t) = R(t) - R_0(t), \; U_1(t) = Q(t) + R_0(t) \qquad (1.14)$$

are the additional actions applied to the object and to the base after a vibration isolation system has been introduced. These can be treated as the control forces.

In some cases, the base can be viewed as a rigid body which is not influenced by force $Q(t)$. The vibration isolator can thus be deemed to be a one-dimensional system which forms force $R(t)$ in accordance with displacements z_A and z_C.

Expressions similar to (1.13) may be obtained also for the multiaxial vibration isolators producing forces in any direction at the attachment points. These absorbers can be considered as an ensemble of three uniaxial vibration isolators, each of which acts in one direction.

In all cases considered above we analysed the action of the vibration protection facility by forming some control forces. The units which produce the control forces may have different structures. *Passive* systems which consist of inertial, elastic and damping elements have been in use for many years in the past. *Active* vibration isolation systems, which involve the use of the active systems and external power sources, are popular nowadays.

The present book is devoted to the analysis of passive and active systems of vibration protection. Considerable attention will be paid to systems which utilize both passive and active elements. Such systems are frequently in use in practice.

Before we proceed to analysing active systems it is necessary to get acquainted with some properties of the object to be protected and the base, both being the "controlled objects" from the perspective of vibration protection systems.

1.2 Dynamic characteristics of the vibration-isolated system and the base

In what follows we assume that the dynamic excitations cause small vibration both in the system to be protected and in the base. The system and the base are considered to be linear systems, for which the main results of the classical theory of small vibration are applicable. We recall here only the basic ideas of this theory, the theory itself is covered adequately in many textbooks, see for example [3] and [32].

Let us assume that the vibration-isolated system (or the base) can be modelled by a holonomic stationary mechanical system with n degrees of freedom, whose position is defined completely by the n−dimensional vector of the generalised coordinates $\mathbf{q}$. The motion is subject to the *generalised forces* $\mathbf{Q}(t)$ (n-dimensional vector) and confined to a small neighbourhood of an equilibrium position $\mathbf{q} = 0$. The *resistance forces* are proportional to the velocities of the material points. The system is assumed to possess a *full dissipation*, i.e. any motion is accompanied by energy dissipation. Under these assumptions, the equations of motion are governed by Lagrange's equations

$$A\ddot{\mathbf{q}} + B\dot{\mathbf{q}} + C\mathbf{q} = \mathbf{Q}(t). \tag{1.15}$$

Here A, B, and C are quadratic $(n \times n)$ symmetric matrices of the positive definite quadratic forms

$$T = \frac{1}{2}A\dot{\mathbf{q}} \cdot \dot{\mathbf{q}},\ \Phi = \frac{1}{2}B\dot{\mathbf{q}} \cdot \dot{\mathbf{q}},\ \Pi = \frac{1}{2}C\mathbf{q} \cdot \mathbf{q}, \tag{1.16}$$

where T, Φ and Π are the kinetic energy, Rayleigh's dissipative function and potential energy, respectively. Here and throughout the rest of the book we assume that the first vector in the scalar product implies a row, whereas the second vector stands for a column.

The roots of the frequency equation

$$\det(C - Ak^2) = 0 \tag{1.17}$$

form the spectrum of the system's *eigenfrequencies*. The following systems of homogeneous algebraic equations

$$(C - Ak_s^2)\mathbf{h}_s = 0 \tag{1.18}$$

determine the *eigenvectors* $\mathbf{h}_1, ..., \mathbf{h}_n$ up to a scalar constant.

The general solution of the nonhomogeneous differential equation (1.15) is known to be the sum of a particular solution and a general solution of the homogeneous differential equation

$$A\ddot{\mathbf{q}} + B\dot{\mathbf{q}} + C\mathbf{q} = 0. \tag{1.19}$$

To obtain a general solution of the latter equation, it is sufficient to have n linearly independent solutions. To this end, it is convenient to take the *impulse responses* $\mathbf{s}_1(t), ..., \mathbf{s}_n(t)$ subject to the initial conditions

$$\mathbf{s}_r(0) = 0, \ \dot{\mathbf{s}}_r(0) = \mathbf{e}_r \ \ (r = 1, ..., n), \tag{1.20}$$

where $\mathbf{e}_r$ is a column whose only nontrivial element, $r - th$, is equal to unity. The matrix

$$S(t) = \|\mathbf{s}_1(t), ..., \mathbf{s}_n(t)\| \tag{1.21}$$

is referred to as the impulse response matrix. The general solution of the homogeneous differential equation (1.19) takes the form

$$\mathbf{q}_h = S(t)\mathbf{c}, \tag{1.22}$$

where $\mathbf{c}$ is a vector of arbitrary constants.

It can be easily verified by direct substitution that the particular solution of eq. (1.15), satisfying the trivial initial conditions $\mathbf{q}_*(0) = 0$ and $\dot{\mathbf{q}}_*(0) = 0$, can be cast in the form

$$\mathbf{q}_*(t) = \int_0^t S(t-\tau)\mathbf{Q}(\tau)\,d\tau. \tag{1.23}$$

Equations (1.22) and (1.23) yield the general solution of eq. (1.15)

$$\mathbf{q}(t) = \mathbf{q}_h + \mathbf{q}_* = S(t)\mathbf{c} + \int_0^t S(t-\tau)\mathbf{Q}(\tau)\,d\tau. \tag{1.24}$$

Substituting the initial conditions $t = 0$, $\mathbf{q} = \mathbf{q}_0$ and considering that $S(0)$ is the unity matrix, yields

$$\mathbf{c} = \mathbf{q}_0 = \mathbf{q}(0),$$

and, hence,

$$\mathbf{q}(t) = S(t)\mathbf{q}(0) + \int_0^t S(t-\tau)\mathbf{Q}(\tau)\,d\tau. \tag{1.25}$$

For systems with full dissipation, the impulse responses $\mathbf{s}_r(t)$ $(r = 1, ..., n)$ tend to zero as $t \to \infty$. Then, the solution (1.24) tends to a *steady-state motion* which does not depend upon the initial conditions. For sufficiently great values of t the motion governed by eq. (1.15) is expected to be very close to the steady-state motion, which can be written in the form

$$\mathbf{q}(t) = \int_{-\infty}^{t} S(t-\tau)\mathbf{Q}(\tau)\, d\tau. \tag{1.26}$$

In what follows, our main interest lies in two types of motion of the system to be protected, namely, steady-state motions caused by *vibrational dynamic excitation* and motions subject to trivial initial conditions, i.e. governed by eq. (1.23) in the case of *impact excitation*.

Equation (1.15) can be written in operational form as

$$(Ap^2 + Bp + C)\mathbf{q} = \mathbf{Q}(t), \tag{1.27}$$

where $Ap^2 + Bp + C$ implies a matrix with the entries being polynomials in the differentiation operator p.

By means of the inverse operator

$$W(p) = (Ap^2 + Bp + C)^{-1}, \tag{1.28}$$

we can represent the solution of eq. (1.27) in the form

$$\mathbf{q}(t) = W(p)\mathbf{Q}(t). \tag{1.29}$$

Function $W(p)$ is the Laplace transform of the function $S(t)$, therefore, the Laplace transforms of the vectors $\mathbf{q}(t)$ and $\mathbf{Q}(t)$ are related as follows

$$\mathbf{q}_L(p) = W(p)\mathbf{Q}_L(p), \tag{1.30}$$

that yields immediately eq. (1.25). Hence, the relation between the input and the output of the system can be written both in form (1.29) and (1.30). In what follows, we consider both equations to be equivalent and use the operational form (1.29). Given the initial conditions or the steady-state motion (1.26), equations (1.29) and (1.30) determine the motion uniquely.

Operator $W(p)$ is referred to as the *transfer function matrix* for the system (1.15), and its entries $w_{sm}(p)$ being rational functions of p are referred to as the *system transfer functions*. Operator $w_{sm}(p)$ relates an $m-th$ component of the vector $\mathbf{Q}(t)$ and an $s-th$ generalised coordinate. If all components of the vector $\mathbf{Q}(t)$, but an $m-th$, vanish, i.e.

$$Q_i(t) = 0 \ \ (i \neq m), Q_m(t) \neq 0,$$

then

$$q_s(t) = w_{sm}(p)Q_m(t). \tag{1.31}$$

Provided that $Q_m(t)$ in eq. (1.31) implies a harmonic process

$$Q_m(t) = Q_{m0}\cos\omega t, \tag{1.32}$$

then, the steady-state motion $q_s(t)$ is also a harmonic process given by

$$q_s(t) = Q_{m0} \left| w_{sm}(i\omega) \right| \cos[\omega t + \arg w_{sm}(i\omega)]. \tag{1.33}$$

The complex-valued function $w_{sm}(i\omega)$ is termed the *frequency response*. Its absolute value which is the ratio of the amplitude $q_s(t)$ to the amplitude $Q_m(t)$ is called the *amplitude-frequency response*, while arg $w_{sm}(i\omega)$ which determines the phase-angle between the output and the input of the system, is called the *phase-frequency response*. Functions $w_{sm}(i\omega)$ are the entries of the frequency response matrix $W(i\omega)$.

By virtue of the known representation for the inverse matrix, we obtain from eq. (1.28)

$$w_{sm}(p) = \frac{W_{sm}(p)}{\det(Ap^2 + Bp + C)}, \tag{1.34}$$

where $W_{sm}(p)$ is the algebraic adjunct of the corresponding element of the matrix $Ap^2 + Bp + C$. Replacing p by $i\omega$ leads to

$$w_{sm}(i\omega) = \frac{W_{sm}(i\omega)}{\det(C - A\omega^2 + Bi\omega)}. \tag{1.35}$$

As a matter of fact, the systems to be protected and their bases are lightly damped mechanical systems, that is, the entries of matrix B are small values. When the excitation frequency ω coincides with one of the eigenfrequencies, e.g. $\omega = k_s$, the absolute value of the denominator in eq. (1.35) is small and, therefore, the value of the amplitude-frequency characteristics becomes large. As follows from eq. (1.33), some small driving forces cause large amplitude vibration, known as the *resonant vibration*. The study of the resonant vibrations of the system is one of the principal problems of vibration protection.

Up to now, we have assumed that the behaviour of the system is governed by Lagrange's equations. In order to derive the equations of motion for a particular system, it is necessary to construct a dynamical model of the system to be protected and the base. This amounts to identifying the system as an assemblage of masses, elastic elements and dampers acting together as a whole. More often that not, such a modelling meets considerable difficulties, for instance, the presence of complex irregular structural members (electronic devices, gyros etc.) complicates matters appreciably. Choosing a physical model, one should take into account the bandwidth of the spectrum of the external excitation. The higher the excitation frequencies, the larger the number of degrees of freedom needed for the correct modelling of the resonant vibration. As modern complex structures are often subjected to rather high frequency loads (up to several thousand Hz), an adequate model turns out to be very complicated, and the problem may be of such large dimension, that a complete solution may not be feasible and the approach of constructing equations of motion becomes impractical.

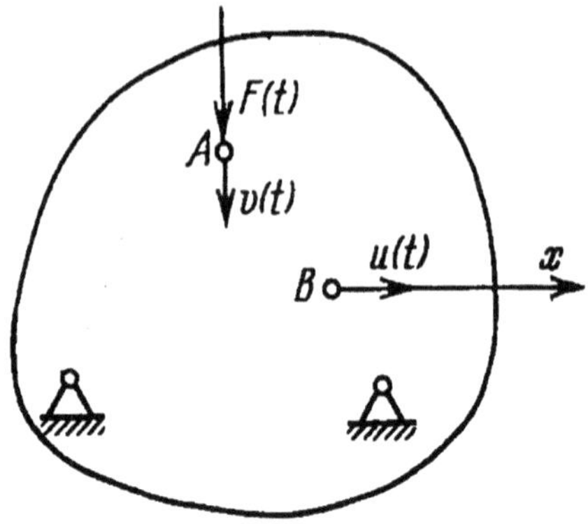

FIGURE 1.7.

In such a case, one has to reject this approach and utilize other approaches based on consideration of the *dynamic characteristics*.

Let a force $F(t)$ be applied to point A, (see Fig. 1.7). Consider the vibrational displacement of a generic point B caused by the above force. Let $u(t)$ denote the projection of this displacement into the direction Bx. The functions $u(t)$ and $F(t)$ are related by means of the operator Φ

$$u(t) = \Phi\{F(t)\}.$$

If the motion of the system is governed by eq. (1.15), one can express the operator Φ in terms of $W(p)$. Indeed, under the assumption that the vibrations are small, the displacement $u(t)$ can be cast as a linear function of the generalised coordinates

$$u = \mathbf{l}_B \cdot \mathbf{q}, \tag{1.36}$$

with $\mathbf{l}_B$ being some constant vector. On the other hand, provided that $v(t)$ implies the displacement of point A in the direction of the force $F(t)$, $v(t)$ is linear in $\mathbf{q}$, i.e.

$$v = \mathbf{l}_A \cdot \mathbf{q}.$$

Let us determine the generalised force $\mathbf{Q}$ due to the force $F(t)$. With this in view, we write the following expression for the virtual work

$$\mathbf{Q} \cdot \delta\mathbf{q} = F\delta v = F\mathbf{l}_A \cdot \delta\mathbf{q}. \tag{1.37}$$

Hence,

$$\mathbf{Q}(t) = F(t)\,\mathbf{l}_A. \tag{1.38}$$

By considering eq. (1.29) we obtain

$$u(t) = \mathbf{l}_B \cdot W(p)\mathbf{l}_A F(t).$$

Taking into account that matrix $W(p)$ is symmetric, we arrive at

$$u(t) = W(p)\mathbf{l}_B \cdot \mathbf{l}_A F(t) = e_{AB}(p)F(t) = e_{BA}(p)F(t)\,. \tag{1.39}$$

Operator $e_{AB}(p)$, relating a force applied to a system point and acting in a prescribed direction to the projection of the displacement of another point on some direction, is named the *dynamic compliance operator* of the system, whereas the corresponding frequency characteristics $e_{AB}(i\omega)$ is called the *dynamic compliance.* We refer to the inverse operators

$$d_{AB}(p) = e_{AB}^{-1}(p) \tag{1.40}$$

as the *dynamic rigidity operator,* and to frequency characteristics $d_{AB}(i\omega)$ as the *dynamic rigidity.* The dynamic compliances and the dynamic rigidities constitute the set of dynamic characteristics. Among them, of particular importance are the characteristics which relate force $F(t)$ applied to a generic point (for example A in Fig. 1.7) and $v(t)$ which is the displacement of the same point in the direction of the above force. These characteristics, $e_A(i\omega)$ and $d_A(i\omega)$, are known as the *dynamic compliance* and the *dynamic rigidity at point A,* respectively.

As it will be shown below, the dynamic characteristics of the system and the base are widely used in the theory of vibration protection. Given a transfer function $W(p)$, the dynamic characteristics can be obtained by means of eqs. (1.39) and (1.40). Another way of establishing the above characteristics is to use test results.

There exist some devices which allow the dynamic rigidities and the dynamic compliances (both the absolute values and the phase angles) to be recorded. Harmonic excitation of the prescribed amplitude and frequency is applied in the chosen points, and the displacement or the velocity is recorded at other points.

Given the force of a single-frequency harmonic excitation

$$F(t) = F_0 \cos\omega t,$$

eq. (1.39) renders the following expression for the steady-state motion

$$u(t) = u_0 \cos(\omega t + \varphi) = F_0 \left| e_{AB}(i\omega) \right| \cos[\omega t + \arg e_{AB}(i\omega)]. \tag{1.41}$$

Thus, the amplitude of the steady-state harmonic vibration is the force amplitude multiplied by the absolute value of the corresponding dynamic compliance. It follows from eq. (1.40) that

$$\left| d_A(i\omega) \right| = \left| e_A(i\omega) \right|^{-1}, \quad \arg d_A(i\omega) = -\arg e_A(i\omega), \tag{1.42}$$

and hence

$$F_0 = u_0 \left| d_A(i\omega) \right|. \tag{1.43}$$

Let us now consider some general properties of the dynamic characteristics.

We assume for simplicity that the generalised coordinates $q_1, ..., q_n$ are the principal coordinates of the system. In this case, the motion of the system is known to be governed by the following equations, cf. [32]

$$\ddot{q}_s + \sum_{m=1}^{n} b_{sm}\dot{q}_m + k_s^2 q_s = Q_s(t) \quad (s = 1, ..., n). \tag{1.44}$$

Here b_{sm} are the entries of the matrix of coefficients B representing the dissipative properties of the system. Let eq. (1.44) be written in vector form as

$$(I_n p^2 + Bp + K)\mathbf{q} = \mathbf{Q}(t), \tag{1.45}$$

where I_n implies a unity matrix, while K denotes the following diagonal matrix

$$K = \operatorname{diag}\left\{k_1^2, ..., k_n^2\right\}.$$

When the system under consideration is weakly damped, the elements b_{sm} are all small. The question arises as to what properties do the dynamic compliances and dynamic rigidities have. As this particular case is typical for the mechanical systems to be protected, we consider first a system with no dissipation.

Letting $B = 0$, we obtain

$$W(p) = (I_n p^2 + K)^{-1} = \operatorname{diag}\left\{(k_1^2 + p^2)^{-1}, ..., (k_n^2 + p^2)^{-1}\right\}, \tag{1.46}$$

so that

$$w_{ss}(p) = (k_s^2 + p^2)^{-1}, \; w_{sm}(p) = 0 \; (s \neq m). \tag{1.47}$$

Equation (1.39) renders

$$e_{AB}(p) = W(p)l_A \cdot l_B = \sum_{s,m=1}^{n} w_{sm}(p) l_{As} \cdot l_{Bm}, \tag{1.48}$$

which leads to the following result

$$e_{AB}(p) = \sum_{s=1}^{n} l_{As} l_{Bs} (k_s^2 + p^2)^{-1} = \frac{\sum\limits_{s=1}^{n} l_{As} l_{Bs} \prod\limits_{m=1}^{n}{}_{s} (k_m^2 + p^2)}{\prod\limits_{r=1}^{n} (k_r^2 + p^2)}. \tag{1.49}$$

Here and in what follows, Π_s implies that the product does not contain the $s-th$ factor. It is easy to prove that the dynamic compliance $e_{AB}(i\omega)$ is real for any ω, whereas its absolute value grows without bound as $\omega \to k_r$. This means that the roots of the equation

$$\prod_{r=1}^{n}(k_r^2 - \omega^2) = 0$$

coinciding with the natural frequencies of the system are its resonant frequencies. However it is much more difficult to make a statement concerning the numerator's roots. This is possible to do only in the case for which the dynamic compliance at the point is considered. In this particular case

$$e_A(i\omega) = \frac{\sum\limits_{s=1}^{n} l_{As}^2 \prod\limits_{m=1}^{n}{}_s (k_m^2 - \omega^2)}{\prod\limits_{r=1}^{n}(k_r^2 - \omega^2)}. \tag{1.50}$$

The numerator of the latter equation is as follows

$$V(i\omega) = \sum_{s=1}^{n} l_{As}^2 \prod_{m=1}^{n}{}_s (k_m^2 - \omega^2).$$

It is easy to verify that $V(0) > 0$ and the value

$$V(ik_s) = \sum_{s=1}^{n} l_{As}^2 \prod_{m=1}^{n}{}_s (k_m^2 - k_s^2)$$

is positive for odd s and negative for even s. It is assumed that (i) k_s is not a multiple eigenfrequency, (ii) its value increases as k increases, and (iii) all $l_{As} \neq 0$. Thus, function $V(i\omega)$ reverses its sign $n-1$ times which implies that it has $n-1$ roots $\varkappa_s$ that lie between the eigenfrequencies of the system

$$k_1 < \varkappa_1 < k_2 < \varkappa_2 < \ldots < \varkappa_{n-1} < k_n. \tag{1.51}$$

When $\omega = \varkappa_s$, the dynamic compliance vanishes. For this reason these frequencies are called *antiresonant frequencies.* When the system undergoes a harmonic excitation of frequency $\varkappa_s$ at point A the displacement amplitude of this point is equal to zero. This leads to a conclusion that the antiresonant frequencies coincide with the eigenfrequencies of the system with an imposed additional constraint which prevents the displacement of point A from moving in the direction of force $F(t)$, see [3]. An analytical expression for this constraint is

$$\mathbf{l}_A \cdot \mathbf{q} = \mathbf{0}. \tag{1.52}$$

If k_s is a multiple eigenfrequency, the numerator and the denominator in (1.50) have a common factor $\left(k_s^2 - \omega^2\right)$. Removing this common factor implies that the system "loses" one eigenfrequency and one antiresonant frequency, the alteration (1.51) being preserved. Provided that $l_{As} = 0$ then vibrations due to the $s-th$ normal mode causes no displacement of point A in the given direction which means that this point is a node of the above mode. In this case removing the common factor $\left(k_s^2 - \omega^2\right)$ in the numerator and the denominator in eq. (1.50) leads to the same conclusions as in the case of a multiple eigenfrequency.

When b_{sm} are not equal to zero and sufficiently small, eqs. (1.48) and (1.35) render

$$e_{AB}(p) = \sum_{s,m=1}^{n} \frac{W_{sm}(p)}{\det(I_n p^2 + Bp + k)} l_{As} l_{Bm}. \tag{1.53}$$

As follows from the linear theory of vibration the roots of the characteristic equation

$$\det(I_n p^2 + Bp + k) = 0 \tag{1.54}$$

are complex-valued and form complex conjugate pairs

$$p_s = \pm\lambda_s i - \mu_s,$$

where μ_s are positive and small while λ_s differs insignificantly from k_s. From the expression for the dynamic compliance

$$\begin{aligned} e_{AB}(i\omega) &= \frac{\sum_{s,m=1}^{n} l_{As} l_{Bm} W_{sm}(i\omega)}{\det(K - I_n\omega^2 + Bi\omega)} \\ &= \frac{\sum_{s,m=1}^{n} l_{As} l_{Bm} W_{sm}(i\omega)}{\prod_{r=1}^{n} (i\omega - i\lambda_r + \mu_r)(i\omega + i\lambda_r + \mu_r)} \\ &= \frac{\sum_{s,m=1}^{n} l_{As} l_{Bm} W_{sm}(i\omega)}{\prod_{r=1}^{n} (\lambda_r^2 + \mu_r^2 + 2\mu_r i\omega - \omega^2)} \end{aligned} \tag{1.55}$$

we can conclude that λ_r are the resonant frequencies. When $\omega = \lambda_r$ the denominator (1.55) contains a small factor $2\mu_r i\omega + \mu_r^2$ which means that the absolute value of the dynamic compliance becomes large.

By considering eq. (1.55) we obtain the following expression for the dynamic compliance at point A

$$e_A(i\omega) = \frac{\sum\limits_{s,m=1}^{n} l_{As} l_{Am} W_{sm}(i\omega)}{\prod\limits_{r=1}^{n} (\lambda_r^2 + \mu_r^2 + 2\mu_r i\omega - \omega^2)}. \tag{1.56}$$

Let us study the denominator of this quotient. To this end, we consider a system which is obtained from the original one by imposing the constraint (1.52). Since

$$\mathbf{q} = W(p)\,\mathbf{Q} = W(p)\,\mathbf{l}_A \cdot F(t),$$

we have

$$\mathbf{l}_A \cdot W(p)\,\mathbf{l}_A \cdot F(t) = 0,$$

which is equivalent to

$$W(p)\,\mathbf{l}_A \cdot \mathbf{l}_A \cdot F(t) = \frac{\sum\limits_{s,m=1}^{n} W_{sm}(p) l_{As} l_{Am}}{\det(I_n p^2 + Bp + k)} F(t) = 0.$$

This means that the following equation

$$V_*(p) = \sum_{s,m=1}^{n} W_{sm}(p) l_{As} l_{Am} = 0 \tag{1.57}$$

is the characteristic equation for the system with the imposed constraint. For this reason, the roots of this equation should be complex-valued numbers

$$p_k = \pm\alpha_k i - \beta_k \;(k = 1, ..., n-1), \tag{1.58}$$

where all $\beta_k > 0$. As shown earlier, $\alpha_k = \varkappa_k$ and $\beta_k = 0$ when $b_{sm} = 0$. For small b_{sm}, the values of α_k must be approximately equal to $\varkappa_k$ whereas β_k must be small. Thus, the dynamic compliance thus can be represented in the form

$$e_A(i\omega) = \varepsilon\,\frac{\prod\limits_{k=1}^{n-1} (\alpha_k^2 + \beta_k^2 + 2\beta_k i\omega - \omega^2)}{\prod\limits_{s=1}^{n} (\lambda_s^2 + \mu_s^2 + 2\mu_s i\omega - \omega^2)}, \tag{1.59}$$

with ε being a constant factor. Provided that the system has no multiple or closely spaced eigenfrequencies, i.e. the condition $k_{s+1}^2 - k_s^2 \gg 2\mu_s k_s$ holds, and the alteration of the eigenfrequencies λ_k and the antiresonant frequencies α_k must be kept.

We introduce the following notation

$$\left.\begin{aligned} \tau_k^2 &= \left(\alpha_k^2+\beta_k^2\right)^{-1}, \; \beta_k = \eta_k/\tau_k, \\ T_s^2 &= \left(\lambda_s^2+\mu_s^2\right)^{-1}, \; \mu_s = \zeta_s/T_s. \end{aligned}\right\} \tag{1.60}$$

Then,

$$\begin{aligned} e_A(i\omega) &= \frac{\prod\limits_{k=1}^{n-1}(\alpha_k^2+\beta_k^2)\prod\limits_{k=1}^{n-1}(1+2\tau_k\eta_k i\omega-\tau_k^2\omega^2)}{\prod\limits_{s=1}^{n}(\lambda_s^2+\mu_s^2)\prod\limits_{s=1}^{n}(1+2T_s\zeta_s i\omega-T_s^2\omega^2)} \\ &= e_A(0)\frac{\prod\limits_{k=1}^{n-1}(1+2\tau_k\eta_k i\omega-\tau_k^2\omega^2)}{\prod\limits_{s=1}^{n}(1+2T_s\zeta_s i\omega-T_s^2\omega^2)}. \end{aligned} \tag{1.61}$$

As β_k^2 and μ_s^2 are values of the second order of smallness, one can take $\tau_k \approx \alpha_k^{-1}$ and $T_s \approx \lambda_s^{-1}$. Moreover, the values of the time constants τ_k and T_s alternate, i.e.

$$T_1 < \tau_1 < T_2 < \tau_2 < \ldots < \tau_{n-1} < T_n. \tag{1.62}$$

Nondimensional positive damping factors η_k and ξ_k are usually smaller than 0,05-0,1. Hence, at frequencies distant from the resonant or antiresonant ones, the following conditions hold

$$\begin{aligned} \left|1-\tau_k^2\omega^2\right| &\gg 2\tau_k\eta_k\omega, \; \left|1-T_s^2\omega^2\right| \gg 2T_s\zeta_s\omega. \\ (k &= 1,\ldots,n-1; \; s=1,\ldots,n) \end{aligned} \tag{1.63}$$

As a first approximation we can take

$$e_A(i\omega) = e_A(0)\,\frac{\prod\limits_{k=1}^{n-1}(1-\tau_k^2\omega^2)}{\prod\limits_{s=1}^{n}(1-T_s^2\omega^2)}, \tag{1.64}$$

which implies that the dissipative properties of the system are completely ignored.

Let

$$\omega = T_r^{-1} + \nu,$$

ν being small. In other words, let frequency ω be close to a resonant frequency $\lambda_r \approx T_r^{-1}$. In this case, the damping should be taken into account in the $r-th$ factor of the denominator (1.61). Assuming that conditions (1.63) hold for any k and any $s \neq r$ we obtain

$$e_A(i\omega) \approx e_A(0) \frac{\prod_{k=1}^{n-1}(1-\tau_k^2 T_r^{-2})}{[2\zeta_r i + 2T_r\nu(\zeta_r i - 1) - T_r^2\nu^2] \prod_{s=1}^{n}{}_r (1-T_s^2 T_r^{-2})}$$
$$= E[2\zeta_r i + 2T_r\nu(\zeta_r i - 1) - T_r^2\nu^2]^{-1}, \tag{1.65}$$

where

$$E = e_A(0) \frac{\prod_{k=1}^{n-1}{}_r (1-\tau_k^2 T_r^{-2})}{\prod_{s=1}^{n}(1-T_s^2 T_r^{-2})}.$$

By considering (1.62) we conclude that $E > 0$ since both the numerator and denominator of the latter equation have $r-1$ negative factors. Therefore,

$$|e_A(iT_r^{-1})| = \frac{E}{2\zeta_r}, \quad \arg e_A(iT_r^{-1}) \approx -\frac{\pi}{2}. \tag{1.66}$$

The bounds of the resonant domain are conventionally given by those values of ω at which the absolute value of the dynamic compliance is smaller than the resonant one by a factor of $\sqrt{2}$. It is easy to verify that this condition renders the following values for these bounds (up to the values of the second order of smallness)

$$\omega_{1,2} = T_r^{-1} \mp T_r^{-1}\zeta_r \tag{1.67}$$

or $\nu_{1,2} = \mp T_r^{-1}\zeta_r$. Substituting these values of ν into eq. (1.65) yields

$$\left.\begin{aligned} &|e_A(i\omega_1)| = |e_A(i\omega_2)| \approx E/4\zeta_r, \\ &\arg e_A(i\omega_1) = -\pi/4, \quad \arg e_A(i\omega_2) = -3\pi/4. \end{aligned}\right\} \tag{1.68}$$

It can be proven by analogy that at the antiresonant frequency

$$e_A(i\tau_r^{-1}) \approx E' \, 2\eta_r i, \tag{1.69}$$

where

$$E' = e_A(0) \frac{\prod_{k=1}^{n-1}{}_r (1-\tau_k^2 \tau_r^{-2})}{\prod_{s=1}^{n}(1-T_s^2 \tau_r^{-2})} < 0.$$

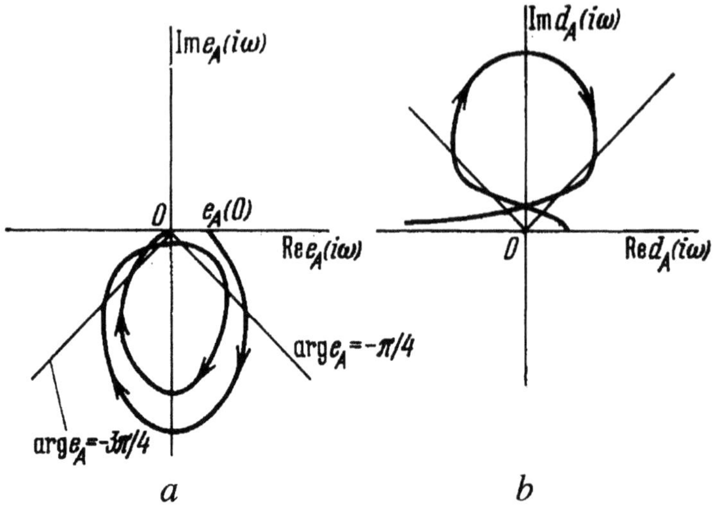

FIGURE 1.8.

Hence,

$$\left|e_A(i\tau_r^{-1})\right| \approx \left|E'\right| \cdot 2\eta_r; \quad \arg e_A(i\tau_r^{-1}) \approx -\frac{\pi}{2}. \tag{1.70}$$

In effect we have obtained a rather clear conception about the dynamic compliance at an arbitrary point of a weakly-damped system. Figure 1.8a displays a typical locus of the function $e_A\,(i\omega)$ while Fig 1.8b displays that of the dynamic compliance $d_A\,(i\omega) = e_A^{-1}\,(i\omega)$. When dependence $e_A\,(i\omega)$ is not available and must be determined experimentally it is sufficient to determine $e_A\,(0)$, the resonant and antiresonant frequencies and the bandwidth of each resonant and antiresonant frequency. This permits a unique determination of all the parameters in eq. (1.61).

In some cases, the object to be protected possesses zero eigenfrequencies. This means that it can move as a free rigid body, and equation for the dynamic compliance somewhat modifies. Assuming for example that there exists a single rigid-body mode, eq. (1.56) takes the following form

$$e_A(i\omega) = \frac{\sum\limits_{s,m=1}^{n} l_{As} l_{Am} W_{sm}(i\omega)}{-\omega^2 \prod\limits_{r=1}^{n-1} (\lambda_r^2 + \mu_r^2 + 2\mu_r i\omega - \omega^2)}. \tag{1.71}$$

We now turn to analysis of operator $e_{AB}\,(p)$ which relates the displacement of point B and the force acting on point A. Accounting for smallness of the dissipation coefficients $b_{sm,}$ eq. (1.53) reduces to

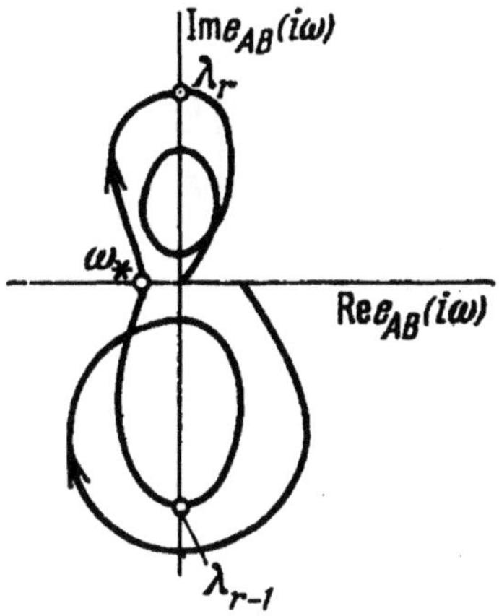

FIGURE 1.9.

$$e_{AB}(p) \approx \sum_{s=1}^{n} l_{As} l_{Bs} \frac{\prod_{m=1}^{n}{}_{s} \left(p^2 + \lambda_m^2 + b_{mm} p\right)}{\prod_{k=1}^{n} \left(p^2 + \lambda_k^2 + b_{kk} p\right)}.$$

Let us now determine the dynamic compliance at frequency $\omega = \lambda_r$. Inserting $p = i\lambda_r$ and neglecting asymptotically small terms yields

$$e_{AB}(i\lambda_r) \approx l_{Ar} l_{Br} \frac{\prod_{m=1}^{n}{}_{r} \left(-\lambda_r^2 + \lambda_m^2 + b_{mm} i\lambda_r\right)}{\prod_{k=1}^{n} \left(-\lambda_r^2 + \lambda_k^2 + b_{kk} i\lambda_r\right)} = \frac{l_{Ar} l_{Br}}{b_{kk} i\lambda_r}.$$

As the normal modes are determined up to an arbitrary constant factor one can assume without loss of generality that $l_{A1} l_{B1} > 0$. Also, let

$$l_{Am} l_{Bm} > 0 \text{ for } m = 1, 2, \ldots r - 1; \; l_{Ar} l_{Br} < 0.$$

Then,

$$\arg e_{AB}\left(i\lambda_m\right) = -\frac{\pi}{2} \text{ for } m \leq r - 1, \;\; \arg e_{AB}\left(i\lambda_r\right) = -\frac{3\pi}{2}.$$

In this case, locus $e_{AB}\left(i\omega\right)$ is as shown in Fig. 1.9. After $r-1$ revolutions in the lower half-plane the locus curve intersects the real axis at frequency ω_* ($\lambda_1 < \omega_* < \lambda_2$) and moves into the upper half-plane. This means that the system loses one antiresonant frequency. Provided that

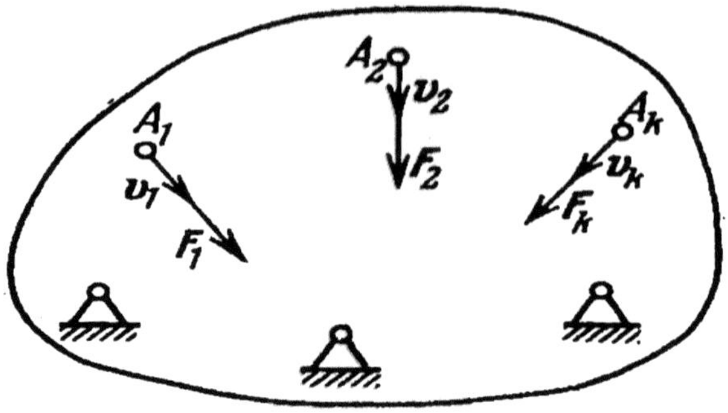

FIGURE 1.10.

$$l_{Am}l_{Bm} < 0 \text{ for } m = r, r+1, ...s; \; l_{A,s+1}l_{B,s+1} < 0,$$

the locus encircles $s - r$ times in the upper half-space and returns to the lower half-space. Hence, the locus form depends crucially upon the number of sign alteration in the series $l_{Am}l_{Bm}$. Provided that A and B coincide, it is apparent that $l_{Am}l_{Bm} = l_{Am}^2 > 0$, and the locus is as shown in Fig. 1.8a.

Let the forces $F_1(t)$, $F_2(t)$, ..., $F_k(t)$ act on the points A_1, A_2, ..., A_k and $v_1(t)$, $v_2(t)$, ..., $v_k(t)$ denote projections of the displacement of the points A_k along the directions of the forces F_s $(s = 1, ..., k)$, see Fig. 1.10. By virtue of the superposition principle which is valid for linear system we obtain

$$v_s(t) = \sum_{m=1}^{k} e_{ms}(p)F_m(t) \; (s = 1, ..., k). \tag{1.72}$$

Here $e_{ms}(p)$ are the dynamic compliance operators relating forces $F_m(t)$ to displacements $v_s(t)$. Relationship (1.72) can be written in vector form as

$$\mathbf{v}(t) = E(p)\mathbf{F}(t), \tag{1.73}$$

where $\mathbf{v}(t)$ and $\mathbf{F}(t)$ are $k-$dimensional vectors of displacement and force, respectively, while $E(p)$ stands for a matrix operator whose entries are $e_{ms}(p)$. This operator is called the *matrix of dynamic compliance operators*, whereas the corresponding frequency characteristic, $E(i\omega)$, is referred to as the *dynamic compliance matrix* at points A_1, A_2,..., A_k. Note that matrix $E(i\omega)$ is symmetric due to eq. (1.39). Its inverse

$$D(i\omega) = E^{-1}(i\omega) \tag{1.74}$$

is called the *dynamic rigidity matrix* at the above points. Thus, for the dynamic rigidity matrix and dynamic compliance matrix to be determined, it is necessary to know the points A_1, A_2,..., A_k and the directions of the forces $F_1(t)$, $F_2(t)$, ..., $F_k(t)$.

Let us express matrix $E(p)$ in terms of $W(p)$. Since

$$v_s = \mathbf{l}_s \cdot \mathbf{q},$$

we obtain

$$\mathbf{v}(t) = L\mathbf{q}, \tag{1.75}$$

where L is a $(k \times n)$ matrix, whose rows are vectors $\mathbf{l}_s$. On the other hand, by equating the elementary works done by forces $\mathbf{F}$ and $\mathbf{Q}$ we have

$$\mathbf{Q} \cdot \delta\mathbf{q} = \mathbf{F} \cdot \delta\mathbf{v} = \mathbf{F} \cdot L\delta\mathbf{q}.$$

This yields

$$\mathbf{Q} = L^T\mathbf{F}, \tag{1.76}$$

where T denotes transpose. From eqs. (1.29), (1.75) and (1.76) one obtains

$$\mathbf{v} = LW(p)\,L^T\mathbf{F},$$

and thus

$$E(p) = LW(p)L^T. \tag{1.77}$$

We consider next two examples.

1.2.1 *Example 1: The dynamic compliance matrix for a free rigid body*

Let us construct the dynamic compliance matrix for a free rigid body exhibiting small vibrations, see Fig. 1.11. Let $\mathbf{s}$ and θ denote respectively the vector of the displacement of the centre of mass and the vector of rotation. The projections of the displacement vector, s_x, s_y and s_z along the principal axis x, y and z and the angles of rotation about the principal axis, θ_x, θ_y and θ_z are chosen as the generalised coordinates. Let $\mathbf{r}_i(x_i, y_i, z_i)$ denote the vector of the point where force $\mathbf{F}_i$ is applied and $\mathbf{n}_i(\alpha_i, \beta_i, \gamma_i)$ denote the unit vector of this force. The standard geometric relations then yield

$$\begin{aligned} v_i &= (\mathbf{s} + \boldsymbol{\theta} \times \mathbf{r}_i) \cdot \mathbf{n}_i = \alpha_i s_x + \beta_i s_y + \gamma_i s_z + \\ &\quad (\gamma_i y_i - \beta_i z_i)\theta_x + (\alpha_i z_i - \gamma_i x_i)\theta_y + +(\beta_i x_i - \alpha_i y_i)\theta_z \; (i = 1, ..., k). \end{aligned} \tag{1.78}$$

Matrix L is then as follows

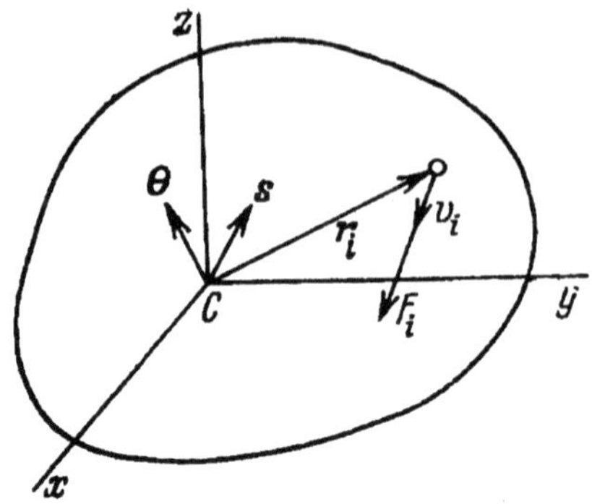

FIGURE 1.11.

$$L = \left\| \begin{matrix} \alpha_1 & \beta_1 & \gamma_1 & \gamma_1 y_1 - \beta_1 z_1 & \alpha_1 z_1 - \gamma_1 x_1 & \beta_1 x_1 - \alpha_1 y_1 \\ \cdots & \cdots & \cdots & \cdots\cdots & \cdots\cdots & \cdots\cdots \\ \alpha_k & \beta_k & \gamma_k & \gamma_k y_k - \beta_k z_k & \alpha_k z_k - \gamma_k x_k & \beta_k x_k - \alpha_k y_k \end{matrix} \right\|. \tag{1.79}$$

Let us construct matrix $W(p)$. In this case, B and C are zero matrices as we are dealing with a free rigid body. For this reason, matrix $W(p)$ has the expression

$$W(p) = (Ap^2)^{-1} = \frac{1}{mp^2} \operatorname{diag}\left\{1, 1, 1, \frac{1}{\rho_x^2}, \frac{1}{\rho_y^2}, \frac{1}{\rho_z^2}\right\}, \tag{1.80}$$

where m is the body mass, while ρ_x, ρ_y and ρ_z are the radii of inertia. Inserting eqs. (1.79) and (1.80) into eq. (1.77) yields the components of matrix $E(p)$

$$\begin{aligned} e_{sm}(p) = {} & \frac{1}{mp^2}\Big[\alpha_s\alpha_m + \beta_s\beta_m + \gamma_s\gamma_m + \frac{1}{\rho_x^2}(\gamma_s y_s - \beta_s z_s)(\gamma_m y_m - \beta_m z_m) \\ & + \frac{1}{\rho_y^2}(\alpha_s z_s - \gamma_s x_s)(\alpha_m z_m - \gamma_m x_m) \\ & + \frac{1}{\rho_z^2}(\beta_s x_s - \alpha_s y_s)(\beta_m x_m - \alpha_m y_m)\Big]. \end{aligned} \tag{1.81}$$

1.2.2 Example 2. The dynamic compliance of a two degree-of-freedom system

We obtain now the dynamic compliances of the system depicted in Fig. 1.12. Letting $v_A = q_1$ and $v_B = q_2$, we derive the equation of motion of the system subject to forces F_A and F_B

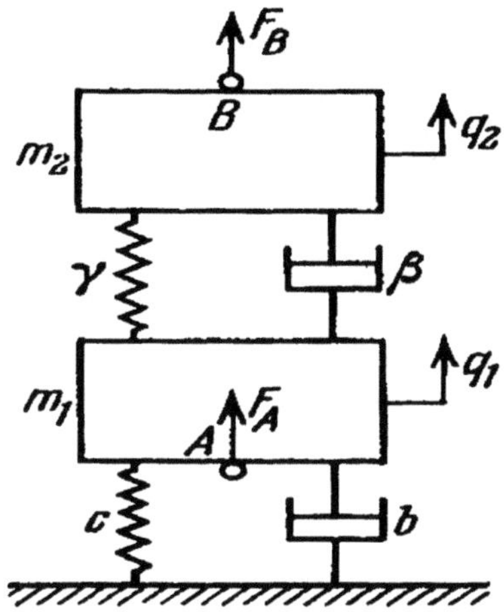

FIGURE 1.12.

$$\left.\begin{aligned} (m_1p+bp+c)q_1+(\beta p+\gamma)(q_1-q_2)&=F_A(t), \\ m_2p^2q_2+(\beta p+\gamma)(q_2-q_1)&=F_B(t). \end{aligned}\right\} \tag{1.82}$$

Solving these equations yields

$$\left.\begin{aligned} q_1&=\Delta^{-1}(p)[(m_2p^2+\beta p+\gamma)F_A+(\beta p+\gamma)F_B], \\ q_2&=\Delta^{-1}(p)[(\beta p+\gamma)F_A+(m_1p^2+bp+\beta p+c+\gamma)F_B], \end{aligned}\right\} \tag{1.83}$$

where

$$\Delta(p)=\begin{vmatrix} m_1p^2+bp+\beta p+c+\gamma & -(\beta p+\gamma) \\ -(\beta p+\gamma) & m_2p^2+\beta p+\gamma \end{vmatrix}. \tag{1.84}$$

Hence,

$$\left.\begin{aligned} e_A(p)=&\ \ \Delta^{-1}(p)(m_2p^2+\beta p+\gamma),\ e_{AB}(p)=\Delta^{-1}(p)(\beta p+\gamma), \\ e_B(p)=&\ \ \Delta^{-1}(p)(m_1p^2+bp+\beta p+c+\gamma). \end{aligned}\right\} \tag{1.85}$$

As the typical properties of functions $e_A\,(i\omega)$ and $e_B\,(i\omega)$ have been studied earlier we analyse the dynamic compliance $e_{AB}\,(i\omega)$ in detail. Let $-\mu_1 \pm i\lambda_1$ and $-\mu_2 \pm i\lambda_2$ be the roots of the characteristic equation of the system

$$\Delta\,(p)=0.$$

Denoting

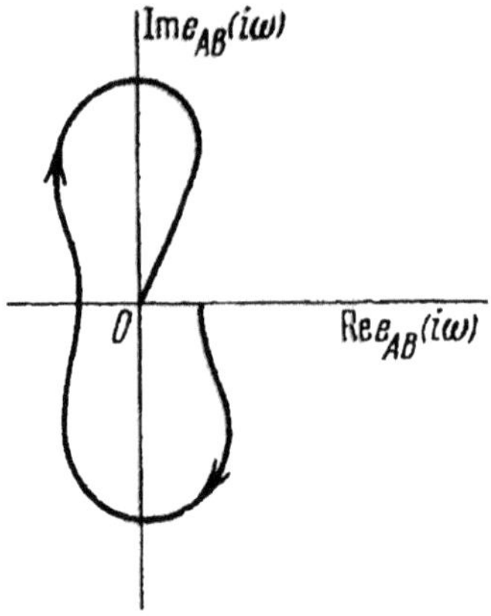

FIGURE 1.13.

$$\frac{\beta}{\gamma} = T, \quad \frac{1}{\lambda_1^2 + \mu_1^2} = T_1^2, \quad \frac{1}{\lambda_2^2 + \mu_2^2} = T_2^2, \; \xi_1 = \mu_1 T_1, \; \xi_2 = \mu_2 T_2,$$

we have

$$e_{AB}(p) = \frac{1 + Tp}{(c+\gamma)(1 + 2T_1\zeta_1 p + T_1^2 p^2)(1 + 2T_2\zeta_2 p + T_2^2 p^2)}. \tag{1.86}$$

The locus of the dynamic compliance $e_{AB}(i\omega)$ for $T_1 > T > T_2$ is displayed in Fig. 1.13.

It is necessary to stress again that the above properties of the dynamic characteristics of mechanical systems manifest themselves only when the system is lightly damped. Moreover, as pointed out above the resonant and antiresonant frequencies must be well separated, and the resonant domain (1.68) must contain no antiresonant frequency. We explain the implication of this requirement for the example system depicted in Fig 1.12. By transforming dynamic compliance $e_A(i\omega)$ to the form (1.61) we find that

$$|e_A(i\omega)| = \frac{\sqrt{(1 - \tau_1^2\omega^2)^2 + 4\eta_1^2\tau_1^2\omega^2}}{\sqrt{[(1 - T_1^2\omega^2)^2 + 4\zeta_1^2 T_1^2\omega^2][(1 - T_2^2\omega^2)^2 + 4\zeta_2^2 T_2^2\omega^2]}}. \tag{1.87}$$

The system exhibits its resonance properties at frequency $\omega = T_1^{-1}$ only if the numerator (1.87) is not small. To ensure the latter, it is necessary to require that the first term in the numerator is much greater than the second term. The condition

$$|1 - \tau_1^2 T_1^{-2}| \gg 2\eta_1\tau_1 T_1^{-1} \tag{1.88}$$

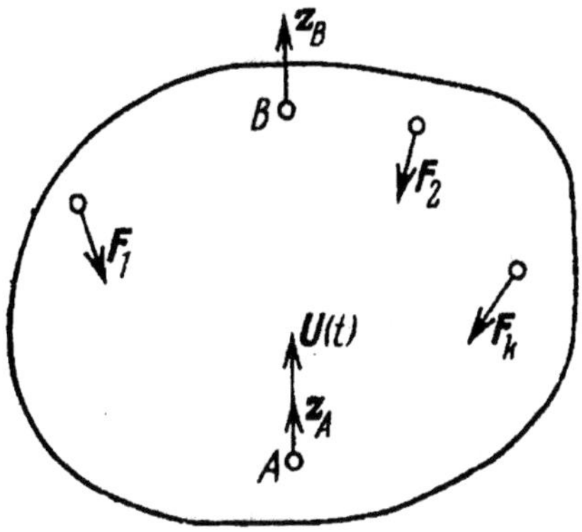

FIGURE 1.14.

can be shown to imply that $\tau_1 T_1^{-1}$ deviates significantly from unity. Analogously, for antiresonance to occur at frequency $\omega = \tau_1^{-1}$, the following conditions

$$\left|1 - \tau_1^{-2} T_1^{-2}\right| \gg 2\zeta_1 \tau_1^{-1} T_1, \quad \left|1 - T_2^2 \tau_1^{-2}\right| \gg 2\zeta_2 T_2 \tau_1^{-1} \tag{1.89}$$

must be satisfied. In a simple case, in which $b = 0$ and $c = 0$ conditions (1.88) and (1.89) can be shown to be equivalent to the following

$$T_1^{-1} \gg \frac{b\,(m_1 + m_2)}{m_2^2}, \quad \tau_1^{-1} \gg \frac{b\,(m_1 + m_2)}{m_2^2}.$$

The latter implies that the resonant and antiresonant phenomena occur only if mass m_2 is not too small.

1.3 Efficiency conditions in the case of harmonic excitation

The most important characteristic of any system of vibration protection is its efficiency. A system of vibration protection is effective when it reduces the vibration transmission to the object. As pointed out earlier, vibration protection systems create control forces which are additional dynamic actions. The performance of a particular vibration protection system is judged by comparing the level of the vibration field before and after the control is introduced. This section is concerned with criteria for establishing the efficiency of some vibration protection systems for the case of a single-frequency harmonic excitation.

1. Assume that the action of a vibration protection system reduces to a single control force $U\,(t)$, cf. Fig. 1.14. Such a situation is observed when a one-dimensional dynamic absorber is used. Another possibility is a uniaxial

vibration isolator whose backward influence on the base is negligible. Let force $U(t)$ be applied at point A of the object in a prescribed direction. In general, the goal of the control is to influence some parameter of the vibration field, for example the displacement z_B of another point B. Let the displacement of this point without control be $\zeta_B(t)$. When the control force $U(t)$ acts the superposition principle can be applied, to give

$$z_B(t) = \zeta_B(t) + e_{AB}(p)U(t), \tag{1.90}$$

with $e_{AB}(p)$ being the dynamic compliance operator. We assume now that the control force due to the vibration protection system relates to displacement $z_B(t)$ by means of a linear operator, i.e.

$$U(t) = -w_z(p)z_B(t). \tag{1.91}$$

In other words, the control is realised by means of feedback, the minus sign implying a negative feedback. Substituting (1.91) into (1.90) yields

$$z_B(t) = [1 + e_{AB}(p)w_z(p)]^{-1}\zeta_B(t). \tag{1.92}$$

The operator

$$k_z(p) = [1 + e_{AB}(p)w_z(p)]^{-1} \tag{1.93}$$

expresses the displacement of point B in the system with control in terms of that in the system without control. This characterizes the efficiency of the vibration protection system completely. Provided that

$$\zeta_B(t) = \zeta_0 \cos(\omega t),$$

eq. (1.92) yields

$$z_B(t) = z_0 \cos(\omega t + \varphi) = \zeta_0 |k_z(i\omega)| \cos[\omega t + \arg k_z(i\omega)]. \tag{1.94}$$

The quantity

$$k_0(\omega) = |k_z(i\omega)| = |1 + e_{AB}(i\omega)w_z(i\omega)|^{-1}, \tag{1.95}$$

which is seen to be equal to the ratio of amplitudes of two harmonic processes $z_B(t)$ and $\zeta_B(t)$ is referred to as the *efficiency factor* at frequency ω. The smaller the factor, the more effective the vibration protection system utilized. The efficiency is frequently described by the following quantity

$$L(\omega) = -20 \log k_0(\omega), \tag{1.96}$$

which characterizes the reduction of vibration amplitudes in decibels.

Consider now a particular case in which the controlled variable is displacement $z_A(t)$ of point A. The displacement's direction is assumed to coincide with the direction of the control force $U(t)$. In this case, $e_{AB}(p)$ coincides with the dynamic compliance operator $e_A(p)$ which leads to

$$k_0(\omega) = |1 + e_A(i\omega)w_z(i\omega)|^{-1}. \tag{1.97}$$

The efficiency condition may be represented in the form

$$k_0(\omega) = \frac{|d_A(i\omega)|}{|d_A(i\omega) + w_z(i\omega)|} < 1, \tag{1.98}$$

wherein $d_A(i\omega) = e_A^{-1}(i\omega)$ is the dynamic rigidity of the object at point A. Condition (1.98) indicates that the control is effective at the resonance frequencies of the object since $|d_A(i\omega)|$ is small at these frequencies. An explanation is that the attachment of a vibration protection system shifts the resonant frequencies of the object. Condition (1.98) is also satisfied at the frequencies at which $|w_z(i\omega)|$ is large. This explains the efficiency of the simple passive dynamic absorber studied in Section 1.1. In general, the efficiency of the vibration protection system can be increased by increasing the feedback gain since this increases $|w_z(i\omega)|$.

2. Let the control force $U(t)$ be due to displacement $z_A(t)$ and the objective of the control be to decrease $z_B(t)$. In this case, an estimation of the control efficiency becomes essentially elaborate. Indeed,

$$z_A(t) = \zeta_A(t) + e_A(p)U(t) = \zeta_A(t) - e_A(p)w_z(p)z_A(t),$$

where $\zeta_A(t)$ denotes displacement of point A without control. Rearranging leads to

$$z_A(t) = [1 + e_A(p)w_z(p)]^{-1}\zeta_A(t). \tag{1.99}$$

On the other hand

$$z_B(t) = \zeta_B(t) + e_{AB}(p)U(t) = \zeta_B(t) - e_{AB}(p)w_z(p)z_A(t). \tag{1.100}$$

Substituting (1.99) into (1.100) we obtain

$$z_B(t) = \zeta_B(t) + e_{AB}(p)w_z(p)[1 + e_A(p)w_z(p)]^{-1}\zeta_A(t). \tag{1.101}$$

Hence, to assess the efficiency of the control, one needs $\zeta_A(t)$ and $\zeta_B(t)$. Provided that the system is subject to harmonic excitations

$$\zeta_A(t) = \zeta_{A0}\cos(\omega t), \quad \zeta_B(t) = \zeta_{B0}\cos(\omega t + \alpha),$$

one obtains

$$z_B(t) = \zeta_{B0}\cos(\omega t + \alpha) - \operatorname{Re} w(i\omega)\,\zeta_{A0}\cos(\omega t) + \operatorname{Im} w(i\omega)\,\zeta_{A0}\sin(\omega t),$$

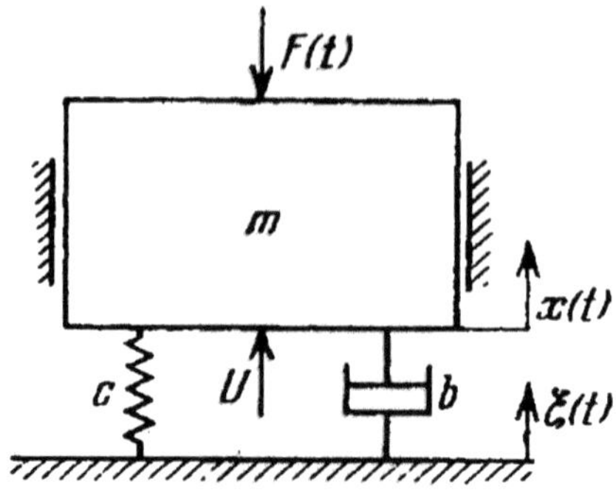

FIGURE 1.15.

where

$$w(i\omega) = e_{AB}(i\omega)w_z(i\omega)[1 + e_A(i\omega)w_z(i\omega)]^{-1}. \tag{1.102}$$

This enables one to determine the amplitude of $\zeta_B(t)$ which is

$$z_{B0} = \sqrt{(\zeta_{A0}\operatorname{Re} w(i\omega) - \zeta_{B0}\cos\alpha)^2 + (\zeta_{A0}\operatorname{Im} w(i\omega) - \zeta_{B0}\sin\alpha)^2}. \tag{1.103}$$

3. Let us apply the equations obtained to deriving the conditions for an efficient active vibration protection system of a single-degree-of-freedom system. The system is assumed to consist of a mass m, an elastic element of stiffness c and a dashpot b, see Fig. 1.15. Let us assume that the mass m is driven by a prescribed force $F(t)$ and the displacement of the base is $\xi(t)$. The objective of the control will be, for example, (i) to reduce the absolute displacement $x(t)$ and (ii) to reduce the force $R(t)$ acting on the base.

To begin with, we assume that the control force $U(t)$ is due to the absolute displacement, i.e.

$$U(t) = -w_x(p)x(t). \tag{1.104}$$

The equations of the motion

$$mp^2x = -(bp + c)y + F(t) - w_x(p)x, \quad x = y + \xi, \tag{1.105}$$

yield

$$x(t) = [mp^2 + bp + c + w_x(p)]^{-1}[F(t) + (bp + c)\xi(t)]. \tag{1.106}$$

Comparing the expressions for $x(t)$ for $w_x(p) \neq 0$ with that for $w_x(p) = 0$ we obtain the following equation for operator $k_x(p)$

$$k_x(p) = \left[1 + \frac{w_x(p)}{mp^2 + bp + c}\right]^{-1}. \tag{1.107}$$

This operator is known to describe the efficiency of the control. The latter expression also coincides with eq. (1.93) because in this particular case

$$e_{AB}(p) = e_A(p) = (mp^2 + bp + c)^{-1}. \tag{1.108}$$

Solving eq. (1.105) for $y\,(t)$ yields

$$y(t) = [mp^2 + bp + c + w_x(p)]^{-1}\{F\,(t) - [w_x(p) + mp^2]\xi(t)\}. \tag{1.109}$$

One can see from the latter equation that the efficiency of the control in terms of the absolute displacement estimated by means of the reduction of the relative displacement depends of the nature of the external excitation. In the case of force excitations alone, which implies that $\xi(t) = 0$, we have

$$y(t) = [mp^2 + bp + c]^{-1}F\,(t) \qquad \text{if } w_x(p) = 0,$$

$$y(t) = [mp^2 + bp + c + w_x(p)]^{-1}F\,(t) \quad \text{if } w_x(p) \neq 0,$$

and, hence,

$$k_{yx}(p) = \frac{mp^2 + bp + c}{mp^2 + bp + c + w_x(p)} = \left[1 + \frac{w_x(p)}{mp^2 + bp + c}\right]^{-1} = k_x(p) \tag{1.110}$$

as anticipated since $x \equiv y$. When the excitation is purely kinematic, i.e. $F\,(t) \equiv 0$, we have

$$y(t) = -mp^2[mp^2 + bp + c]^{-1}\xi\,(t) \qquad \text{if } w_x(p) = 0,$$

$$y(t) = -[mp^2 + w_x(p)][mp^2 + bp + c + w_x(p)]^{-1}\xi\,(t) \quad \text{if } w_x(p) \neq 0,$$

which gives

$$k_{yx}(p) = \left[1 + \frac{w_x(p)}{mp^2}\right]\left[1 + \frac{w_x(p)}{mp^2 + bp + c}\right]^{-1}. \tag{1.111}$$

Let us now determine the force acting on the base under the assumption that the base is subject to the control force $U\,(t)$ and the forces in the spring and the dashpot. In other words, an active system of vibration protection is mounted between the object and the base and acts with equal forces on both the object and the base. In this case,

$$\begin{aligned} R(t) &= (bp + c)y + w_x(p)x = \\ &= \frac{bp + c + w_x(p)}{mp^2 + bp + c + w_x(p)}F\,(t) - \frac{mp^2(bp + c)}{mp^2 + bp + c + w_x(p)}\xi(t). \end{aligned} \tag{1.112}$$

Thus, the reduction of the system reaction after the control has been introduced depends on the nature of the dynamic excitation, too. Provided that $\xi(t) = 0$, we have

$$\begin{aligned} k_R(p) &= \frac{[bp + c + w_x(p)](mp^2 + bp + c)}{[mp^2 + bp + c + w_x(p)](bp + c)} = \\ &= \left[1 + \frac{w_x(p)}{bp + c}\right]\left[1 + \frac{w_x(p)}{mp^2 + bp + c}\right]^{-1}. \end{aligned} \tag{1.113}$$

When the excitation is kinematic, we have

$$k_R(p) = \left[1 + \frac{w_x(p)}{mp^2 + bp + c}\right]^{-1} = k_x(p). \tag{1.114}$$

The efficiency of the control due to relative displacement

$$U(t) = -w_y(p)y(t) \tag{1.115}$$

can be analysed by analogy. The result is

$$k_y(p) = \left[1 + \frac{w_y(p)}{mp^2 + bp + c}\right]^{-1}, \tag{1.116}$$

$$k_{xy}(p) = \left[1 + \frac{w_y(p)}{bp + c}\right]\left[1 + \frac{w_y(p)}{mp^2 + bp + c}\right]^{-1}, \text{ if } F(t) = 0, \tag{1.117}$$

$$k_{xy}(p) = \left[1 + \frac{w_y(p)}{mp^2 + bp + c}\right]^{-1}, \text{ if } \xi(t) = 0, \tag{1.118}$$

$$k_R(p) = k_{xy}(p) \text{ if } F(t) = 0 \tag{1.119}$$

$$k_R(p) = \left[1 + \frac{w_y(p)}{mp^2 + bp + c}\right]^{-1}, \text{ if } \xi(t) = 0, \tag{1.120}$$

4. Let us consider now a multidimensional system which consists of s one-dimensional elements attached to the object at the points $A_1, ..., A_s$, see Fig. 1.16. These elements are assumed to form the control forces $U_1, U_2, ...,$ U_s which will be treated as vector $\mathbf{U}(t)$. The objective of the control is to reduce the displacements of points $B_1, B_2, ..., B_k$. The control vector $\mathbf{U}$ is related to the displacements $z_{B1}, z_{B2}, ..., z_{Bk}$ by the $s \times k$ transfer function matrix $W_z(p)$

$$\mathbf{U}(t) = -W_z(p)\mathbf{z}_B, \tag{1.121}$$

where $\mathbf{z}_B$ denotes the vector of displacement of points $B_1, B_2, ..., B_k$. Let $E_{AB}(p)$ be the matrix of dynamic compliance operator whose entries $e_{lm}(p)$ $(l = 1, ..., s, m = 1, ..., k)$ relate the displacement of point B_m with the force applied at point A_l. Then

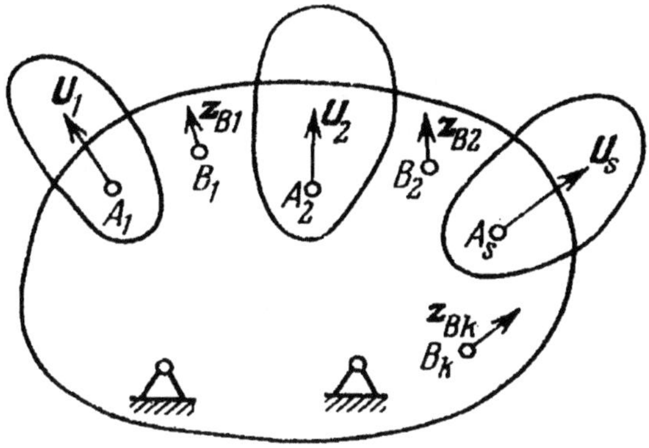

FIGURE 1.16.

$$\mathbf{z}_B = \boldsymbol{\zeta}_B + E_{AB}(p)\mathbf{U}(t) = \boldsymbol{\zeta}_B - E_{AB}(p)W_z(p)\mathbf{z}_B, \tag{1.122}$$

where $\boldsymbol{\zeta}_B$ denotes the vector of displacements of points $B_1, B_2, ..., B_k$ of the object without mounted elements of vibration protection. From eq. (1.122) one obtains

$$\mathbf{z}_B = [I_k + E_{AB}(p)W_z(p)]^{-1}\boldsymbol{\zeta}_B. \tag{1.123}$$

The following matrix operator

$$K(p) = [I_k + E_{AB}(p)W_z(p)]^{-1} = \{k_{uv}(p)\}_{u,v=1}^{k} \tag{1.124}$$

relating the vectors $\mathbf{z}_B$ and $\boldsymbol{\zeta}_B$ describes the efficiency of the control. The efficiency at frequency ω can be determined as the norm of the following matrix

$$K(i\omega) = [I_k + E_{AB}(i\omega)W_z(i\omega)]^{-1}. \tag{1.125}$$

Let us consider one of the possible norms of the matrix.

Provided that the components of the vector $\boldsymbol{\zeta}_B(t)$ are harmonic functions, i.e.

$$\zeta_r(t) = \zeta_{r0}\cos(\omega t + \alpha_r), \ (r = 1, ..., k),$$

eq. (1.123) results in

$$\begin{aligned} z_r(t) &= \sum_{m=1}^{k} |k_{rm}(i\omega)|\,\zeta_{m0}\cos[\omega t + \alpha_m + \arg k_{rm}(i\omega)] \quad (1.126)\\ (r &= 1, ..., k). \end{aligned}$$

It is apparent that the following inequality holds

$$z_{r0} \leqslant \sum_{m=1}^{s} |k_{rm}(i\omega)| \, \zeta_{m0} \; (r = 1, ..., k) \tag{1.127}$$

for the amplitudes z_{r0} of the harmonic processes $z_r(t)$ regardless of the phases α_m. We add together all these equations, to get

$$\sum_{r=1}^{k} z_{r0} \leqslant \sum_{m=1}^{s} \zeta_{m0} \sum_{r=1}^{k} |k_{rm}(i\omega)| \, . \tag{1.128}$$

Provided that the following inequality

$$\sum_{r=1}^{k} |k_{rm}(i\omega)| < 1 \tag{1.129}$$

holds for any m, eq. (1.128) yields

$$\sum_{r=1}^{k} z_{r0} \leqslant \sum_{m=1}^{k} \zeta_{m0}. \tag{1.130}$$

This means that the system of vibration protection reduces the sum of the vibration amplitudes at the points wherein its elements are attached. This allows the quantity

$$k_0(\omega) = \sup_{m} \sum_{r=1}^{k} |k_{rm}(i\omega)| \tag{1.131}$$

to be chosen as an efficiency factor for the system of vibration protection. Condition $k_0(\omega) < 1$ provides performance in the above sense for any values of ζ_{r0} and α_r. Therefore, in order to prove the efficiency at frequency ω it is sufficient to find the column of matrix (1.125) with the maximal sum of its components.

In the particular case in which points B_m coincide with points A_l the matrix $E_{AB}(p)$ is simply the matrix of dynamic compliance operators at point A, i.e. $E_{AB}(p) = E_A(p)$. Then we have

$$K(p) = [I_k + E_{AB}(p)W_z(p)]^{-1} = [D_A(p) + W_z(p)]^{-1} D_A(p), \tag{1.132}$$

where $D_A(p) = E_A(p)^{-1}$ stands for the matrix of dynamic rigidity operators.

5. Up until now, we have considered vibration protection systems which produce some additional control forces which vanish if the feedback does not exist. The problem of the efficiency of the vibration isolator which replaces the rigid connection of the object and the base must be studied from a somewhat other perspective.

In order to analyse the efficiency of a one-dimensional vibration isolator we consider first a free object subject to dynamic force excitations $F_1, ...,$

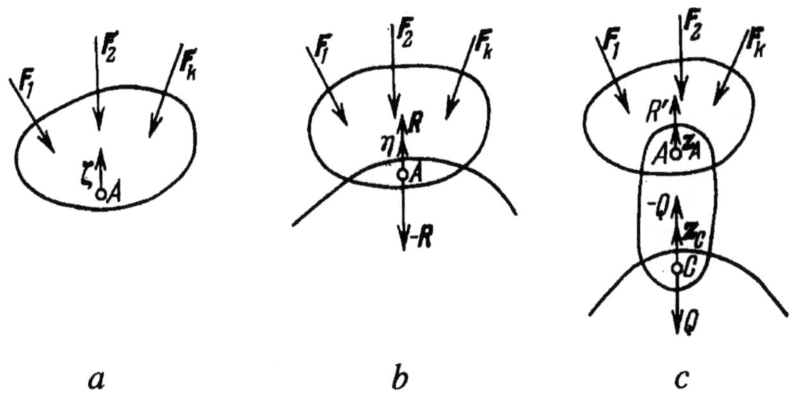

FIGURE 1.17.

F_k, see Fig. 1.17a. Let $\zeta(t)$ denote the displacement of point A in the direction of the vibration isolator's reaction. Let us consider the same object which is now attached to the base at point A, cf. Fig. 1.17 b. The latter Figure shows two forces, namely $R(t)$ and $-R(t)$ acting on the object and the base, respectively. The superposition principle allows the displacement of point A to be determined as follows

$$\eta(t) = \zeta(t) + e_A(p)R(t). \tag{1.133}$$

On the other hand, point A belongs to the base which implies that

$$\eta(t) = -e_0(p)R(t), \tag{1.134}$$

$e_0(p)$ being the dynamic compliance operator at the attachment point. From eqs. (1.133) and (1.134) we obtain

$$\zeta(t) = -[e_0(p) + e_A(p)]R(t). \tag{1.135}$$

Let us assume that a linear vibration isolator is placed between the base and the object, see Fig. 1.17c. Let $z_A(t)$ and $R'(t)$ stand for the displacement and the reaction of the vibration isolator at point A, respectively, while $z_C(t)$ and $Q(t)$ denote the displacement of point C of the base and the force acting at this point, respectively. It is clear that

$$z_A = \zeta(t) + e_A(p)R'(t), \tag{1.136}$$

$$z_C = e_0(p)Q(t). \tag{1.137}$$

By using eqs. (1.13) we obtain

$$\left.\begin{aligned} R' &= \gamma_{11}(p)z_A + \gamma_{12}(p)z_C, \\ Q &= \gamma_{21}(p)z_A + \gamma_{22}(p)z_C. \end{aligned}\right\} \tag{1.138}$$

Removing $R'(t)$, z_A, z_C and ζ from eqs. (1.135)-(1.138) we have

$$
\begin{aligned}
Q(t) &= \\
&- \frac{[e_A(p)+e_0(p)]\gamma_{21}(p)}{[1-\gamma_{22}(p)e_0(p)][1-\gamma_{11}(p)e_A(p)]-\gamma_{12}(p)\gamma_{21}(p)e_0(p)e_A(p)} R(t) \\
&= k_Q(p)R(t).
\end{aligned}
\tag{1.139}
$$

Operator $k_Q(p)$, relating forces $Q(t)$ and $R(t)$, describes the efficiency of the vibration isolator as a facility which decreases the force acting on the base. As $z_C(p)$ is related to $R(t)$ and $Q(t)$ by means of eqs. (1.134) and (1.137), it is clear that the above operator determines the efficiency of the vibration isolator due to the displacement of base point C.

One can easily obtain from eq. (1.139) an expression for the operator which characterizes the relative efficiency caused by replacing one vibration isolator by another one. Provided that $\gamma_{ij}^0(p)$ and $\gamma_{ij}(p)$ are the operators describing the properties of these isolators, we have

$$
\varkappa_Q(p) = \frac{k_Q(p)}{k_Q^0(p)} = \frac{\gamma_{21}[(1-\gamma_{22}^0 e_0)(1-\gamma_{11}^0 e_A)-\gamma_{12}^0\gamma_{21}^0 e_A e_0]}{\gamma_{21}^0[(1-\gamma_{22}e_0)(1-\gamma_{11}e_A)-\gamma_{12}\gamma_{21}e_A e_0]}. \tag{1.140}
$$

Consider now a particular case, namely, the vibration isolator is assumed to obey the following law

$$
R' = -Q = -w(p)(z_A - z_C). \tag{1.141}
$$

(For example, $w(p) = c + bp$ for a simplest passive vibration isolator). In this case,

$$
\gamma_{11}(p) = \gamma_{22}(p) = -w(p), \ \gamma_{12}(p) = \gamma_{21}(p) = w(p). \tag{1.142}
$$

Upon substituting eq. (1.142) into eq. (1.139) we obtain

$$
k_Q(p) = -\frac{e_A(p)+e_0(p)}{e_A(p)+e_0(p)+w^{-1}(p)}. \tag{1.143}
$$

6. Let us consider now a uniaxial vibration isolator. We consider first the motion of the base without the attached object and denote the displacement of point C of the base as $\psi(t)$, see Fig. 1.18a. The displacement of the same point after the object has been rigidly attached to the base, cf. Fig. 1.18b, can be found from the following equations

$$
\left.
\begin{aligned}
\zeta(t) &= \psi(t) - e_0(p)R(t), \\
\zeta(t) &= e_C(p)R(t),
\end{aligned}
\right\}
\tag{1.144}
$$

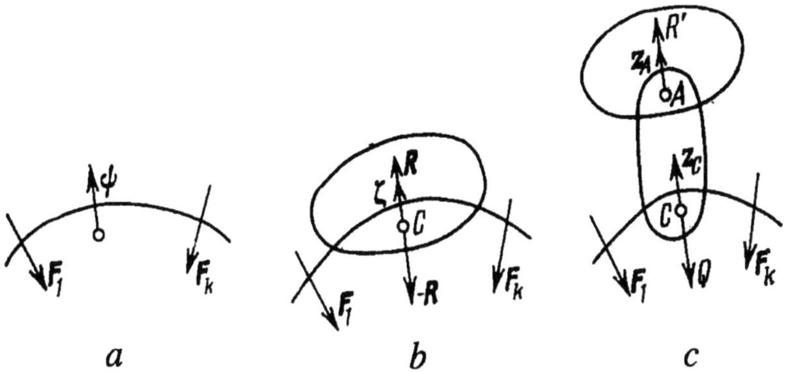

FIGURE 1.18.

with $e_C(p)$ being the operator of dynamic compliance at the attachment point. Eliminating $R(t)$ yields

$$\zeta(t) = e_A(p)[e_A(p) + e_0(p)]^{-1}\psi(t). \tag{1.145}$$

Provided that a vibration isolator is attached, see Fig. 1.18c, we have

$$\left.\begin{aligned} z_A(t) &= e_A(p)R'(t), \\ z_C(t) &= \psi(t) + e_0(p)Q(t). \end{aligned}\right\} \tag{1.146}$$

Removing z_C, ψ, R' and Q from eqs. (1.145), (1.146) and (1.138) leads to

$$\begin{aligned} &z_A(t) = \\ &\frac{\gamma_{12}(p)[e_A(p) + e_0(p)]}{[1-\gamma_{11}(p)e_A(p)][1-\gamma_{22}(p)e_0(p)] - \gamma_{12}(p)\gamma_{21}(p)e_0(p)e_A(p)}\zeta(t) \\ &= k_z(p)\zeta(t). \end{aligned} \tag{1.147}$$

Operator $k_z(p)$, relating the displacements of point A before and after attachment of the vibration isolator, describes its efficiency. If relations (1.142) are satisfied, we get

$$k_z(p) = \frac{e_A(p) + e_0(p)}{e_A(p) + e_0(p) + w^{-1}(p)}. \tag{1.148}$$

7. Let us establish now the efficiency conditions for the simplest vibration isolator for which $w(p) = c + bp$. Equation (1.148) yields

$$k_z(p) = \frac{w(p)}{d(p) + w(p)}, \tag{1.149}$$

where

$$d(p) = [e_A(p) + e_0(p)]^{-1}. \tag{1.150}$$

Let us find the efficiency factor at frequency ω. If

$$d\,(i\omega) = \mu\,(\omega) + i\nu\,(\omega)\,,$$

then,

$$k_0(\omega) = |k_z(i\omega)| = \sqrt{\frac{c^2 + b^2\omega^2}{[\mu(\omega) + c]^2 + [\nu(\omega) + b\omega]^2}}. \tag{1.151}$$

Provided that the damping in the object, the base and in the vibration isolator is small, factor k_0 reaches its maximum (which implies the minimum efficiency) at those frequencies at which

$$\mu\,(\omega_r) + c = 0.$$

These are the resonant frequencies of the system consisting of the object, the base and the vibration isolator. In order to satisfy the condition $k_0(\omega_r) < 1$ it is necessary to require that the following condition

$$c^2 + b^2\omega^2 < [\nu(\omega) + b\omega]^2$$

holds. The latter condition is equivalent to the following one

$$c^2 < \nu^2(\omega) + 2\nu(\omega)b\omega. \tag{1.152}$$

As shown in Section 1.2, the imaginary part of the dynamic compliance for any mechanical system is negative, that is, the imaginary part of the corresponding dynamic rigidity is positive. For this reason, condition (1.152) holds, if

$$c < \nu\,(\omega)\,.$$

This means that for the performance of the simplest vibration isolator in a broad band containing the resonant frequencies of the object, it is sufficient that its stiffness is smaller than the imaginary part of $d\,(i\omega)$ at each resonant frequency. In practice, the stiffness cannot be vanishingly small meaning that the effectiveness of the vibration protection for lightly-damped objects in a broad-band frequency domain is difficult.

8. Let us study now the efficiency of a vibration isolator with an intermediate mass, cf. Fig. 1.6b. As follows from eqs. (1.12) and (1.13)

$$\left.\begin{aligned} \gamma_{11} = \gamma_{22} = &\ -(c + bp)(mp^2 + bp + c)/\left(mp^2 + 2bp + 2c\right), \\ \gamma_{21} = \gamma_{21} = &\ (bp + c)^2/\left(mp^2 + 2bp + 2c\right). \end{aligned}\right\} \tag{1.153}$$

Let us compare this vibration isolator with a vibration isolator composed of the same simplest cascades in series but without any intermediate mass. The corresponding expressions for the operators γ_{ij}^0 can be obtained from eq. (1.153) letting $m = 0$. The result is

$$\gamma_{11}^0 = \gamma_{22}^0 = -\frac{1}{2}(bp + c), \; \gamma_{12}^0 = \gamma_{21}^0 = \frac{1}{2}(bp + c). \tag{1.154}$$

Substituting (1.153) and (1.154) into (1.140) and assuming a rigid base, $e_0 = 0$, yields

$$\begin{aligned} \varkappa_Q(p) &= \frac{(bp + c)(2d_A + bp + c)}{(bp + c)(2d_A + bp + c) + mp^2(d_A + bp + c)} \\ &= \left[1 + \frac{mp^2(bp + c + d_A)}{(bp + c)(2d_A + bp + c)}\right]^{-1}, \end{aligned} \tag{1.155}$$

where $d_A = e_A^{-1}(p)$. The efficiency condition at frequency ω is given by

$$\varkappa_Q(i\omega) = \left|1 + \frac{m\omega^2\left[bi\omega + c + d_A(i\omega)\right]}{(bi\omega + c)\left[bi\omega + c + 2d_A(i\omega)\right]}\right|^{-1} < 1. \tag{1.156}$$

Rearranging, we obtain

$$m > 2\frac{c(c + \mu)(c + 2\mu) + (c + \mu)b^2\omega^2 + 2c\nu(b\omega + \nu)}{\omega^2[(c + \mu)^2 + (b\omega + \nu)^2]}, \tag{1.157}$$

where $\mu(\omega) = \mathrm{Re}\; d_A(i\omega)$ and $\nu(\omega) = \mathrm{Im}\; d_A(i\omega)$. In the case of small dissipation, the right hand side of the latter equation is maximal at those frequencies at which

$$c + \mu(\omega_p) = 0.$$

By virtue of eq. (1.157) we obtain at these frequencies

$$m > \frac{4c\nu(\omega_p)}{\omega_p^2[b\omega_p + \nu(\omega_p)]}. \tag{1.158}$$

This condition holds provided that

$$m > \frac{4c}{\omega_p^2}.$$

Figure 1.19 shows a vibration isolator with a simplest dynamic vibration absorber attached to the intermediate mass. Let m_1, γ and β denote respectively the mass, the spring stiffness and the damping factor of the absorber. It is easy to establish that the force produced by the absorber is

$$U(t) = \frac{m_1p^2(\gamma + \beta p)}{m_1p^2 + \beta p + \gamma} z(t). \tag{1.159}$$

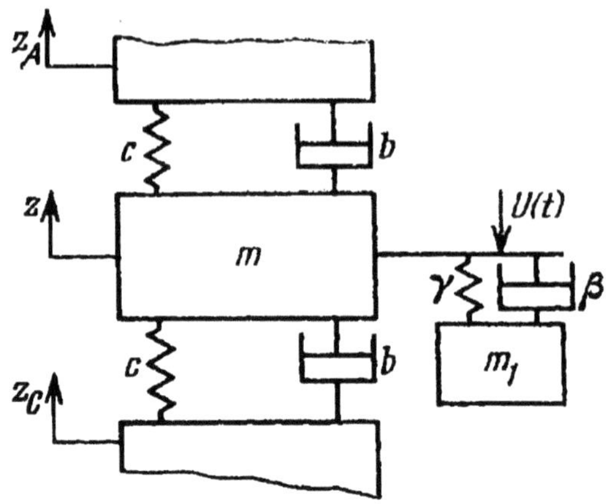

FIGURE 1.19.

Clearly, this force acts on the intermediate mass. Let us estimate the absorber efficiency. To this end, we construct the governing equation for the intermediate mass which is as follows

$$mp^2 z = P - S - U, \tag{1.160}$$

where P and S are forces applied to this mass from the upper and lower cascades of the vibration isolators, respectively. Substituting eq. (1.160) into (1.159), yields

$$S = P - mp^2 \left[1 + \frac{m_1(\gamma + \beta p)}{m(m_1 p^2 + \beta p + \gamma)}\right] z. \tag{1.161}$$

This means that the attachment of the vibration absorber is equivalent to replacement of mass m in the equation of motion by the following operator

$$m_*(p) = m + \frac{m_1(\gamma + \beta p)}{m_1 p^2 + \beta p + \gamma}. \tag{1.162}$$

The additional "complex-valued mass"

$$\Delta m = m_*(iw) - m = \frac{m_1(\gamma + \beta i\omega)}{\gamma - m_1\omega^2 + \beta i\omega}$$

provides maximum effect at frequency $\omega_1 = \sqrt{\gamma/m_1}$. Also, $\beta = 0$ implies an infinite additional mass at the above frequency. At this frequency the absolute value of the equivalent mass is given by

$$|m_1(i\omega)| = \left|m + m_1 - \frac{m_1^2\omega_1}{\beta} i\right| = \sqrt{(m + m_1)^2 + m_1^2 \frac{\gamma^2}{\beta^2\omega_1^2}} =$$

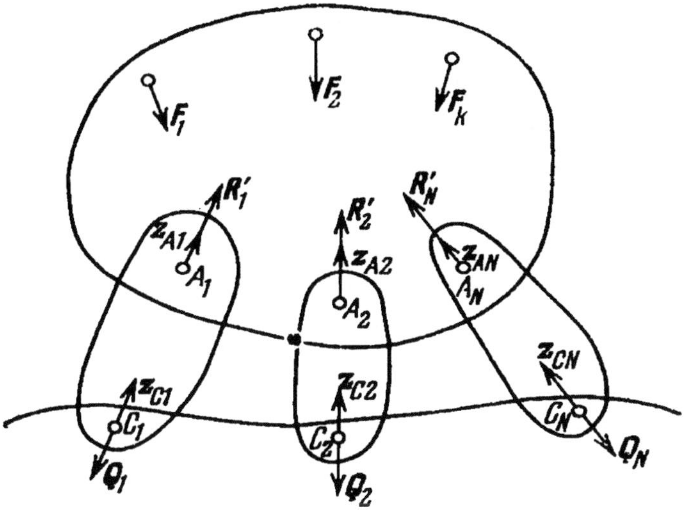

FIGURE 1.20.

$$= \sqrt{(m+m_1)^2 + m_1^2\left(\frac{m_1\omega_1}{\beta}\right)^2}. \tag{1.163}$$

When $m_1 \ll m$, the reduction effect is considerable only if $\gamma \gg \beta\omega_1$, i.e. if $m_1\omega_1 \gg \beta$.

9. We consider now a system which consists of N uniaxial vibration isolators attached to the object at points $A_1, ..., A_N$ and to the base to be isolated at points $C_1, ..., C_N$, see Fig. 1.20. Let $\boldsymbol{\zeta}(t)$ denote the vector composed of projections of the displacements of points A_k $(k = 1, ..., N)$ of the free object onto the directions of the vibration isolators' reactions. Let $\boldsymbol{\eta}(t)$ denote those of the same object rigidly attached to the base. If $\mathbf{R}(t)$ stands for the vector of reaction in the latter case, one obtains

$$\boldsymbol{\eta}(t) = E_A(p)\mathbf{R}(t) + \boldsymbol{\zeta}(t), \tag{1.164}$$

where $E_A(p)$ is the matrix of the dynamic compliance operator of the object at the attachment points. The displacements of points A_k and C_k coincide in the case of the rigid attachment. Let $\mathbf{R}(t)$ denote the vector of forces acting on the base which results in

$$\boldsymbol{\eta}(t) = -E_0(p)\mathbf{R}(t). \tag{1.165}$$

Here $E_0(p)$ is the matrix of the dynamic compliance operator of the base at points C_k. Equations (1.164) and (1.165) give

$$\boldsymbol{\zeta}(t) = -[E_0(p) + E_A(p)]\mathbf{R}(t). \tag{1.166}$$

Let the vibration isolators be mounted between the base and the object. Let $\mathbf{z}_A(t)$ and $\mathbf{z}_C(t)$ denote the displacement vectors at points A_k and C_k, respectively, while $\mathbf{R}'$ and $\mathbf{Q}$ denote the vectors of forces applied to the object and to the base at the above points, respectively. Then

$$\left.\begin{aligned} \mathbf{z}_A &= \boldsymbol{\zeta}(t) + E_A(p)\mathbf{R}'(t)\,, \\ \mathbf{z}_C &= E_0(p)\mathbf{Q}(t)\,. \end{aligned}\right\} \tag{1.167}$$

Relationship (1.138) holds for any uniaxial vibration isolator. The vector form of this relationship is as follows

$$\left.\begin{aligned} \mathbf{R}' &= \Gamma_{11}(p)\mathbf{z}_A + \Gamma_{12}(p)\mathbf{z}_C\,, \\ \mathbf{Q} &= \Gamma_{21}(p)\mathbf{z}_A + \Gamma_{22}(p)\mathbf{z}_C\,, \end{aligned}\right\} \tag{1.168}$$

where $\Gamma_{ij}(p)$ $(i,j,=1,2)$ are diagonal $N \times N$ matrices, operators $\gamma_{ij}(p)$ being their entries. Let us determine the operator which relates vectors $\mathbf{Q}(t)$ and $\mathbf{R}(t)$. It follows from eqs. (1.165) and (1.167) that

$$\mathbf{Q}(t) = E_0^{-1}(p)\mathbf{z}_C(t)\,, \quad \mathbf{R}(t) = -E_0^{-1}(p)\boldsymbol{\eta}(t)\,. \tag{1.169}$$

Thus, it is sufficient to determine the operator relating $\mathbf{z}_C(t)$ and $\boldsymbol{\eta}(t)$, and then reverse its sign.

Substituting eq. (1.168) into eq. (1.167) yields

$$\left.\begin{aligned} \mathbf{z}_A &= \boldsymbol{\zeta} + E_A(p)[\Gamma_{11}(p)\mathbf{z}_A + \Gamma_{12}(p)\mathbf{z}_C], \\ \mathbf{z}_C &= E_0[\Gamma_{21}(p)\mathbf{z}_A + \Gamma_{22}(p)\mathbf{z}_C]. \end{aligned}\right\} \tag{1.170}$$

One obtains from these equations that

$$\mathbf{z}_C = \{(I - E_A\Gamma_{11})(E_0\Gamma_{21})^{-1}(I - E_0\Gamma_{22}) - E_A\Gamma_{12}\}^{-1}\boldsymbol{\zeta}. \tag{1.171}$$

Expressing $\boldsymbol{\zeta}$ in terms of $\boldsymbol{\eta}$, using eqs. (1.165) and (1.166) and substituting the result into eq. (1.171) gives

$$\mathbf{z}_C = \{(I - E_A\Gamma_{11})(E_0\Gamma_{21})^{-1}(I - E_0\Gamma_{22}) - E_A\Gamma_{12}\}^{-1}(E_0 + E_A)E_0^{-1}\boldsymbol{\eta}. \tag{1.172}$$

Accounting for eq. (1.169) we obtain

$$\mathbf{Q}(t) = K_Q(p)\mathbf{R}(t)\,,$$

wherein

$$K_Q(p) = -\{(I - E_A\Gamma_{11})(E_0\Gamma_{21})^{-1}(I - E_0\Gamma_{22}) - E_A\Gamma_{12}\}^{-1}(E_0 + E_A)E_0^{-1}. \tag{1.173}$$

This operator relates the vector of forces acting on the protected base with that in the case of rigid attachment. The efficiency of the system of vibration protection is also defined by the norm of matrix $K_Q(i\omega)$.

Provided that all vibration isolators are governed by eq. (1.141), we have

$$\begin{aligned} \Gamma_{11}(p) &= \Gamma_{22}(p) = -W(p), \; \Gamma_{12}(p) = \Gamma_{21}(p) = W(p), \\ \mathbf{R}' &= -\mathbf{Q}, \end{aligned} \qquad (1.174)$$

where $W(p)$ is a diagonal matrix with components $w^{(k)}(p)$ $(k = 1, ...N)$. Moreover,

$$K_Q(p) = -\{(I + E_A W)W^{-1}E_0^{-1}(I + E_0 W) - E_A W\}^{-1}(E_0 + E_A)E_0^{-1}.$$

Expanding the latter expression yields

$$K_Q(p) = -[W^{-1}(p) + E_0(p) + E_A(p)]^{-1}[E_A(p) + E_0(p)]. \qquad (1.175)$$

10. To estimate the efficiency of the vibration protection system which consists of a number of vibration isolators it is necessary to compare the displacement vectors for rigid and vibration-protected attachments. When the attachment is rigid, the following equations, which are analogous to eqs. (1.144), are valid

$$\left. \begin{aligned} \boldsymbol{\zeta}(t) &= \boldsymbol{\psi}(t) - E_0(p)\mathbf{R}(t), \\ \boldsymbol{\zeta}(t) &= E_A(p)\mathbf{R}(t), \end{aligned} \right\} \qquad (1.176)$$

with $\boldsymbol{\psi}(t)$ being the vector of displacement of points C_k in the absense of the object. Equation (1.176) yields

$$\boldsymbol{\psi}(t) = [E_0(p) + E_A(p)]E_A^{-1}(p)\boldsymbol{\zeta}(t). \qquad (1.177)$$

For a vibration-protected system

$$\left. \begin{aligned} \mathbf{z}_A &= E_A(p)\mathbf{R}'(t), \\ \mathbf{z}_C &= \boldsymbol{\psi}(t) + E_0(p)\mathbf{Q}(t). \end{aligned} \right\} \qquad (1.178)$$

By virtue of eqs. (1.168) and (1.174)-(1.176) one can express $\mathbf{z}_A$ in terms of $\boldsymbol{\zeta}$, giving

$$\begin{aligned} &\mathbf{z}_A(t) = \\ &\{(I - E_0\Gamma_{22})(I - E_A\Gamma_{11})^{-1}E_A\Gamma_{12} - E_0\Gamma_{21}\}^{-1}(E_0 + E_A)E_A^{-1}\boldsymbol{\zeta} \\ &= K_z(p)\boldsymbol{\zeta}(t). \end{aligned} \qquad (1.179)$$

Provided that eq. (1.174) holds, one obtains

$$K_z(p) = E_A(p)[E_0(p)+E_A(p)+W^{-1}(p)]^{-1}[E_A(p)+E_0(p)]E_A^{-1}(p). \quad (1.180)$$

The efficiency conditions at frequency ω can be established in the form of conditions imposed on the norm of matrix $K_z(i\omega)$.

1.4 Efficiency in case of polyharmonic, random and non-stationary excitations

As shown in the previous Section, the efficiency of any one-dimensional facility of vibration protection (vibration absorber or vibration isolator) may be described by an operator $k(p)$, the latter relating the parameters of the vibration field of concern, e.g. displacement or force, before and after this facility has been mounted. The relevant formulae (1.92), (1.139) and (1.147) may be written in a general form as follows

$$z'(t) = K(p)z(t), \quad (1.181)$$

where $z(t)$ and $z'(t)$ denote one of the above parameters of the system without and with a vibration protection facility, respectively. A similar dependence

$$\mathbf{z}'(t) = K(p)\mathbf{z}(t), \quad (1.182)$$

replacing eqs. (1.123), (1.171) and (1.179) may be derived for multidimensional systems.

As has been shown above, the efficiency factor can be introduced for the system driven harmonically at frequency ω. The efficiency factor is a value which characterizes the efficiency of the vibration protection. Either the absolute value of $k(i\omega)$ or the norm of the matrix $K(i\omega)$ can be taken as this factor.

It is known however that the vibration excitations in actual systems are usually not single-frequency harmonic processes. More often, they can be modelled by polyharmonic or random functions of time. Moreover, dealing with engineering systems, one has to work with non-stationary processes caused by impacts and transient processes. The intent of this Section is to introduce some efficiency criteria for vibration protection facilities which are applicable for the above cases. The reader is assumed to be familiar with the fundamental concepts of the theory of random functions, as well as with those of Fourier and Laplace transformations.

Let the dynamic excitations be some polyharmonic functions of time with frequencies $\omega_1, \omega_2, ..., \omega_r$ and the vibration protection facility be a linear one-dimensional one. The steady-state responses $z(t)$ and $z'(t)$ will be some polyharmonic processes, too. Namely, provided that

$$z(t) = \sum_{m=1}^{r} a_m \cos(\omega_m t + \psi_m), \tag{1.183}$$

the superposition principle results in

$$z'(t) = \sum_{m=1}^{r} |k(i\omega_m)| \, a_m \cos[\omega_m t + \psi_m + \arg k(i\omega_m)]. \tag{1.184}$$

A vibration protection facility can be considered as efficient for each harmonic if the amplitude of each harmonic in $z'(t)$ is smaller than that in $z(t)$. We thus come to the following efficiency criterion

$$k_0 = \sup_m |k(i\omega_m)| < 1. \tag{1.185}$$

Provided that this condition is satisfied, the vibration protection facility is efficient for any harmonic.

It is however possible to introduce another efficiency criteria. An important characteristic of the polyharmonic process is known to be the root-mean-square. The variances of processes (1.183) and (1.184) are, see [65]

$$\sigma_z^2 = \frac{1}{2} \sum_{m=1}^{r} a_m^2, \quad \sigma_{z'}^2 = \frac{1}{2} \sum_{m=1}^{r} |k(i\omega_m)|^2 \, a_m^2. \tag{1.186}$$

The vibration protection facility is called *efficient with respect to the root-mean-square* if

$$k_\sigma = \frac{\sigma_{z'}}{\sigma_z} < 1. \tag{1.187}$$

The system which is effective with respect to each harmonic is apparently effective in the root-mean-square, the converse statement being wrong.

Assume now that the dynamic excitations are some stationary random processes which implies the processes $z(t)$ and $z'(t)$ are stationary random processes, too. The spectral densities of these processes, $S_z(\omega)$ and $S_{z'}(\omega)$ are related as follows, see [128]

$$S_{z'}(\omega) = |k(i\omega)|^2 \, S_z(\omega). \tag{1.188}$$

The vibration protection facility is *effective with respect to the spectral density* if for all frequencies at which $S_z(\omega) \neq 0$ the following condition holds

$$S_{z'}(\omega) < S_z(\omega).$$

This leads to the efficiency condition in the form

$$k_s = \max_{\omega} |k(i\omega)| < 1. \tag{1.189}$$

The mathematic expectations and the variances of $z\,(t)$ and $z'\,(t)$ are given by

$$m_{z'} = k\,(0)\, m_z, \tag{1.190}$$

$$\sigma_z^2 = \frac{1}{2\pi} \int_{-\infty}^{+\infty} S_z(\omega) d\omega,\ \sigma_{z'}^2 = \frac{1}{2\pi} \int_{-\infty}^{+\infty} |k(i\omega)|^2\, S_z(\omega) d\omega. \tag{1.191}$$

Provided that

$$m_{z'} < m_z$$

or

$$\sigma_z^2 < \sigma_{z'}^2$$

the protection facility is *effective with respect to the mathematic expectation or mean square*, respectively.

It is worth noting that contrary to the other criteria introduced the efficiency with respect to the root-mean-square depends not only upon the properties of the vibration protection system but also upon the parameters of the excitation.

Let us consider now a system subject to an impact excitation and study the motion under the trivial initial condition. Let $z\,(t)$ and $z'\,(t)$ denote the vibration field parameter to be reduced before and after a vibration protection facility has been mounted, respectively. Clearly, $z\,(t)$ and $z'\,(t)$ are non-stationary processes. An important property of non-stationary processes are frequently their peak values

$$h_z = \max_t |z\,(t)|\,,\ h_{z'} = \max_t |z'\,(t)|\,. \tag{1.192}$$

If $h_z < h_{z'}$, the vibration protection system is effective with respect to the peak values. This efficiency criterion is difficult to relate to both function $k\,(p)$ and the impact parameters. Nonetheless, one has to make use of this criterion since reducing the peak values is the principal goal in the impact protection problem.

The following quotient

$$k_J = \frac{J'}{J}, \tag{1.193}$$

where

$$J = \int_0^{+\infty} z^2(t)\,dt, \quad J' = \int_0^{+\infty} [z'(t)]^2\,dt$$

may serve as another efficiency criterion. If $k_J < 1$, the system is effective with respect to the integral over the square value of the parameter. For the Fourier transforms

$$Z(\omega) = \int_0^{+\infty} z(t)\,e^{-i\omega t}dt, \quad Z'(\omega) = \int_0^{+\infty} z'(t)\,e^{-i\omega t}dt = k(i\omega)\,Z(\omega),$$

Parseval's theorem, [140], yields

$$J(\omega) = \frac{1}{\pi}\int_0^{+\infty} |Z(\omega)|^2\,d\omega, \quad J'(\omega) = \frac{1}{\pi}\int_0^{+\infty} |Z'(\omega)|^2\,d\omega.$$

These relations show that the system which is effective with respect to the spectral density ($k(i\omega) < 1$) is effective with respect to the integral over the square value, too.

The above criteria may be generalised to multidimensional systems. If the norm of matrix $K(i\omega)$ is due to eq. (1.131), the following condition

$$N_K = \sup_m k_0^2(\omega_m) < 1 \tag{1.194}$$

provides the efficiency with respect to each harmonic while the condition

$$N_s = \max_\omega k_0(\omega) < 1 \quad (0 < \omega < \infty)$$

ensures the efficiency with respect to root-mean-square.

2
Linear active systems

2.1 Structure and elements of active systems

2.1.1 Transfer functions of active systems

As mentioned in Chapter 1, the purpose of any vibration protection system is to produce control forces, i.e. the forces which act on the object to be protected and change its vibration field. Systems generating a single control force are referred to as *one-dimensional* while systems generating a number of control forces are called *multidimensional.*

Figure 2.1 shows the block diagram of a multidimensional active vibration protection system. Following from the ideas of control theory, the object to be protected is termed a control object or simply an object and in some cases the object may include passive vibration protection systems. The parameters of the vibration fields, (absolute and relative displacements, velocities, accelerations at some points, forces and stresses in the structural elements etc.), external forces or kinematic excitations play the role of an input to the *controller.* In the first case, the active system is a closed-loop system possessing control due to deviations while in the second case an active system has a control due to disturbances. The input variables are measured by *sensors* and transformed into a convenient form. The information from sensors is transformed and acts as an input to a compensator. The output of the compensator is amplified by an amplifier and acts as an input for the *actuator*, the latter forming the control force. Additional circuits, depicted by dotted lines in Fig. 2.1, may exist as well. Vibration field parameters can directly influence the actuator, in other words, the control

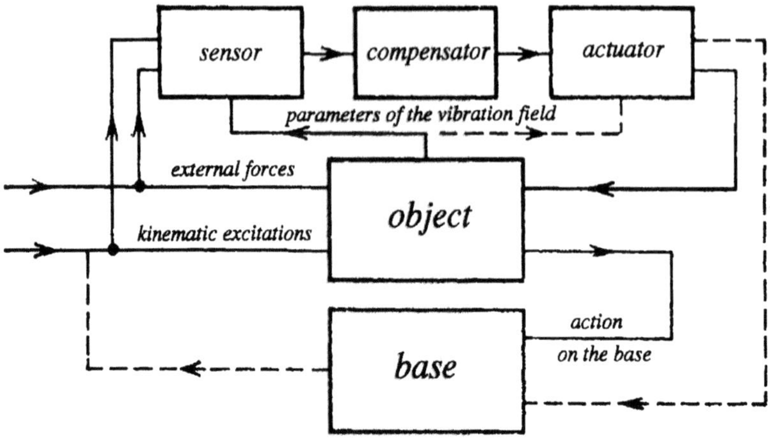

FIGURE 2.1.

forces may depend on the motion of the object. In addition to this, the actuators which isolate the object act not only on the object but also on the base changing the kinematic excitations. Analogously, in the case when the base is to be protected from the vibrating object the control forces act not only on the base but also on the object. These additional circuits are displayed in Fig. 2.1 for the example of object isolators.

If the system is linear then each of its elements can be described by a transfer function. Figure 2.2 shows the block diagram of a linear multidimensional system, where $\mathbf{F}(t)$ and $\boldsymbol{\xi}(t)$ are the vectors of the forces and kinematic excitations, respectively, and $\mathbf{z}(t)$ denotes the vector of the vibration field parameters measured. Transfer function matrices W_{sF}, $W_{s\xi}$ and W_{sz} characterise the properties of the corresponding linear sensors. These matrices are diagonal provided that each component of the vectors $\mathbf{F}(t)$, $\boldsymbol{\xi}(t)$ and $\mathbf{z}(t)$ is measured independently. In some cases, for instance, when the angles of rotation are reconstructed in terms of measured displacements of particular points, these matrices may have a more complicated structure. The outputs of the sensors are signals which form a vector $\boldsymbol{\sigma}$. These signals are summed, transformed (e.g. integrated or differentiated) and amplified within a compensator, all these linear transformations being describes by matrix $W_c(p)$. Vector $\mathbf{u}$ which is the compensator output acts as the input to the actuator, furthermore, $\mathbf{z}(t)$ may acts directly as an additional input to the actuator. Transfer function $W_a(p)$ relates the input variables of the actuator to the control forces $U(t)$ acting on the object (or base) with the transfer function $W_0(p)$ and forces $R(t)$ applied to the base (or object). Transfer function $W_1(p)$ relates the excitation acting on the base to the kinematic excitations.

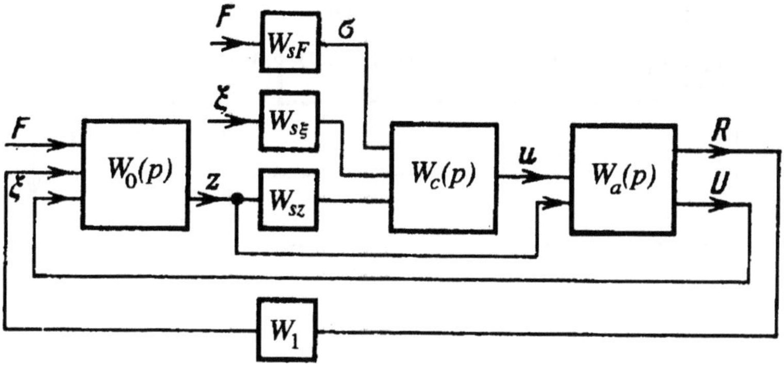

FIGURE 2.2.

Figure 2.3 displays a block diagram of a one-dimensional system whose control force $U(t)$ is formed by a single output $z(t)$ and the base is assumed to be rigid. The governing equation of motion for the object is given by

$$z = \sum_{i=1}^{k} w_{0i}^{(F)}(p)F_i + \sum_{j=1}^{s} w_{0j}^{(\xi)}(p)\xi_j + w_0^{(U)}(p)U, \tag{2.1}$$

where F_i and ξ_j are components of the dynamic excitations, and $w_{0i}^{(F)}$, $w_{0j}^{(\xi)}$ and $w_0^{(U)}$ are the rows of the transfer functions of the object. The transfer function relating $U(t)$ and $z(t)$ is as follows

$$\begin{aligned} U &= w_a^{(u)}(p)u + w_a^{(z)}(p)z \\ &= [w_a^{(u)}(p)w_c(p)w_{sz}(p) + w_a^{(z)}(p)]z = -w_z(p)z. \end{aligned} \tag{2.2}$$

In what follows, we refer to $w_z(p)$ as the *feedback transfer function.* A minus sign appears in eq. (2.2) since feedbacks are ordinarily negative. Inserting eq. (2.2) into (2.1) yields

$$z = \sum_{i=1}^{k} w_{0i}^{(F)}(p)F_i + \sum_{j=1}^{s} w_{0j}^{(\xi)}(p)\xi_j - w_0^{(U)}(p)w_z(p)z. \tag{2.3}$$

The function

$$\Phi(p) = w_0^{(U)}(p)w_z(p) \tag{2.4}$$

is treated as the transfer function of an open-loop system. It follows from eq. (2.3) that

$$z = [1 + \Phi(p)]^{-1} \left[\sum_{i=1}^{k} w_{0i}^{(F)}(p)F_i + \sum_{j=1}^{s} w_{0j}^{(\xi)}(p)\xi_j \right]. \tag{2.5}$$

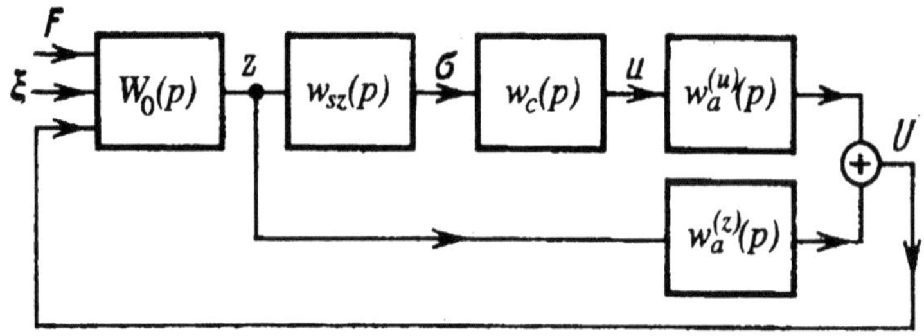

FIGURE 2.3.

The function

$$k_z(p) = [1 + \Phi(p)]^{-1} \tag{2.6}$$

is the transfer function of a closed-loop system whereas the following functions

$$w_i^{(F)}(p) = k_z(p) w_{0i}^{(F)}(p), \; w_j^{(\xi)}(p) = k_z(p) w_{0j}^{(\xi)}(p) \tag{2.7}$$

are the system transfer functions due to corresponding disturbances.

If there were no active facility in the system, the following expression for variable z would hold

$$z = \sum_{i=1}^{k} w_{0i}^{(F)}(p) F_i + \sum_{j=1}^{s} w_{0j}^{(\xi)}(p) \xi_j, \tag{2.8}$$

which could be obtained from eq. (2.1) by letting $U = 0$. Comparing eqs. (2.8) and (2.5) we find that operator $k_z(p)$, which determines the change in the output variable z after introducing a controller, characterises the active facility efficiency with respect to this variable. As shown in Section 1.3 transfer function $w_0^{(U)}$ coincides with one of the dynamic compliance operators provided that the control object is the mechanical system treated in Section 1.2.

We consider now the dynamic characteristics of the elements utilised in active systems. Taking into account that detailed descriptions of these and their properties is adequately covered in the vast literature on this topic, see e.g. [124], [106] and [126], we limit ourselves to a brief overview of the foundations.

2.1.2 *Sensors*

Measurement of relative displacements

Any electric displacement gauge, e.g. potentiometer, strain-gauge, inductive or capacitive one etc., can be used for measuring displacement of a

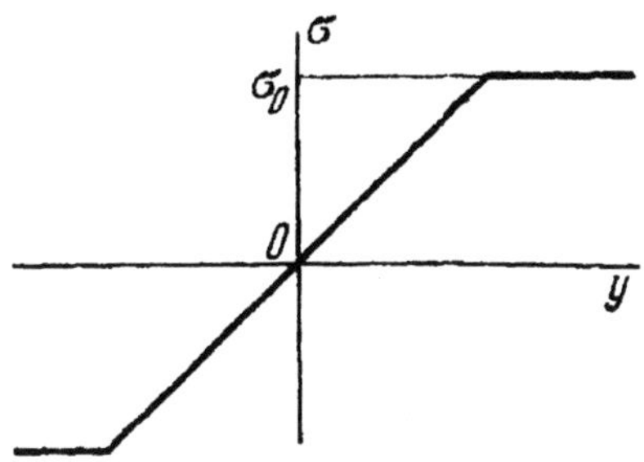

FIGURE 2.4.

particular point of the object relative to the base or any other point. Relative displacement results in a change in a certain parameter of the electric circuit, e.g. resistance, inductance or capacitance, which in turn changes the output variable, say, voltage or current.

Displacement gauges are commonly used in vibration protection systems for measuring processes with relatively low frequencies. In this case, elastic deformations of the gauge component are usually negligible which allows the gauge to be viewed as a massless element with the following characteristic

$$\sigma = f(y),$$

where y is the displacement to be measured and σ is the output variable. As a rule, function $f(y)$ is linear for small y such that

$$\sigma = k_y y. \tag{2.9}$$

Nonlinearity of the characteristic is to be taken into account for considerable relative displacements which are typical for resonant regimes. The dependence $\sigma(y)$ often has the form shown in Fig. 2.4, where the output value is bounded and can not exceed σ_0.

Measurement of absolute displacements

Absolute displacements and absolute accelerations are measured by means of transducers of seismic type. Design solutions are very different, see to this aim [48], however in all cases the output electric signal depends on displacement of the "seismic" mass relative to the transducer housing, see Fig 2.5, this dependence being linear for small displacements. It is easy to show that the relative displacement of mass m_s is related to the absolute displacement of the housing as follows

$$\Delta(t) = \frac{m_s p^2}{m_s p^2 + b_s p + c_s} x(t) = \frac{T_s^2 p^2}{T_s p^2 + 2\zeta_s T_s p + 1} x(t), \tag{2.10}$$

where c_s and b_s are the rigidity and the resistance factor, respectively, while T_s and ζ_s are the time constant and the damping factor, respectively,

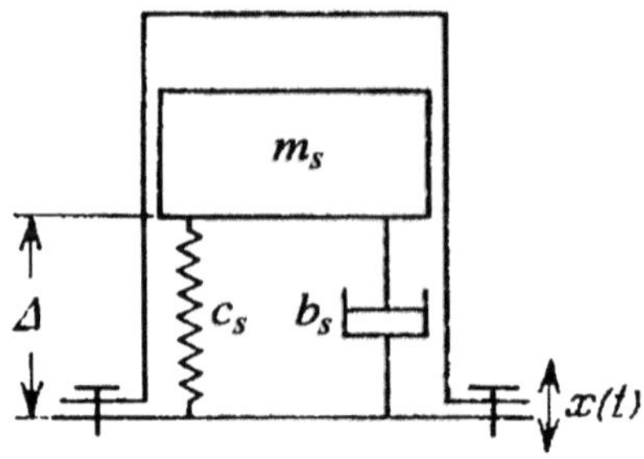

FIGURE 2.5.

$$T_s = \sqrt{\frac{m_s}{c_s}}, \ \zeta_s = \frac{b_s}{2\sqrt{c_s m_s}}, \ k_s = k_\sigma \frac{m_s}{c_s}. \tag{2.11}$$

Therefore, the transfer function of the transducer measuring absolute displacement is given by

$$w_{sx}(p) = \frac{k_\sigma T_s^2 p^2}{T_s^2 p^2 + 2\zeta_s T_s p + 1} = \frac{k_s p^2}{T_s^2 p^2 + 2\zeta_s T_s p + 1}. \tag{2.12}$$

Here k_σ denotes the proportionality coefficient between the output signal σ and the seismic mass displacement $\Delta(t)$.

For the accelerometer which measures absolute acceleration $\ddot{x}(t)$ we obtain

$$w_{s\ddot{x}}(p) = \frac{k_s}{T_s^2 p^2 + 2\zeta_s T_s p + 1}. \tag{2.13}$$

Let us recall that, due to the properties of the resonance characteristics $|w_{sx}(i\omega)|$ and $|w_{s\ddot{x}}(i\omega)|$, a transducer measuring absolute displacement can only be used for measuring high frequency processes $\left(\omega > 6\pi T_s^{-1}\right)$ while an accelerometer is used at relatively low frequencies $\left(\omega < 2\pi T_s^{-1}\right)$. For this reason, more often than not accelerometers are utilised in active systems.

Measurement of rotation angles

In some cases, there is a need to measure the rotation of the object which is treated as a rigid body. Relative rotations can be measured indirectly, for this purpose, the relative displacements at several points of the object are measured and then divided by the distance between these points. Devices for measuring the relative displacements can be deemed as massless. For small rotation angles, one gets

$$\sigma = k_\varphi \varphi. \tag{2.14}$$

For measuring absolute angles of rotation and accelerations, torsional vibration gauges are used. Their transfer functions are similar to those given by eqs. (2.12) and (2.13). Gyroscopic transducers, which are single-degree-of-freedom gyros with an elastic element for compensating gyroscopic moment and a damper, are used for measuring low-frequency angles of rotation. The transfer function of these transducers may be represented in the following form

$$w_{g\theta} = \frac{k_g p}{T_g^2 p^2 + 2\zeta_g T_g p + 1}. \qquad (2.15)$$

Here

$$\sigma = w_{g\theta}(p)\,\theta,$$

with θ denoting the absolute angle of rotation. The sensitivity of the transducer is dependent on the gain k_g which is proportional to the angular velocity of the gyrowheel.

Measurement of forces and stresses

In order to measure the forces acting on the objects or the forces in vibrating components of the object diverse load gauges are used. They are usually strain-gauge transducers and may be considered as being massless linear elements for which

$$w_{lF}(p) = k_F = const. \qquad (2.16)$$

The transducers measuring stresses possess the same characteristic.

2.1.3 Compensators and amplifiers

Integrators and differentiators

Signals are frequently integrated and differentiated in control systems. Integration of signal proportional to relative displacement is performed for example for creating astatic systems, while differentiating is utilised for introducing damping into the system.

RC-circuits are used as integrators and differentiators, [126]. Figure 2.6 shows a simple *RC*-circuit with the following transfer function

$$w_c(p) = \frac{R}{R + (Cp)^{-1}} = \frac{RCp}{1 + RCp} = \frac{T_c p}{1 + T_c p}, \qquad (2.17)$$

where $T_c = RC$ is the time constant. If the signal frequency ω satisfies the condition $T_c\omega \ll 1$, this circuit is close to being ideally integrating. Thus, *RC*-circuits differentiate low frequency signals.

The transfer function of the circuit in Fig. 2.6 is given by

$$w_c(p) = \frac{(Cp)^{-1}}{R + (Cp)^{-1}} = \frac{1}{1 + CRp} = \frac{1}{1 + T_c p}. \qquad (2.18)$$

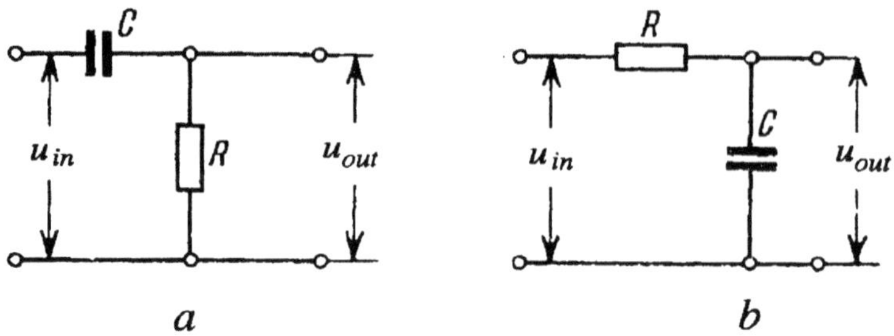

FIGURE 2.6.

If $T_c\omega \gg 1$ this circuit becomes an integrating one. It is clear that the time constant T_c of the integrator must increase with decreasing signal frequency. For this reason, when integrating very low frequency signals which is often the case in vibration protection systems, it is expedient to use electronic DC amplifiers with capacitive feedback. For large gains, this amplifier is an ideal integrator with the following transfer function

$$w_c(p) = -(T_c p)^{-1}. \tag{2.19}$$

Notice that in the case of harmonic input, the phase of a differentiator output leads the input phase by $\pi/2$ whereas the phase of an integrator output lags behind the input phase by $\pi/2$. This is the reason for referring to integrators and differentiators as lead and delay circuits, respectively.

Active vibration protection systems can also utilise some mechanical devices which combine the functions of sensors and compensators. Such a mechanical facility consisting of a spring c and a linear damper b is shown in Fig. 2.7. Values of σ and σ_1 are related to $y\,(t)$ by the following expressions

$$\sigma \quad = \quad \frac{bp}{c+bp}y = \frac{T_c p}{1+T_c p}y,$$

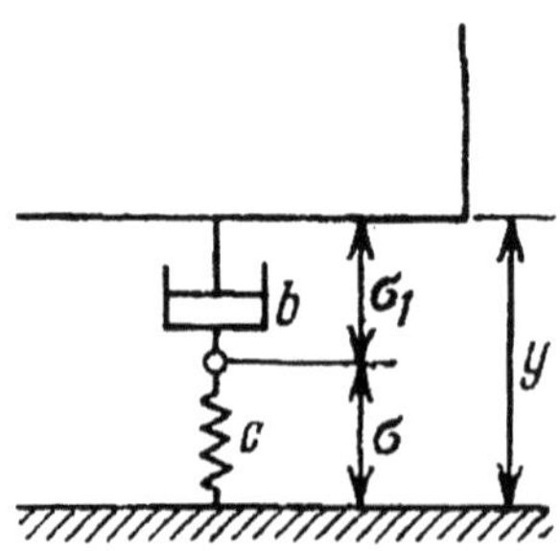

FIGURE 2.7.

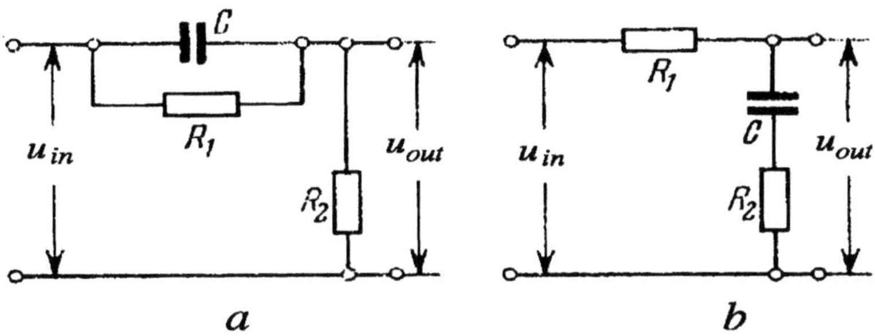

FIGURE 2.8.

$$\sigma_1 = \frac{bp}{c+bp}y = \frac{1}{1+T_c p}y, \tag{2.20}$$

where $T_c = b/c$. Therefore, this system can be used as a differentiator and as an integrator.

Frequency filters

In active systems there is often a need to avoid feedback signals whose frequencies lie in a certain frequency band. In this case, frequency filters are used. Figure 2.8a shows a RC-circuit which is a high frequency filter. The output voltage u_{out} is related to the input voltage u_{in} as follows

$$u_{out} = \frac{R_2}{R_2 + (R_1^{-1} + Cp)^{-1}} u_{in} = \frac{R_2}{R_1 + R_2} \frac{1 + T_1 p}{1 + T_2 p} u_{in} = w_c(p) u_{in}, \tag{2.21}$$

where $T_1 = CR_1$ and $T_2 = CR_1R_2/(R_1 + R_2)$. If $R_1 \gg R_2$, then, the system does not pass signals with frequencies smaller than T_1^{-1}. If the frequency of the signal is sufficiently high, then

$$|w_c(i\omega)| \approx \frac{R_2}{R_1} \cdot \frac{T_1\omega}{T_2\omega} \approx 1.$$

Figure 2.8b shows a circuit which is a low frequency filter. Here

$$u_{out} = \frac{R_2 + (Cp)^{-1}}{R_1 + R_2 + (Cp)^{-1}} u_{in} = \frac{1 + T_1 p}{1 + T_2 p} u_{in} = w_c(p) u_{in}, \tag{2.22}$$

where $T_1 = CR_2$ and $T_2 = C(R_1 + R_2)$. If $R_1 \gg R_2$, the system does not pass the signals with frequencies considerably higher than T_2^{-1}. Mechanical filters may be designed as well. For example, we obtain for the system depicted in Fig. 2.9 that

$$\sigma = \frac{b_1 p + c}{(b + b_1)p + c} z = \frac{1 + T_2 p}{1 + T_1 p} z, \tag{2.23}$$

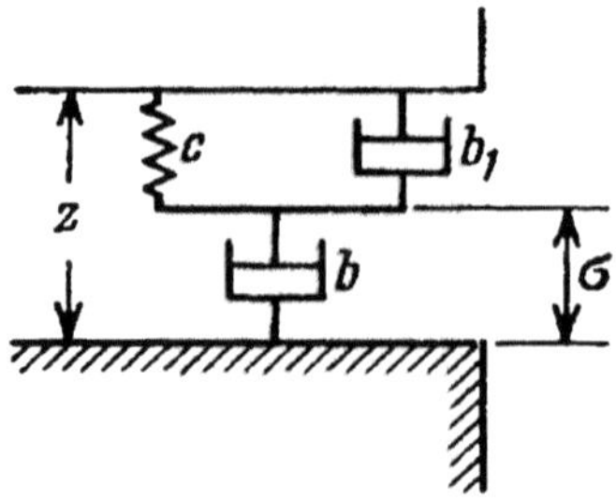

FIGURE 2.9.

where $T_1 = (b + b_1)/c$ and $T_2 = b_1/c$. If $b \gg b_1$, then $T_1 \gg T_2$ and filter (2.23) does not pass high frequencies.

Amplifiers

DC and AC amplifiers are used to amplify electric signals. At the frequencies relevant to active vibration protection systems the amplifiers may be considered as being non-inertial with the following transfer function

$$w_c(p) = k. \tag{2.24}$$

Mechanic, pneumatic, hydraulic and pneumomechanic amplifiers can be involved in active systems as well.

2.1.4 Actuators

The actuators form the control forces, and, for this reason, they are the most important elements of active systems. It is the actuator which determines a particular type of active vibration protection systems. Depending upon the type of structure, there exist *hydraulic, pneumatic, electromechanical* and *electromagnetic* systems.

A governing law for load $U(t)$ depends not only on the input signal $u(t)$ but also on the motion of the object. Thus, in addition to the envisaged feedback, there appears an additional link between the object and the actuator described by transfer function $w_a^{(z)}(p)$ in eq. (2.2).

In the majority of cases, actuators are nonlinear systems with complex dynamic characteristics. A linearisation of these characteristics is feasible under some simplifying assumptions. The applicability of linear models is bounded and dependent on the validity of basic assumptions.

In this Chapter we study linearised characteristics of actuators while account of some nonlinear factors is considered in Chapter 5.

Equations for dynamic characteristics of the most popular actuators are derived in what follows.

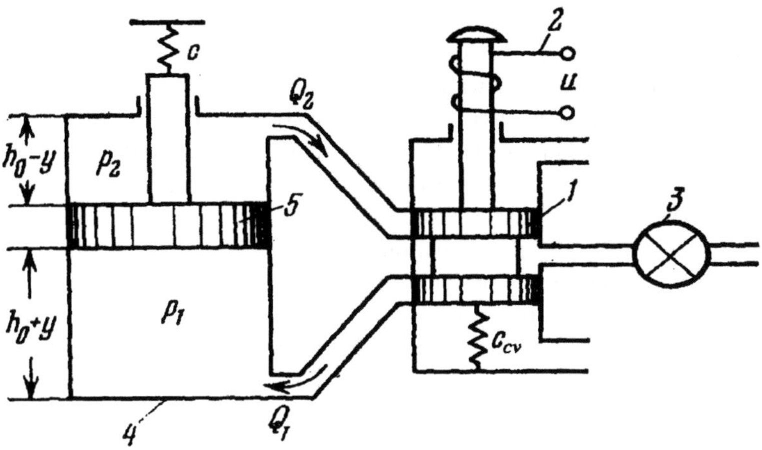

FIGURE 2.10.

Hydraulic actuators

One of the simplest hydraulic actuators is schematically displayed in Fig. 2.10. Displacement of control valve 1 is due to electromagnet 2 which in turn is activated by the input signal $u(t)$. This facility, in which the control valve motion is due to an electric signal, is called *electrohydraulic*. Displacement of the control valve causes the working fluid to enter into one of the chambers of cylinder 4, the other chamber being connected via the control valve to an outlet discharge pipe. The pressure in the working fluid is provided by pump 3. The pressure difference in the chambers produces a force $U_*(t)$ which acts on piston 5 and the object connected to the piston. The relationship between the input voltage $u(t)$ and the control valve displacement $v(t)$ can be cast in the form

$$v = \frac{k_v}{m_{cv}p^2 + b_{cv}p + c_{cv}}u = \frac{k_v}{c_{cv}}\frac{1}{T_{cv}^2 p^2 + 2\zeta_{cv}T_{cv}p + 1}u, \qquad (2.25)$$

where m_{cv}, c_{cv} and b_{cv} are the mass of the control valve, the rigidity of its elastic element (e.g. rigidity of the compressive working fluid) and the resistance factor, respectively, and k_v denotes the factor relating the electromagnet force to the input voltage. It is worthwhile mentioning that, in general, any electromagnet is a vibratory system. However, its time constant is ordinarily much smaller than T_{cv}, and thus can be neglected. Dependence v versus u may deviate from that given by eq. (2.25). For instance, in [106] it is suggested that a control valve facility consisting of an ideal delay circuit, non-inertial linear amplifier and a limiter in series should be adopted.

We proceed now to deriving the piston dynamic characteristic which relates the force acting on the piston to the piston displacement. To this aim, we consider a dynamic model of the system which takes into account compressibility of the working fluid and leakage via the clearance between the piston and the cylinder wall.

For the time being we assume that the working fluid enters the lower cylinder chamber whereas the upper chamber is connected to the outlet discharge pipe. Pressure in the lower and upper chambers is denoted by p_1 and p_2, respectively, S is the piston area, h_0 is the length of each chamber in the middle position of the piston, y is the piston displacement, with $y = 0$ denoting the middle position of the piston, V_1 and V_2 are chambers' volumes, Q_1 is the volume flow rate of fluid entering the lower chamber, and Q_2 is the volume flow rate of fluid leaving the upper chamber. It is clear that

$$V_1 = S(h_0 + y), \quad V_2 = S(h_0 - y), \tag{2.26}$$

and hence

$$\dot{V}_1 = -\dot{V}_2 = S\dot{y}. \tag{2.27}$$

A relative decrease in volume for a compressive fluid may be considered to be proportional to an increase in pressure

$$\Delta V = -\mu V \Delta p, \tag{2.28}$$

where μ designates the compressibility factor. If pressure and volume are time-dependent, it follows from eq. (2.28) that

$$\dot{V} = -\mu V \dot{p}. \tag{2.29}$$

Let us assume that the fluid leakage is proportional to the difference in pressure in the cylinder chambers

$$Q_l = \varkappa(p_1 - p_2), \tag{2.30}$$

where Q_l denotes the volume flow rate through clearance and $\varkappa$ is the leakage factor. Considering eqs. (2.29) and (2.30) yields

$$\left.\begin{aligned} \dot{V}_1 &= Q_1 - \mu V_1 \dot{p}_1 - \varkappa(p_1 - p_2), \\ \dot{V}_2 &= -Q_2 - \mu V_2 \dot{p}_2 + \varkappa(p_1 - p_2). \end{aligned}\right\} \tag{2.31}$$

Substituting eqs. (2.26) and (2.27) into (2.31) we obtain the following expressions for the volume flow rates

$$\left.\begin{aligned} Q_1 &= S\dot{y} + \mu S(h_0 + y)\dot{p}_1 + \varkappa(p_1 - p_2), \\ Q_2 &= S\dot{y} - \mu S(h_0 - y)\dot{p}_2 + \varkappa(p_1 - p_2). \end{aligned}\right\} \tag{2.32}$$

On the other hand, the value of volume flow rate through the control valve depends on the control valve displacement and on the difference of the input and output pressures.

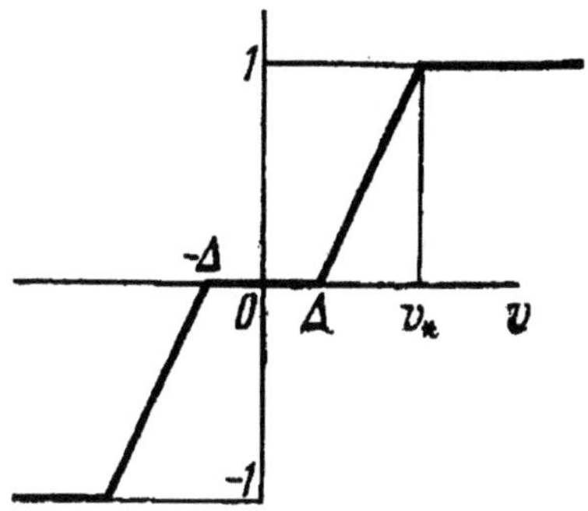

FIGURE 2.11.

Since considerable pressure differences are typical for hydraulic actuators it is common practice to assume that the volume flow rates Q_1 and Q_2 depend only on the control valve position v. The following relation is conventionally presumed

$$Q_i = f(v)\, Q_{\max} \ (i = 1, 2). \tag{2.33}$$

Here $f(v)$ is a certain, in general nonlinear function schematically shown in Fig. 2.11. In this Figure, 2Δ denotes the value of the control valve overlap which is equal to the difference between the sizes of the control valve port and the port of the cylinder in which the control valve slides. The maximum volume flow rates corresponding to the fully open port is denoted in eq. (2.33) as $Q_{\max}$. Provided that the control valve overlap is sufficiently small and $v_{\max}$ does not exceed v_* which corresponds to "saturation" in Fig. 2.11, dependence (2.33) can be treated as being linear, i.e.

$$Q_i = k_v v Q_{\max} \ (i = 1, 2). \tag{2.34}$$

Inserting eq. (2.33) into (2.32) we obtain a system of nonlinear equations relating pressures p_1 and p_2 to the piston displacement $y(t)$ and the control valve displacement $v(t)$

$$\left.\begin{aligned} Q_{\max} f(v) &= S\dot{y} + \mu S(h_0 + y)\dot{p}_1 + \varkappa(p_1 - p_2), \\ Q_{\max} f(v) &= S\dot{y} - \mu S(h_0 - y)\dot{p}_2 + \varkappa(p_1 - p_2). \end{aligned}\right\} \tag{2.35}$$

These equations can be slightly simplified if we adopt that in the regime under consideration the piston exhibits small vibrations about the middle position, i.e. $|y| \ll h_0$. Summing the equations in (2.35) yields

$$\frac{\mu h_0}{2}\dot{U}_* + \frac{\varkappa}{S}U_* = Q_{\max} f(v) - S\dot{y}, \tag{2.36}$$

where $U_* = S(p_1 - p_2)$ is the force acting on the piston. If function $f(v)$ can be linearised we obtain from eq. (2.36) the following linearised dynamic characteristic for the working cylinder

$$\frac{\mu h_0}{2}\dot{U}_* + \frac{\varkappa}{S}U_* = Q_{\max}\dot{k}_v v - S\dot{y}. \tag{2.37}$$

When the control valve is at rest in the middle position, i.e. $v = 0$, and no leakage is observed, $\varkappa = 0$, it follows from eq. (2.37) that

$$U_* = -\frac{2S}{\mu h_0}y. \tag{2.38}$$

Thus, the working fluid produces a restoring force and, in a certain sense, is equivalent to an elastic element with the following rigidity

$$c_f = \frac{2S}{\mu h_0}. \tag{2.39}$$

The dynamic characteristic of the working cylinder is significantly simplified if the fluid compressibility and the leakage through the piston clearance are neglected. Indeed, in this case $\mu = 0$ and $\varkappa = 0$, and eqs. (2.36) and (2.37) yield

$$\dot{y} = \frac{Q_{\max}}{S}f(v) \tag{2.40}$$

or

$$\dot{y} = \frac{Q_{\max}}{S}k_v v. \tag{2.41}$$

In this case, the actuator provides rather a law of the piston motion than a control force. The actuator motion is therefore independent of the properties of the control object and depends only on the control valve motion.

Using the simplified dynamic characteristics (2.40) and (2.41) is admissible for the case of an approximate investigation of low frequency processes. It follows from eq. (2.41) that the relative displacement of the piston is given by

$$y_p = \frac{Q_{\max}k_v}{Sp}v \quad \left(p \equiv \frac{d}{dt}\right). \tag{2.42}$$

In case of high frequency vibrations of the control valve, the amplitude of the relative vibration of the piston becomes small and the hydraulic facility, due to its mechanical properties, acts as a rigid element connecting the object and the base. For this reason, the properties of the vibration protection system at high frequencies turn out to be insufficient. In order to improve these properties, an elastic element is usually installed between the piston and the object. Denoting the rigidity of this element as c and the relative displacement of the object as y we obtain the following equation for the control force

$$U(t) = -c(y - y_p) = -cy + \frac{cQ_{\max}k_v}{Sp}v. \tag{2.43}$$

If any additional elastic element is absent, then, at high frequencies, it is necessary to account for the rigidity of the fluid in the cylinder given by eq. (2.39). It is easy to see that eq. (2.43) is derived from eq. (2.37) by substituting $\varkappa = 0$ and c from eq. (2.39).

More accurate calculations demand taking into account the dependences of the pressures in the cylinder chambers on volume flow rates Q_1 and Q_2. The pressure loss due local resistance in the control valve is assumed to be proportional to the square of the volume flow rate and depends upon the control valve position whereas the pressure loss in the pipes is linear to the volume flow rate.

Equation (2.37) determines the force acting on the piston. In order to obtain the control force $U(t)$, we account for the piston inertia force and the resistance force, to get

$$U(t) = U_*(t) - m_p\ddot{x}_p - b_p\dot{y}. \tag{2.44}$$

Here m_p denotes the equivalent mass of the piston and elements attached to the piston, $\ddot{x}_p$ is the absolute acceleration of the piston, and b_p is a factor taking into account the resistance force. A dry friction force proportional to $\operatorname{sign}\dot{y}$ is considered in some cases. As mentioned above, a hydraulic isolator produces not only a control force $U(t)$ applied to the object to be protected, but also a force acting on the base. Clearly, the latter force is equal to $-U(t)$ when the cylinder housing is attached to the base.

The investigation performed does not take into account certain phenomena occurring in the hydraulic systems, like an input pressure pulsation and wave processes in pipelines. These phenomena deserve some attention in the practical design of active systems, see [31], however their analysis is rather sophisticated and beyond the scope of the present book.

It is also worth noticing some shortcoming of hydraulic facilities. For example, instability of the working fluid properties depending on temperature and the presence of the air in the working fluid. The latter phenomenon increases drastically the fluid compressibility and has considerable influence on the dynamic characteristic of the system.

Pneumatic actuators

Pneumatic actuators may have very different design solutions. In some cases, the actuator is a two-way cylinder, pressure in its chambers being controlled by a control valve or switch-gear. In other cases, chambers with air flow are used, the pressure being changed by means of diverse air-flow throttles. In addition to the working chambers whose volume is controlled by a moveable element, e.g. a piston or a membrane, pneumatic systems also use constant volume chambers linked via throttles to working chambers.

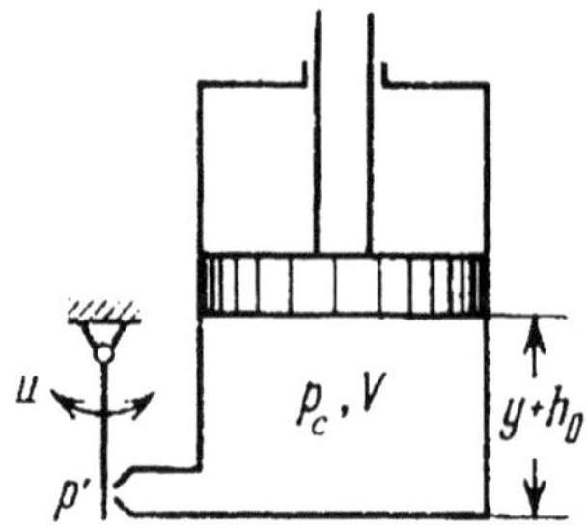

FIGURE 2.12.

We consider a pneumatic working chamber bounded by a piston of area S, Fig. 2.12. Let y denote the piston displacement relative to the initial position h_0. The input throttle has cross-sectional area f which is controlled by the flap displacement $u(t)$. The pressure p' before the throttle may vary, for example in the cases when the throttle is mounted between two chambers.

Consistent with the conventional practice of design of pneumatic systems, see e.g. [142], the following assumptions are made:

(i) The influence of mass density of the working body on the flow rate is negligible for small pressure differences.

(ii) The dynamic flow rate characteristics relating varying input and output throttle pressures coincide with the static characteristics calculated under assumptions of constant pressures at any time instant.

(iii) The pressure is equal at any point on the chamber at any instant of time.

Under these assumptions the mass flow rate of the working body, i.e. gas, via the throttle depends on the position of the flap and on the input and output pressures

$$G = G(p', p_c, u), \tag{2.45}$$

where p_c stands for the chamber pressure.

The mass flow rate characteristic depends on the pressure ratio p_c/p'. If this ratio is larger than some critical value (which is 0,53 for air), the outflow is called under-critical otherwise it is called over-critical. In the case of an under-critical outflow, the mass flow rate is determined by the following formula

$$G = \zeta f(u)\, g \sqrt{2\rho' p' \left[\frac{k}{k-1} \left(\frac{p_c}{p'} \right)^{2/k} - \left(\frac{p_c}{p'} \right)^{(k+1)/k} \right]}. \tag{2.46}$$

Here ζ is a flow rate factor which varies between 0,75 and 0,9 depending upon the throttle design, $f(u)$ is the cross-sectional area of the throttle, ρ' is the mass density of the gas before the throttle, k is the ratio of the specific heat at constant pressure to the specific heat at constant volume ($k = 1,4$ for air). In the case of an over-critical outflow, the mass flow rate does not depend upon the pressure ratio and is given by

$$G = \zeta f(u)\, g \sqrt{2\rho' p' \frac{k}{k+1} \left(\frac{2}{k+1}\right)^{2/(k-1)}}. \tag{2.47}$$

Equations (2.46) and (2.47) correspond to an adiabatic outflow which is normally observed for throttles of pneumatic systems. As a rule, process of filling the working chamber is accompanied by intensive heat exchange and may be deemed as being isothermal. In this case, the gas equation

$$p_c V = \frac{M}{\mu} RT, \tag{2.48}$$

where M is mass of the gas in the chamber, μ the molecular mass, T is the absolute temperature and R is the gas constant, yields

$$\frac{d}{dt}(p_c V) = \frac{RT}{\mu}\frac{dM}{dt} = \frac{RT}{\mu g} G(p', p_c, u). \tag{2.49}$$

Taking into account that

$$V = S\,(h_0 + y)\,, \;\; \dot{V} = S\dot{y}$$

we obtain

$$\dot{p}_c S(h_0 + y) + p_c \dot{S} y = \frac{RT}{\mu g} G(p', p_c, u). \tag{2.50}$$

Provided that $p'(t)$ and $u(t)$ are prescribed one can determine $p_c(t)$ from this equation. Linearisation of this equation is feasible if small displacements about some stationary position are considered, i.e. y and $\dot{y}$ are small and p is close to a constant value p_0. In this case, eq. (2.50) can be written as follows

$$Sh_0 \dot{p}_c + Sp_0 \dot{y} = \frac{RT}{\mu g} \left\{ G(p', p_0, 0) + \left(\frac{\partial G}{\partial u}\right)_0 u + \left(\frac{\partial G}{\partial p}\right)_0 (p_c - p_0) \right\}. \tag{2.51}$$

Assuming that in the stationary state $p' = p_0$, and, thus, $G = 0$, eq. (2.51) yields

$$h_0 \dot{U}_* + U_0 \dot{y} = \frac{RT}{\mu g S} \left(\frac{\partial G}{\partial p_c}\right)_0 (U_* - U_0) + \frac{RT}{\mu g} \left(\frac{\partial G}{\partial u}\right)_0 u. \tag{2.52}$$

Here $U_* = Sp_c$ is the pressure force acting on the piston, and $U_0 = Sp_0$. Introducing the time constant

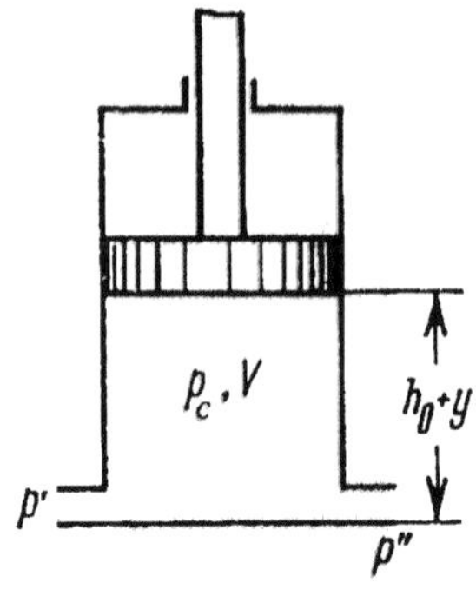

FIGURE 2.13.

$$T_a = \frac{h_0 \mu g S}{RT\left(\frac{\partial G}{\partial p_c}\right)_0}, \tag{2.53}$$

we arrive at the following equation

$$T_a \Delta \dot{U} + \Delta U = \frac{S\left(\frac{\partial G}{\partial u}\right)_0}{\left(\frac{\partial G}{\partial p_c}\right)_0} u - \frac{U_0 \mu g S}{RT\left(\frac{\partial G}{\partial p_c}\right)_0} \dot{y}, \tag{2.54}$$

where $\Delta U = U_* - U_0$. Equation (2.54) must be solved together with the equation of motion for the object to be protected.

Let us consider now a chamber with the running working body depicted in Fig. 2.13. This chamber has two throttles, an input throttle and an output throttle. Denoting their mass flow rates respectively as G' and G'' we obtain

$$G' = G(p', p_c, u), \; G'' = G(p_c, p'', u), \tag{2.55}$$

where p' is the pressure before the input throttle and p'' is the pressure after the output throttle. The control signal $u\,(t)$ is assumed to coincide for both throttles. In this case, eq. (2.50) takes the form

$$\dot{p}_c S(h_0 + y) + p_c S \dot{y} = \frac{RT}{\mu g}[G(p', p_c, u) - G(p_c, p'', u)]. \tag{2.56}$$

Linearising this equation about the following stationary state

$$u = u_0, \; p_c = p_0,$$

which implies that $G' = G''$, yields the following equation which is analogous to eq. (2.54)

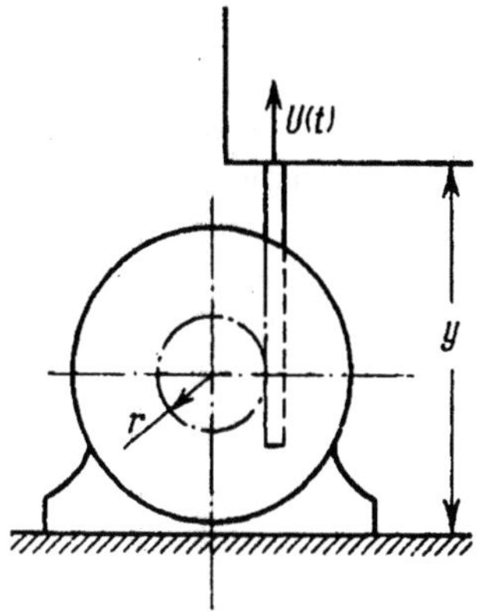

FIGURE 2.14.

$$T_a \Delta \dot{U} + \Delta U =$$

$$\frac{S\left[\left(\frac{\partial G'}{\partial u}\right)_0 - \left(\frac{\partial G''}{\partial u}\right)_0\right]}{\left(\frac{\partial G'}{\partial p_c}\right)_0 - \left(\frac{\partial G''}{\partial p_c}\right)_0} u - \frac{U_0 \mu g S}{RT\left[\left(\frac{\partial G'}{\partial p_c}\right)_0 - \left(\frac{\partial G''}{\partial p_c}\right)_0\right]} \dot{y}. \tag{2.57}$$

Here

$$T_a = \frac{S h_0 V \mu g}{RT}\left[\left(\frac{\partial G''}{\partial p_c}\right)_0 - \left(\frac{\partial G'}{\partial p_c}\right)_0\right]^{-1}. \tag{2.58}$$

If the chamber volume is constant, the dynamic characteristic of the chamber is obtained by letting $\dot{y} = 0$ in eqs. (2.50) and (2.56). When a number of chambers are connected one must solve a system of equations in which the working pressures play the parts of unknown variables.

Electromechanical actuators

DC and AC motors are actuators in electromechanical systems. The rotor rotation is ordinarily transmitted to the object so that the velocity of the object $\dot{y}$ relative to the motor housing is proportional to the rotor angular velocity Ω, see Fig 2.14, while the control force $U(t)$ is proportional to the load torque moment M_a, i.e.

$$\dot{y} = r\Omega, \; M_l = rU. \tag{2.59}$$

Here r is a proportionality factor depending on the gear parameters. The governing equation for the rotor is as follows

$$J\dot{\Omega} = M_t - \beta\Omega - M_l, \tag{2.60}$$

where J is the equivalent moment of inertia of all masses connected to the rotor, β is the coefficient of viscous resistance and M_t is the motor torque. The relationship between the torque and angular velocity is given by the so called working characteristic of the motor. As the motor mounted on a vibration protection system vibrates, one has to use the dynamic operation characteristic for the motor which considers the inertia of the electromagnetic processes, [17], [132] and [138]. The dynamic characteristic of a DC-motor is given by

$$(T_e p + 1)M_t = M_0 - \mu\Omega. \tag{2.61}$$

Here T_e is the electric time constant of the motor which is equal to the ratio of the armature inductance L_a to the resistance of the armature R_a. Further, M_0 and μ denote the starting torque and the characteristic slope for which the following relationships hold

$$M_0 = k_1 \Phi_f u_a, \ \mu = k_2 \Phi_f^2, \tag{2.62}$$

where Φ_f denotes the flux, u_a is the armature voltage, and k_1 and k_2 are factors whose values are dependent on the particular type of the motor. Control of the motor can be performed by changing the armature voltage or the flux. Therefore, u_a and Φ_f may be viewed as the input variables of the actuator. Substituting M_t from eq. (2.60) into (2.61) and taking into account eqs. (2.59) and (2.62) yields

$$(T_e p + 1)\left[(Jp + \beta)\frac{py}{r} + rU\right] = k_1 \Phi_f u_a - k_2 \Phi_f^2 \frac{py}{r}.$$

After a little algebra, we arrive at the following equation

$$U = \frac{k_1 \Phi_f}{r(T_e p + 1)} u_a - \frac{[T_e J p^2 + (J + T_e \beta)p + \beta + k_2 \Phi_f^2]p}{r^2 (T_e p + 1)} y. \tag{2.63}$$

This expression relates the control force U with the object displacement y and the input variables. This equation must be supplemented with the equation of motion which relates y and $U(t)$. Characteristic (2.63) is linear provided that $\Phi_f = 0$.

The dynamic characteristic for an AC-motor differs from (2.61) by the presence of a harmonic component. As shown for example in [106], the dynamic characteristic for a two-phase asynchronous motor is

$$[T_e p + 1]M_t = \delta u_{in} - \mu\Omega + k\Omega \cos 2\nu t, \tag{2.64}$$

where δ, μ and k are factors depending on the motor parameters, u_{in} is the input voltage and ν is the frequency of alternating current.

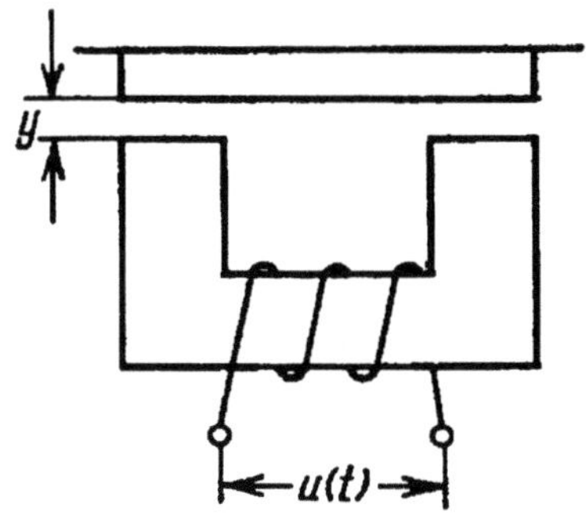

FIGURE 2.15.

Electromagnetic actuators

An electromagnet can be utilised as a facility for providing a force. The simplest scheme of an actuator involving an electromagnet is shown in Fig. 2.15. The input variable is voltage $u(t)$. The electromagnet housing is mounted on the base whereas the armature is attached to the object to be protected from vibration. Relative displacement of the object changes the gap between the core and the armature which in turn changes the electric circuit inductance L, the latter becoming a nonlinear function of the air gap y. The force produced by an electromagnet is proportional to the square of current and nonlinearly depends on the air gap. In general, the following equation holds, see [57]

$$U = \mu_0 S (Ni)^2 \varphi(y). \tag{2.65}$$

Here μ_0 is the magnetic permeability, S is the cross-sectional area of the armature, i is the current and N is the number of windings. Further, $\varphi()$ is a nonlinear function of its argument, for example $\varphi(y) = (y)^{-2}$ if the magnetic field is homogeneous. The current is governed by the following differential equation

$$\frac{d}{dt}[L(y)i] = u(t) - ri, \tag{2.66}$$

where r is the resistance. The system of equations (2.65) and (2.66) defines the dynamic characteristic of the system, i.e. the dependence of the control force $U(t)$ on $u(t)$ and $y(t)$.

2.2 One-dimensional linear active systems

Let us investigate the efficiency of one-dimensional active systems under the assumption that the system to be protected is linear and possesses weak dissipation. As shown in Section 1.3, the transfer function of a closed-loop one-dimensional system can be represented in the form of eq. (1.93), i.e.

$$k_z(p) = [1 + e_{AB}(p) w_z(p)]^{-1}.$$

Let us remember that in this equation $w_z(p)$ denotes the controller transfer function and $e_{AB}(p)$ stands for the dynamic compliance operator which can be represented in the form of eq. (1.53) for the class of systems under consideration. Displacements z_A and z_B are related to the vector of principal coordinates $\mathbf{q}$ as follows

$$z_A = \mathbf{l}_A \cdot \mathbf{q},\ z_B = \mathbf{l}_B \cdot \mathbf{q}, \tag{2.67}$$

where the components of vectors $\mathbf{l}_A$ and $\mathbf{l}_B$ are l_{As} and l_{Bm}, respectively, see (1.53).

Let us write the equations of motion in the principal coordinates in the form of eq. (1.44). Taking into account eq. (1.38) we obtain

$$\ddot{q}_s + \sum_{m=1}^{n} b_{sm}\dot{q}_m + k_s^2 q_s = Q_s(t) + l_{As}U(t) \quad (s = 1, ..., n), \tag{2.68}$$

where $Q_s(t)$ are generalised forces corresponding to the force excitations. Let $l_{Ar} = 0$, i.e. let point A be a nodal point of the $r-th$ normal mode, then the control force $U(t)$ does not explicitly appear in the $r-th$ equation, and its influence on the $r-th$ coordinate is only in terms of small dissipative terms $b_{rm}\dot{q}_m$. If these dissipative terms were absent, the system would *not be completely controllable* due to the terminology of Kalman, [49]. Since the dissipative forces are small, system (2.68) for $l_{Ar} = 0$ is *weakly controllable* in coordinate q_r. In other words, in this case the control force will weakly influence the system vibrations due to this coordinate.

Substituting z_B from eq. (2.67) into eq. (1.91) yields

$$U(t) = -w_z(p)\mathbf{l}_B \cdot \mathbf{q} = -\sum_{m=1}^{n} l_{Bm} w_z(p) q_m. \tag{2.69}$$

If, say, $l_{Br} = 0$, i.e. if point B is a nodal point of the $r-th$ normal mode, then coordinate q_r does not explicitly appear in eq. (2.69). In this case, the influence of q_r on forming the control force $U(t)$ manifests itself only implicitly, i.e. in terms of the dissipative terms in eq. (2.68) which ensures coupling coordinate q_r with another principal coordinates. If dissipation were absent, the system under the condition $l_{Br} = 0$ would *not be completely observable.* In the case of weak dissipation, the systems is *weakly observable* in coordinate q_r.

While designing vibration protection systems one often deals with not completely or weakly controllable and observable systems. If, for example, a control force acts along the symmetry axis of a mechanical system, all asymmetric normal modes of this systems are not controllable from the above control force. A system which is not controllable or not observable in q_r is ineffective at frequencies of external actions close to the $r-th$ eigenfrequency.

If $l_{Ar} \neq 0$ for a certain r, but $l_{Br} = 0$, then introducing an active feedback may lead to amplification of the $r-th$ normal mode. To prevent this, it is sufficient to provide coincident points A and B, i.e. to apply the control force to that point of the mechanical system, whose vibration must be reduced. Having a control point which is coincident with the observed point is often expedient from the perspective of design since this allows a vibration protection facility to be compactly designed.

Next, we consider in detail the case when

$$k_z(p) = [1 + e_A(p)w_z(p)]^{-1}, \tag{2.70}$$

where $e_A(p)$ is given by eq. (1.61) in which time constants T_s and τ_k meet condition (1.62).

The efficiency of a one-dimensional control system with transfer function (2.70) is convenient to study by analysing the properties of the corresponding open-loop system having the following transfer function

$$\Phi_z(p) = e_A(p)w_z(p) = \frac{1 - k_z(p)}{k_z(p)}. \tag{2.71}$$

The efficiency condition of the active facility at frequency ω

$$k_0(\omega) = |k_z(i\omega)| = |1 + \Phi_z(i\omega)|^{-1} < 1 \tag{2.72}$$

is satisfied if

$$|1 + \Phi_z(i\omega)| > 1. \tag{2.73}$$

Since

$$\Phi_z(i\omega) = |\Phi_z(i\omega)| (\cos\varphi + i\sin\varphi),$$

where $\varphi = \arg\Phi_z(i\omega)$, condition (2.73) is transformed to the following form

$$(1 + |\Phi_z(i\omega)|\cos\varphi)^2 + |\Phi_z(i\omega)|^2 \sin^2\varphi > 1,$$

or

$$|\Phi_z(i\omega)| > -2\cos\varphi. \tag{2.74}$$

This form of the efficiency condition is linked to the properties of the open-loop system. It follows from eq. (2.74) that the sufficient condition for efficiency is given by

$$|\Phi_z(i\omega)| > 2. \tag{2.75}$$

On the other hand, if

$$|\Phi_z(i\omega)| < 1, \tag{2.76}$$

the system is efficient only if

$$\cos\varphi < -\frac{1}{2},$$

whereas the value of $k_0\,(\omega)$ may lie within the following bounds

$$0,5 \leq k_0\,(\omega) < \infty,$$

depending upon the value of phase φ.

Therefore, the amplitude of point A can not be reduced more than by a factor of 2. Taking this fact into account we realise that control under condition (2.76) is ineffective. The efficiency of active feedback increases with the growth of $|\Phi_z(i\omega)|$. If $|\Phi_z(i\omega)| \gg 1$, we can take approximately that

$$k_0(\omega) \approx |\Phi_z(i\omega)|^{-1}. \tag{2.77}$$

It is convenient to carry out an investigation of the efficiency of active vibration protection system by means of graphic-analytical methods. One of these methods is the polar plot of the frequency characteristic $\Phi(i\omega)$, Fig. 2.16. All points in the polar plot outside of the circle of radius 1 with the centre at point $-1+i0$ satisfy condition (2.72). This means that control is effective at the frequencies corresponding to these points. The polar plot allows us to find the value of $k_0(\omega)$ which is equal to the inverse of the length of the line segment AB between point $-1+i0$ and the corresponding point of the polar plot.

The second method of analysis of efficiency is the Bode plot for $\Phi_z(i\omega)$. It actually consists of two plots, namely, gain versus frequency and phase versus frequency. Let us recall that the following expression

$$L = 20\log|\Phi(i\omega)|, \tag{2.78}$$

concerned with plotting logarithmic frequency characteristics for $\Phi_z(i\omega)$ considered as a function of $\log\omega$ is referred to as the *logarithmic amplitude characteristic* for the open-loop system, or LAC for short. The factor 20 indicates that L is measured in decibels. Dependence

$$\varphi = \arg\Phi(i\omega) \tag{2.79}$$

on $\log\omega$ is referred to as the *logarithmic phase characteristic*, or LPC for short. The transfer function of the system can be usually represented as a product of transfer functions of the following simple elements:

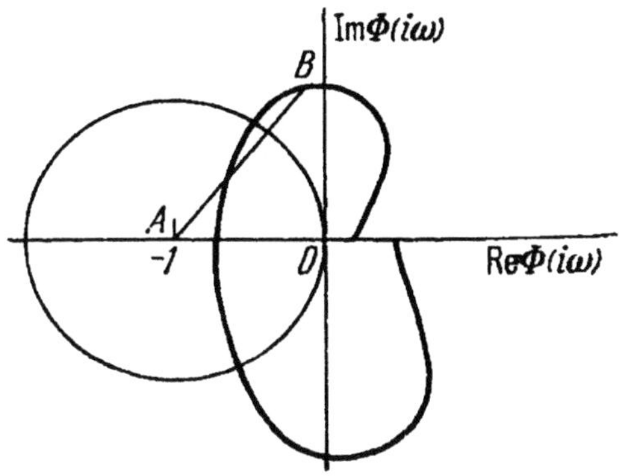

FIGURE 2.16.

(i) aperiodic element

$$w(p) = \frac{k}{Tp+1}, \tag{2.80}$$

(ii) vibratory element

$$w(p) = \frac{k}{T^2p^2 + 2\zeta Tp + 1}, \tag{2.81}$$

(iii) integrator and differentiator

$$w(p) = \frac{k}{p}, \ w(p) = kp, \tag{2.82}$$

and (iv) first order element (also known as proportional-derivative element) and second order element

$$w(p) = k(Tp+1), \ w(p) = k(T^2p^2 + 2\zeta Tp + 1). \tag{2.83}$$

LAC and LPC of the system can be determined as sums of LAC and LPC of the simple elements. On the other hand, LAC and LPC of the simple elements can be approximately plotted in the form of broken lines (so called asymptotic logarithmic characteristics). All this simplifies considerably when plotting logarithmic characteristics.

The methods of plotting these characteristics are beyond the scope of the present book. They can be found in any textbook on control, see, for example, [137], [106] and [101]. Some peculiarities of the methods will be shown with the help of the following particular examples.

Let us consider a system having transfer function (2.70). Active feedback in this system forms a control force proportional to the absolute displacement of point A, i.e.

$$w_z(p) = k. \tag{2.84}$$

In principle, this control can be realised by means of passive elements. To this end, it is sufficient, for instance, to connect point A by means of an elastic element of rigidity k to a fixed point. However, in actual systems there exists no fixed point, and, thus, this solution is unacceptable. Using active elements enables one to realise a control with a transfer function which is close to being constant at low frequencies. At higher frequencies one must realise that the feedback elements are not ideal and consider the real characteristics studied in Section 1.5. The influence of the latter obstacle on the system dynamics is investigated in what follows.

Considering eq. (1.61) we obtain

$$\Phi_z(p) = ke_A(p) = ke_A(0)\frac{\prod_{k=1}^{n-1}(\tau_k^2p^2 + 2\tau_k\eta_kp + 1)}{\prod_{s=0}^{n-1}(T_s^2p^2 + 2T_s\zeta_sp + 1)}. \tag{2.85}$$

Hence,

$$L = 20\log|\Phi_z(i\omega)| = 20\{\log e_A(0) + \log k +$$
$$\sum_{k=1}^{n-1}\log\sqrt{(1-\tau_k^2\omega^2)^2 + 4\tau_k^2\eta_k^2\omega^2} - \sum_{s=0}^{n-1}\log\sqrt{(1-T_s^2\omega^2)^2 + 4T_s^2\zeta_s^2\omega^2}\Big\}. \tag{2.86}$$

For an approximate plot one takes into account that

$$\log\sqrt{(1-T^2\omega^2)^2 + 4T^2\zeta^2\omega^2} \approx \begin{cases} 0 \text{ if } \omega \ll T^{-1}, \\ 2\log T_\omega \text{ if } \omega \gg T^{-1}. \end{cases} \tag{2.87}$$

By means of these approximations an asymptotic LAC is plotted. The latter is a broken line, the break point being referred to as the *corner frequency*. In the case under consideration, the corner frequency coincides with resonant frequencies $\omega_{rs} = T_s^{-1}$ and antiresonant frequencies $\omega_{ak} = \tau_k^{-1}$. The asymptotic LAC corresponding to transfer function (2.85) is shown in Fig. 2.17 with a dashed line. This is horizontal between τ_k^{-1} and T_k^{-1} while between τ_{k+1}^{-1} and T_k^{-1} this goes down with a slope -40 db/decade which implies that L reduces by 40 db when ω is increased by a factor of 10. The asymptotic LAC only needs an improvement in the vicinity of the corner frequencies. At frequency $\omega = \tau_k^{-1}$ we have

$$20\log\sqrt{(1-\tau_k^2\omega^2)^2 + 4\tau_k^2\eta_k^2\omega^2} = 20\log(2\eta_k).$$

Since the corresponding term in eq. (2.86) is positive, and $2\eta_k < 1$, a discrepancy exists between the LAC and the asymptotic LAC, see Fig. 2.17. It can be shown by analogy that L must be increased by $-20\log(2\eta_k)$ at

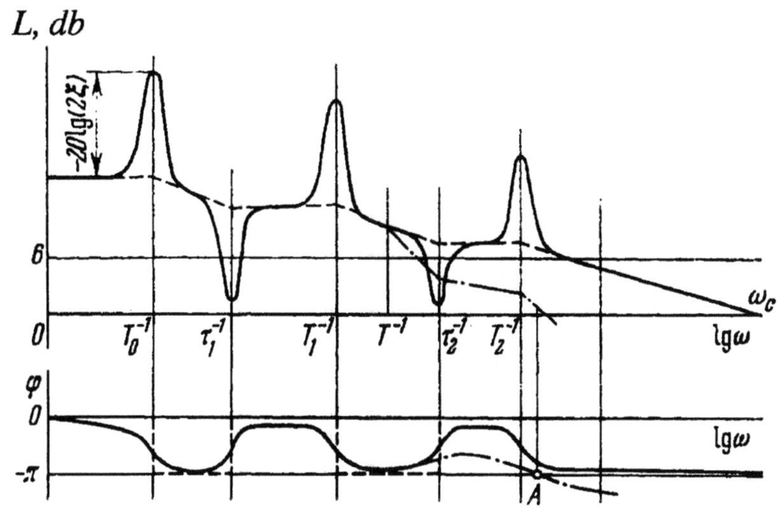

FIGURE 2.17.

$\omega = T_s^{-1}$. The result is a smooth LAC as shown in Fig. 2.17 with a solid line. The LAC intersect the abscissa axis at point $\omega = \omega_c$ which is termed the *crossover frequency.* It is clear that at this frequency

$$|\Phi(i\omega_c)| = 1.$$

Conditions (2.75) and (2.76) are equivalent to the following ones

$$L = 20\log|\Phi(i\omega_c)| > 20\log 2 \approx 6\,(db)$$

$$L < 20\log 1 = 0.$$

Therefore, active feedback can be deemed as being effective if $L > 6db$, while it can be viewed as ineffective if $L < 0$. If $-6db < L < 6db$, the efficiency depends on value of phase φ.

Figure 2.17 shows that in the case considered the active facility is ineffective at high frequencies exceeding the crossover frequency. The LAC decreases at antiresonant frequencies which may lead to inefficiency of the system at frequencies close to τ_1^{-1}, τ_2^{-1} and so forth.

More accurate information about the efficiency of active feedback can be obtained by investigating the LPC of the Bode plot, Fig. 2.17. In the case under consideration the LPC coincides with the phase characteristic of the dynamic compliance $e_A(i\omega)$ which was studied in detail in Section 1.2. As frequency increases, the phase varies between 0 and $-\pi$ and is equal to $-\pi/2$ at $\omega = \tau_k^{-1}$ and T_s^{-1}. The same result can be obtained directly from expression (2.85). Indeed

$$\begin{aligned}\varphi &= \arg\Phi_z(i\omega) = \sum_{k=1}^{n-1}\arg(1-\tau_k^2\omega^2+2\tau_k\eta_k i\omega) \\ &-\sum_{s=0}^{n-1}\arg(1-T_s^2\omega^2+2T_s\zeta_s i\omega).\end{aligned} \tag{2.88}$$

To plot an approximate LPC one can take that

$$\arg(1-T_s^2\omega^2+2T_s\zeta_s i\omega) = \begin{cases} 0 \text{ if } \omega \ll T^{-1}, \\ \pi/2 \text{ if } \omega = T^{-1}, \\ 2\log T_\omega \text{ if } \omega \gg T^{-1}. \end{cases} \tag{2.89}$$

From this equation it is easy to obtain the asymptotic phase characteristic shown in Fig. 2.17 with a dashed line. Refinement of the asymptotic phase characteristic results in the plot depicted with a solid line. On the parts

$$0 \le \omega \le T_0^{-1},\ \tau_k^{-1} \le \omega \le T_k^{-1}$$

the phase values are as follows

$$-\frac{\pi}{2} \le \varphi \le 0.$$

Here $\cos\varphi > 0$ and the efficiency condition (2.74) is satisfied for any value of $|\Phi_z(i\omega)|$. It goes without saying that the efficiency is low for small absolute values of the frequency characteristic, but nevertheless $k_0 < 1$. On the parts

$$T_k^{-1} \le \omega \le \tau_{k+1}^{-1}$$

the phase varies between $-\pi/2$ and $-\pi$. Here $\cos\varphi < 0$ and for condition (2.74) to be satisfied it is necessary that the absolute value of $\Phi_z(i\omega)$ is sufficiently large.

Increasing the feedback gain leads to an increase in the feedback efficiency. It follows from eq. (2.86) that increasing k by a factor of 10 leads to increasing values of L of 20 db at any frequency ω. In reality, however, an increase in feedback gain is limited by a condition for the system stability which is the subject of the next Section.

A velocity-proportional control force, i.e. control force proportional to $\dot{z}_A$, is also utilised in active systems. In this case

$$w_z(p) = k_1 p. \tag{2.90}$$

It is known, that in passive systems under kinematic excitation it is feasible to design a damping which is proportional to the velocity of deflection of elastic elements of an isolator or a dynamic absorber. Such a type of

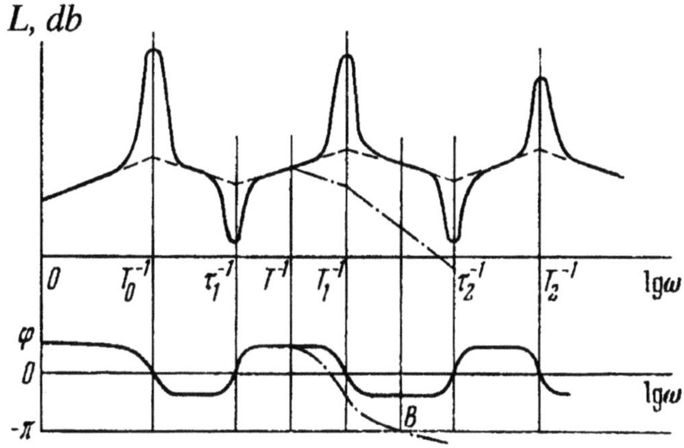

FIGURE 2.18.

damping reduces resonant amplitudes but worsens the properties of the vibration protection system in high frequency domain. Using active elements one can design damping proportional to the absolute velocity which turns out to be useful at all frequencies.

Inserting eq. (2.90) into (2.71) yields

$$\Phi_z(p) = k_1 p e_A(p) = k_1 p e_A(0) \frac{\prod_{k=1}^{n-1} (\tau_k^2 p^2 + 2\tau_k \eta_k p + 1)}{\prod_{s=0}^{n-1} (T_s^2 p^2 + 2T_s \zeta_s p + 1)}. \tag{2.91}$$

This expressions differs from eq. (2.85) only in a factor p. Introducing this factor leads to the appearance of an additional term equal to $20 \log \omega$ in dependence $L(\log \omega)$ and to an additional component which is $\pi/2$ in the LPC. Figure 2.18 displays the Bode plot for transfer function given by eq. (2.91). Similar to the previous case, the efficiency region increases as feedback gain k_1 increases. It is clear from eq. (2.91) that $|\Phi_z(i\omega)| \to 0$ with the growth of ω. For this reason, at high frequencies the system is no longer efficient, however a rather high crossover frequency may be provided.

Next we consider a control force proportional to acceleration of point A, i.e.

$$w_z(p) = k_2 p^2. \tag{2.92}$$

In principle, this control force is realised by attaching a lumped mass k_2 to the object. But attaching an additional mass increases the object weight, and therefore, static deflections of the isolators. Active feedback enables the control force (2.92) to be realised without increasing the weight of the object, which is an evident advantage. Inserting eq. (2.92) into (2.71) we obtain

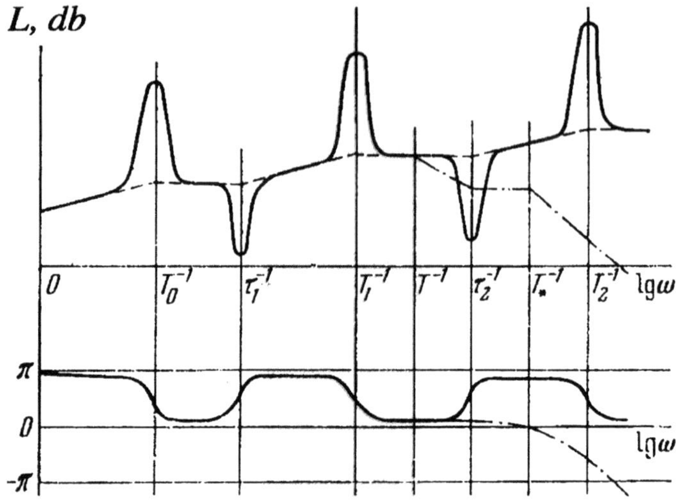

FIGURE 2.19.

$$\Phi_z(p) = k_2 p^2 e_A(p) = k_2 p^2 e_A(0) \frac{\prod_{k=1}^{n-1} (\tau_k^2 p^2 + 2\tau_k \eta_k p + 1)}{\prod_{s=0}^{n-1} (T_s^2 p^2 + 2T_s \zeta_s p + 1)}. \tag{2.93}$$

The corresponding Bode plot is shown in Fig. 2.19. As frequency increases, the value of L grows which in turn implies an increase in the feedback efficiency.

In all cases considered, the efficiency of the control decreases at frequencies τ_k^{-1} which correspond to antiresonant frequencies of the object. To compensate for the decreasing LAC at antiresonant frequency τ_l^{-1}, it suffices to introduce in the feedback circuit a vibratory element with the following transfer function

$$w_v(p) = \frac{k_v}{T^2 p^2 + 2T\zeta p + 1}. \tag{2.94}$$

The values of T and ζ must be as close to the values of τ_l and η_l as possible. After the vibratory element has been introduced we obtain

$$\Phi_z(p) = w_z(p) w_v(p) e_A(0) \frac{\prod_{k=1}^{n-1} (\tau_k^2 p^2 + 2\tau_k \eta_k p + 1)}{\prod_{s=0}^{n-1} (T_s^2 p^2 + 2T_s \zeta_s p + 1)}, \tag{2.95}$$

where $w_z(p)$ is given by one of eqs. (2.84), (2.90) and (2.92). If the relationships

$$T = \tau_i, \ \zeta = \eta_i, \tag{2.96}$$

hold exactly, the $l-th$ component in the numerator in (2.95) and the denominator in eq. (2.94) cancel and, thus, the antiresonant "trough" disappears. Let us analyse the situation in which conditions (2.96) are approximately met.

Let

$$T = \tau_i + \varepsilon, \tag{2.97}$$

where

$$\varepsilon T^{-1} \ll 1, \ \varepsilon \tau_l^{-1} \ll 1.$$

Let us estimate the following expression

$$h = \frac{1 - \tau_l^2 \omega^2 + 2\tau_l \eta_l i\omega}{1 - T^2\omega^2 + 2T\zeta i\omega} \tag{2.98}$$

at $\omega = \tau_l^{-1}$ and $\omega = T^{-1}$. Substituting $\omega = \tau_l^{-1}$ into eq. (2.98) yields

$$\begin{aligned} h\left(\tau_l^{-1}\right) &= \frac{2\eta_l i}{1 - (\tau_l + \varepsilon)^2 \tau_l^{-2} + 2(\tau_l + \varepsilon)\zeta i \tau_l^{-1}} \\ &= \frac{2\eta_l i}{-2\varepsilon\tau_l^{-1} - \varepsilon^2 \tau_l^{-2} + 2\zeta i + 2\zeta i \varepsilon \tau_l^{-1}}. \end{aligned}$$

Assuming that ζ and $\varepsilon\tau_l^{-1}$ have the same order of smallness and neglecting values of higher order of smallness we obtain

$$h\left(\tau_l^{-1}\right) \approx \frac{\eta_1}{\zeta + \varepsilon\tau_l^{-1} + \varepsilon\tau_l^{-1} i}$$

or

$$\left|h\left(\tau_l^{-1}\right)\right| \approx \frac{\eta_l}{\zeta + 2\varepsilon\tau_l^{-1}}.$$

It is clear that this not a small value. This can be somewhat smaller or larger than η_l/ζ depending on the sign of ε. Analogously, at $\omega = T^{-1}$, we get

$$h\left(T^{-1}\right) = \frac{\eta_l - \varepsilon T^{-1} - \varepsilon T^{-1} i}{\zeta}.$$

Therefore, when the values of T and τ_l are close, an antiresonance "trough" at frequency τ_l^{-1} disappears and resonance at frequency T^{-1} does not occur. In this case, an antiresonance of the object is said to be compensated. The introduced control force $w_z(p)$ can be said to be efficient at frequency τ_l^{-1} since absolute value of $\Phi_z(i\omega)$ increases significantly. Introducing a vibratory element (2.94) into the feedback also changes the LPC of the system. At frequencies much greater than T^{-1}, the phase $\arg\Phi_z(i\omega)$ decreases by π. It will be shown below that this phase change affects the stability.

Notice that the following transfer function

$$w(p) = k_2 p^2 \cdot w_k(p) = \frac{k_2 k_k p^2}{T^2 p^2 + 2T\zeta p + 1} \tag{2.99}$$

can be realised with the help of a simple dynamic vibration absorber tuned to frequency T^{-1}. Using active elements permits one to design a self-tuning dynamic vibration absorber.

Let us consider now a system with a control force produced by integrating displacement, i.e.

$$w_z(p) = \frac{k_4}{p}. \tag{2.100}$$

Substituting eq. (2.100) into (1.92) yields

$$z_B(t) = \frac{p}{p + k_4 e_{AB}(p)} \zeta_B(t) = k(p)\zeta_B(t). \tag{2.101}$$

Provided that the object is subjected to a static load, then ζ_B is the static displacement of point B and is therefore time-independent

$$\zeta_B = const.$$

It follows from eq. (2.101) that in a steady-state regime

$$z_B(t) = k(0)\zeta_B = 0, \tag{2.102}$$

i.e. introducing feedback of the type (2.100) eliminates the static displacement of point B. The vibration protection system is thus astatic.

Astatic vibration protection systems which can be realised only by means of active elements are widely used in various branches of technology. A few examples of these are listed in what follows.

(i) There exists a necessity to protect items of equipments mounted aboard of aircraft and spacecraft moving with considerable constant, so-called linear, accelerations. So, for example, constant accelerations can reach several tens of g during a flight of a rocket. If the system is not astatic, the constant accelerations cause considerable displacements of a protected object relative to its base resulting in the object being pressed against the isolator stops which implies that the vibration protection facility no longer functions.

(ii) Ship engines are known to be sources of relatively low frequency (5-6 Hz) vibrational excitations. In order to protect the hull from these vibrations, it is necessary to design extremely soft isolation systems with eigenfrequencies of about 2-3 Hz. But mounting the engine on soft isolators is inadmissible since this leads to unacceptable relative displacements during roll. Only astatic active facilities enable this problem to be solved.

(iii) A similar problem arises in the machine-tool industry. Mounting of precision machinery on soft isolators is usually not admitted as this leads to a skewness of the machine body due to changes in the static loads, displacement of the table or carriage etc. Astatic supports allow automatic levelling of the machine body.

Let us analyse the efficiency condition in the case of control forces due to displacement integration. Inserting eq. (2.100) into (2.71) yields

$$\Phi_z(p) = \frac{k_4}{p} e_A(p) = \frac{k_4 e_A(0)}{p} \frac{\prod_{k=1}^{n-1} (\tau_k^2 p^2 + 2\tau_k \eta_k p + 1)}{\prod_{s=0}^{n-1} (T_s^2 p^2 + 2T_s \zeta_s p + 1)}. \tag{2.103}$$

Figure 2.20 shows the Bode plot for transfer function (2.103). The asymptotic LAC consists of alternating parts with slopes -20 db/decade and -60 db/decade while the phase varies between $-\pi/2$ and $-\pi$. Increasing the slope of the LAC in comparison with the case of the proportional control force leads to a considerable decrease in the crossover frequency. Hence, the control turns out to be efficient only at low frequencies. At first glance it seems that the crossover frequency may be increased by increasing gain k_4 since, similar to the above cases, a tenfold increase in k_4 increases L by 20 db and shifts ω_c about a half decade to the right (i.e. ω_c is increased by a factor of $\sqrt{10}$). But, as it will be shown in the next Section, increasing the feedback gain is constrained by the conditions of system stability. In practice, an integrator is often replaced by an aperiodic element having large time constant. The properties of this control are very close to those of control by integration.

In the problem of protecting aircraft equipment one has to deal with objects whose absolute acceleration contains a large constant component and a vibration component. In this case, using control forces proportional to the third derivative of the absolute displacement

$$w_z(p) = k_5 p^3 \tag{2.104}$$

may be expedient. Such control forces do not react on the constant component of acceleration and favour a decrease in the vibration component.

In all examples considered we studied systems with ideal feedback transfer functions. Transfer functions of real active systems have elements with inertia and for this reason differ from those of idealised systems. For example, the transfer function of a facility consisting of an accelerometer, an amplifier, a control valve and a hydraulic cylinder can be obtained as a product of transfer functions (2.12), (2.24), (2.25) and transfer function

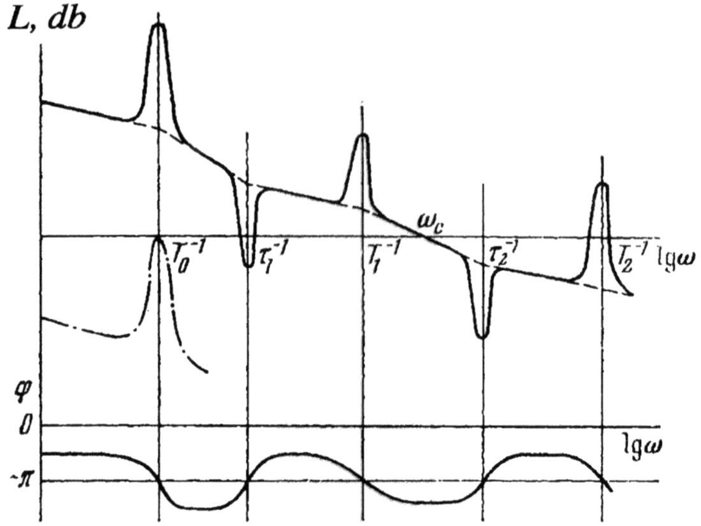

FIGURE 2.20.

corresponding to (2.37). Also, we ignore the additional feedback describing the direct action of the object motion on the control force by letting $w_a^{(z)}(p) = 0$ in Fig. 2.3 . Then

$$w_z(p) = \frac{k_s p^2 \cdot k k_v Q_{\max} \cdot k_v' \cdot S}{(T_s^2 p^2 + 2\zeta_s T_s p + 1) c_v (T_v^2 p^2 + 2\zeta_v T_v p + 1)(T_a p + 1)\varkappa}, \quad (2.105)$$

where

$$T_a = \frac{\mu h_0 S}{2\varkappa}.$$

Therefore, accounting for the fact that the elements of the feedback circuit are not ideal means that additional components, i.e. polynomials of first and second order, appear in the denominator of transfer function $w_z(p)$. In turn, this means that these polynomials appear in the denominator of expression for $\Phi_z(p)$, too. In some cases, those components may also appear in the numerator of $w_z(p)$, however, the order of the polynomial in the denominator must exceed that in the numerator, so that $|w_z(i\omega)| \to 0$ with the growth of ω.

Let us investigate the influence of the elements non-ideality on the form of the LAC and the LPC of the open-loop system. Let, for example, the element form control forces proportional to displacement due to the following transfer function

$$w_z(p) = \frac{k}{Tp+1}. \quad (2.106)$$

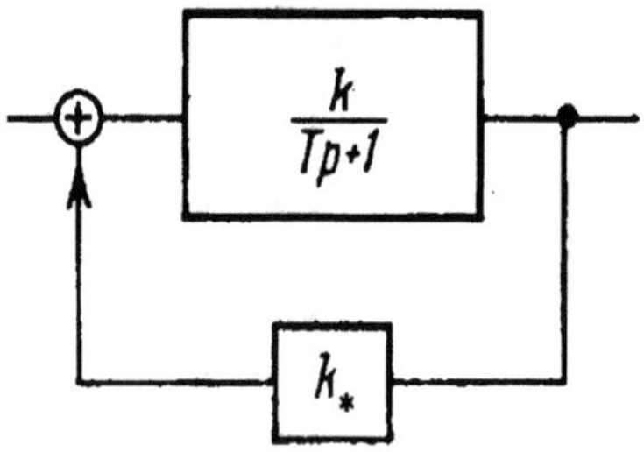

FIGURE 2.21.

This transfer functions differs from that given by eq. (2.84) in polynomial $Tp+1$ in the denominator. In this case, the form of the LAC and the LPC at frequencies smaller than T^{-1} does not change. At frequencies higher than T^{-1} the slope of all parts of the LAC decreases by values of -20 db/decade, the asymptotic LAC being shown in Fig. 2.17 with a dash-dot line. Clearly, the crossover frequency is decreased which in turn implies a reduction in the frequency range of efficient control. Consistent with this, LPC is also changed, namely, at frequencies higher than T^{-1} the value of φ decreases, and at frequencies $\omega \gg T^{-1}$ a change in the phase reaches $-\pi/2$, see Fig. 2.17.

Similar changes take place for cases in which non-ideal elements are used in the feedback forming control forces proportional to velocity or acceleration. The dash-dot lines in Fig. 2.18 represent asymptotic LAC and LPC for the following transfer function

$$w_z(p) = \frac{k_1 p}{T^2p^2 + 2T\zeta p + 1}.$$

Similar to the previous case, a considerable decrease in the crossover frequency is observed here. The phase is decreased by π at $\omega \gg T^{-1}$. Fig. 2.19 displays LAC and LPC for the following transfer function

$$w_z(p) = \frac{k_2 p^2}{(T^2p^2 + 2T\zeta p + 1)(T_* p + 1)} \quad (T > T_*).$$

The presence of non-ideal elements leads to the appearance of a crossover frequency whereas the phase decreases by $3\pi/2$ at high frequencies.

Thus, control efficiency significantly deteriorates because of non-ideal elements in the feedback circuit. If $T_{\max}$ denotes the maximum time constant in the denominator of a transfer function, then deterioration of the efficiency manifests itself at frequencies $\omega > T_{\max}^{-1}$.

To overcome this shortcoming, a number of methods for correction of the feedback transfer function are proposed. One of these methods, namely, the compensation method implies introducing those polynomials in the numerator of $w_z(p)$ which cancel the corresponding polynomials in the denominator. For instance, by introducing an additional element which is a filter with the following transfer function

$$w_1(p) = \frac{Tp+1}{T_1 p+1}$$

into feedback having transfer function (2.106) we obtain feedback with the transfer function

$$w_z'(p) = \frac{k(Tp+1)}{(Tp+1)(T_1 p+1)} = \frac{k}{T_1 p+1},$$

where time constant $T_1 \ll T$. However, it is known, see e.g. [50], that such a compensation is admissible only in those cases when the system is stable and possesses minimum-phase transfer function, i.e. all roots and poles lie in the left half-plane. In addition to this, cancelling polynomials in numerator and denominator means that the transfer function does not completely reflect the dynamic properties of the system, the latter becoming not completely observable. Finally, note that compensating actuators with inertia is especially difficult. As mentioned in Section 2.1, these elements actually possess nonlinear properties, whereas their transfer functions are obtained by means of linearised dynamic characteristics. Any change in the operation regime will lead to a change in transfer system parameters, and, thus, to violation of the compensation condition.

Another way of decreasing the time constant of an element is to apply an additional feedback to this element. Figure 2.21 shows an aperiodic element of the type (2.106) with applied feedback having a gain k_*. In this case

$$w_z^*(p) = \frac{w_z(p)}{1+k_* w_z(p)} = \frac{k}{Tp+1+kk_*} = \frac{\dfrac{k}{1+kk_*}}{\dfrac{T}{1+kk_*}p+1}. \tag{2.107}$$

Therefore, introducing feedback reduces the time constant of the element by a factor of $(1+kk_*)$, the total gain also being changed.

Applying a correction, we can slightly increase the crossover frequency, and therefore, broaden the frequency band of efficient control. Nonetheless, the efficiency domain must be bounded in order to ensure system stability.

If point A where the control force is applied does not coincide with the observed point B, the efficiency conditions slightly change. Also in this case it is convenient to investigate the efficiency of control by analysing the properties of the transfer function for open-loop systems

$$\Phi(p) = e_{AB}(p) w_z(p), \tag{2.108}$$

where $e_{AB}(p)$ denotes the dynamic compliance operator (1.53). Figure 2.22 shows some typical forms of polar plot $\Phi(i\omega)$ for systems with the following transfer functions of feedback:

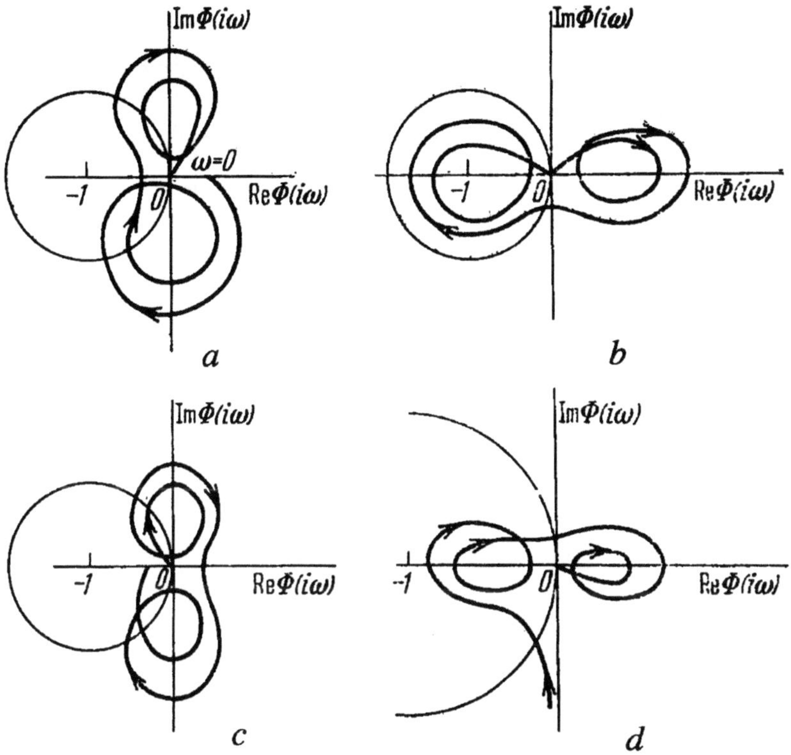

FIGURE 2.22.

(i) Fig. 2.22a for $w_z(p) = k$,
(ii) Fig. 2.22b for $w_z(p) = k_1 p$,
(iii) Fig. 2.22c for $w_z(p) = k_2 p^2$,
(iv) Fig. 2.22d for $w_z(p) = k_4 p^{-1}$.

Also circles of unit radius are plotted there. As shown above, the parts of polar plots outside of these circles correspond to the frequencies at which the control is efficient.

Let us consider some examples. The block diagram of an active isolator with control force due to absolute acceleration is depicted in Fig. 2.23. The object to be protected has one degree-of-freedom and its transfer function is given by

$$e_A(p) = (mp^2 + bp + c)^{-1} = k_0(T_0^2 p^2 + 2\zeta_0 T_0 p + 1)^{-1}, \tag{2.109}$$

where

$$k_0 = c^{-1},\ T_0^2 = mc^{-1},\ \zeta_0 = \frac{b}{2\sqrt{cm}}.$$

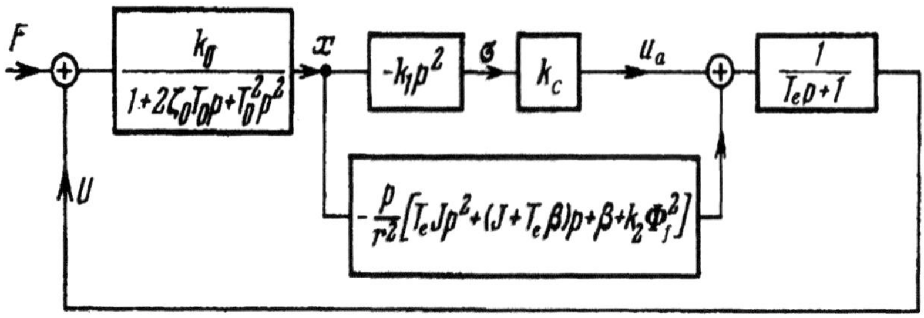

FIGURE 2.23.

Let an AC-motor having dynamic operation characteristic (2.63) be an actuator in the system. The transfer function of the open-loop system is as follows

$$\begin{aligned}\Phi_x(p) &= \frac{-k_0}{T_0^2p^2+2\zeta_0T_0p+1}\left[\frac{k_1k_2p^2}{T_ep+1}+\right.\\ &\left.+\frac{T_eJp^2+(J+T_e\beta)p+\beta+k_2\Phi_f^2}{r^2(T_ep+1)}p\right]\\ &= -\frac{k_0p[T_eJp^2+(J+T_e\beta+k_1k_cr^2)p+\beta+k_2\Phi_f^2]}{r^2(T_0^2p^2+2\zeta_0T_0p+1)(T_ep+1)}\\ &= \frac{kp(T_1p+1)(T_2p+1)}{(T_0^2p^2+2\zeta_0T_0p+1)(T_ep+1)}. \qquad (2.110)\end{aligned}$$

Here

$$\begin{aligned}k &= \frac{k_0(\beta+k_2\Phi_f^2)}{r^2},\\ T_{1,2} &= \frac{J+T_e\beta+k_1k_cr^2\pm\sqrt{(J+T_e\beta+k_1k_cr^2)^2-4JT_e(\beta+k_2\Phi_f^2)}}{2(\beta+k_2\Phi_f^2)}.\end{aligned} \qquad (2.111)$$

Let us estimate possible values of the time constants. Time constant T is smaller than the period of free vibration of the object to be protected by a factor of 2π. The electromechanical system under consideration is normally applied to control those objects with very low eigenfrequencies (about 1,5-2 Hz).

Values of T_1 and T_2 depend on the values of electromechanical and electric time constants of the motor. The electromechanical time constant

$$T_m = \frac{J}{\beta+\mu} = \frac{J}{\beta+k_2\Phi_f^2}$$

is usually much greater than the electric time constant T_e. Assuming that T_e is small yields

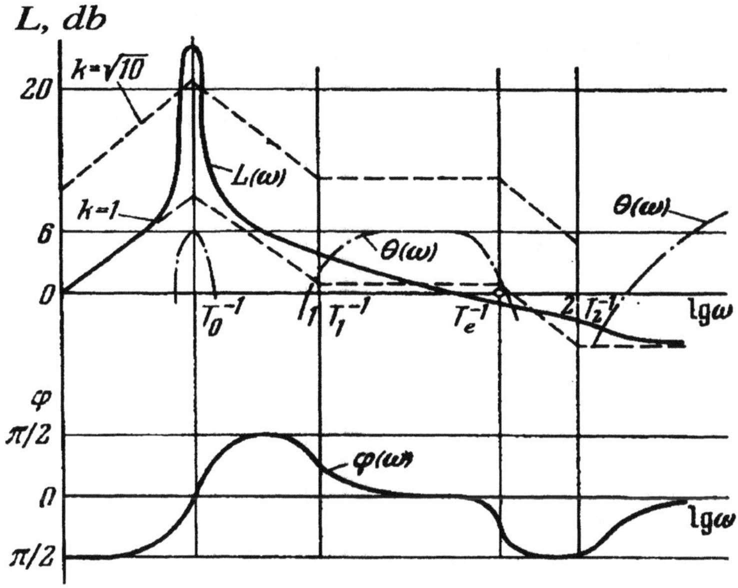

FIGURE 2.24.

$$T_1 \approx \frac{J + k_1 k_c r^2}{\beta + \mu} > T_m, \; T_2 \approx T_e \frac{J}{J + T_e \beta + k_1 k_c r^2} < T_e.$$

Hence, we can adopt that $T_1 \gg T_2$. Usually $T_m \approx 0,05 - 0,1$ s whereas $T_e \approx 0,01 - 0,05$ s. Taking these estimations into account let us adopt the following values: $T_0 = 0,3$ s, $T_1 = 0,1$ s, $T_e = 0,02$ s and $T_2 = 0,01$ s.

We also take that $\zeta_0 = 0,1$ which is a typical value of damping factor for an object mounted on passive isolators. Let us also adopt, for the time being, the total gain k to be equal to unity.

Figure 2.24 shows the Bode plot for the system under consideration. The asymptotic LAC is shown with a dashed line. By means of line $L(\omega) = 6$ db we determine the frequency domain in which condition (2.75) is satisfied. From Fig. 2.24 this domain is seen to be relatively narrow ($2 \text{ s}^{-1} < \omega < 85 \text{ s}^{-1}$). In order to increase the efficiency domain, it is sufficient to increase gain k. Figure 2.24 also displays an asymptotic LAC for $k = \sqrt{10}$, in this case the frequency domain for which $L(\omega) > 6$ db broadens ($1 \text{ s}^{-1} < \omega < 85 \text{ s}^{-1}$). More accurate values for the boundaries of the efficiency domain are obtained from condition (2.74) which can be recast in the following form

$$\theta(\omega) = 6 + 20 \log \cos \varphi < L(\omega).$$

A plot of $\theta(\omega)$ is shown in Fig. 2.24 with a dash-dot line.

We consider next an active system having a control force produced by integration of the relative displacement y of the object. As already mentioned above these astatic systems are used in various branches of technology.

Figure 2.25 shows a block diagram of an astatic system having an electrohydraulic actuator whose dynamic characteristic is given by eq. (2.37). In the latter expression

$$k_a = \frac{Q_{\max} k_v S}{\varkappa}, \; k_1 = \frac{S}{k_v Q_{\max}}, \; T_a = \frac{\mu h_0 S}{2\varkappa}.$$

Parameters k_{cv}, T_1 and ζ characterise the dynamic properties of the control valve.

The transfer function of the open-loop system is given by

$$\begin{aligned} \Phi(p) &= \frac{k_0}{T_0^2 p^2 + 2\zeta_0 T_0 p + 1} \left[\frac{k_c \cdot k_{cv}}{p(T_1^2 p^2 + 2\zeta_1 T_1 p + 1)} + k_1 p \right] \frac{k_a}{T_a p + 1} \\ &= \frac{k(\alpha T_1^2 p^4 + 2\zeta_1 T_1 \alpha p^3 + \alpha p^2 + 1)}{p(T_0^2 p^2 + 2\zeta_0 T_0 p + 1)(T_a p + 1)(T_1^2 p^2 + 2\zeta_1 T_1 p + 1)}, \end{aligned} \tag{2.112}$$

where $k = k_0 k_a k_c k_{cv}$ and $\alpha = k_1 / k_c k_{cv}$.

Let us take the following numerical values for parameters in eq. (2.112):

$$\begin{aligned} T_0 &= 0,3\, s,\ \zeta_0 = 0,1,\ T_1 = 0,01\, s,\ \zeta_1 = 0,5, \\ \alpha &= 0,04\, s^2,\ T_a = 0,05\, s. \end{aligned}$$

In this case

$$\Phi(p) = \frac{k(T_3^2 p^2 - 2\zeta_3 T_3 p + 1)(T_4^2 p^2 + 2\zeta_4 T_4 p + 1)}{p(T_0^2 p^2 + 2\zeta_0 T_0 p + 1)(T_a p + 1)(T_1^2 p^2 + 2\zeta_1 T_1 p + 1)}, \tag{2.113}$$

where $T_3 = 0,2\, s$, $\zeta_3 = 0,025$, $T_4 \approx T_1 = 0,01\, s$, $\zeta_4 \approx \zeta_1 = 0,5$. Cancelling the polynomials in the numerator and denominator yields

$$\Phi(p) = \frac{k(T_3^2 p^2 - 2\zeta_3 T_3 p + 1)}{p(T_0^2 p^2 + 2\zeta_0 T_0 p + 1)(T_a p + 1)}. \tag{2.114}$$

Notice that the system obtained is nonminimum-phase. The explanation for this is that the system in Fig. 2.25 has more than one loop.

Figure 2.26 shows the Bode plot of an open-loop system having $k = 1$. The system is efficient only at low frequencies, namely where $\log \omega < 10$.

We proceed now to the problem of active vibration protection in the two-degree-of-freedom system shown in Fig. 2.27. This dynamic model corresponds to a two-stage vibration protection system with an intermediate mass m_1. Two-stage passive systems are widely used in shipbuilding, and

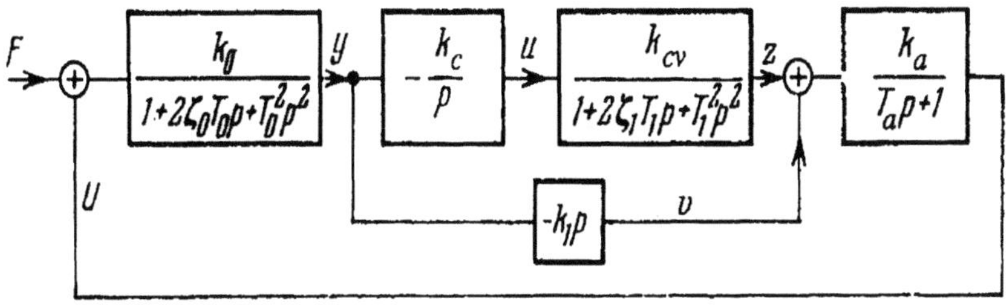

FIGURE 2.25.

their efficiency has been studied in Chapter 1. Here we investigate the influence of the active feedback forming control force $U(t)$. The same model is applied to analyse a one-stage system of vibration protection of an object consisting of two masses m_1 and m_2 connected by an elastic and a dissipative element.

We presume that the objective of vibration protection is to reduce vibrations of the upper mass caused by a kinematic excitation $\xi(t)$, the control force acting on the lower mass, Fig. 2.27. In other words, the active element is mounted on the lower stage. We consider two variants of active systems. In the first case, control force $U(t)$ is formed from the displacement of the upper mass

$$U = -kw(p)z_B. \tag{2.115}$$

The transfer function of the closed-loop system is determined by expression (1.93) in which $w_z(p) = kw(p)$. By virtue of eqs. (1.84) and (1.85) the expression for $e_{AB}(p)$ takes the following form

$$e_{AB}(p) = \frac{\beta p + \gamma}{(m_1p^2 + bp + \beta p + \gamma + c)(m_2p^2 + \beta p + \gamma) - (\beta p + \gamma)^2}.$$

In the second case, the control force is due to displacement of the lower mass

$$U = -kw(p)z_A. \tag{2.116}$$

As vibrations of the upper and lower masses are coupled, there is the following relationship

$$z_B = \frac{\beta p + \gamma}{m_2p^2 + \beta p + \gamma} z_A,$$

which is not affected by introducing control. The control efficiency is thus completely determined by the displacement of the lower mass which corresponds to transfer function (2.70). In this case $w_z(p) = kw(p)$, and

$$e_A(p) = \frac{m_2p^2 + \beta p + \gamma}{(m_1p^2 + bp + \beta p + \gamma + c)(m_2p^2 + \beta p + \gamma) - (\beta p + \gamma)^2}$$

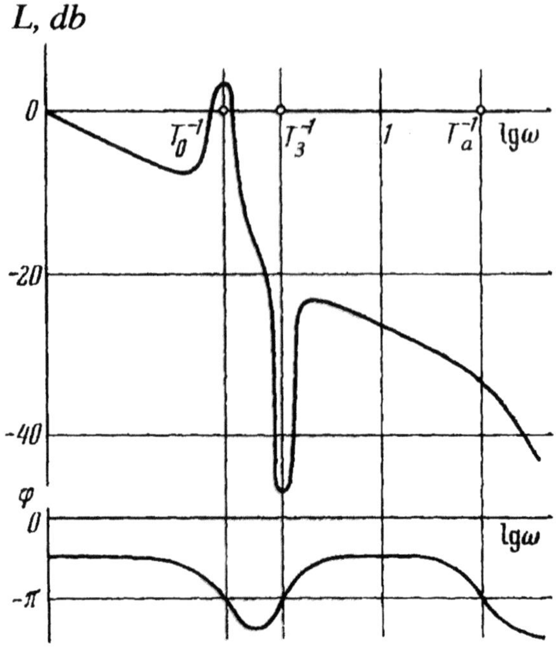

FIGURE 2.26.

in accordance with (1.85).

Let us assume that the main elements of the active feedback are an accelerator and an actuator, for example, a pneumatic facility, which as a first approximation may be considered as an aperiodic element. In this case

$$w(p) = \frac{p^2}{Tp+1}, \tag{2.117}$$

where T is the actuator time constant. Representing the feedback transfer function in the form (2.117) we neglect the fact that all other elements are not ideal. The question whether this is admissible will be discussed in what follows.

Next we construct transfer functions of the open-loop elements. Under control due to displacement of the lower mass, we have

$$\Phi_1(p) = \frac{kp^2}{Tp+1} e_A(p) = \frac{kp^2(m_2p^2 + \beta p + \gamma)}{(Tp+1)\Delta(p)}, \tag{2.118}$$

where $\Delta(p)$ is given by eq. (1.84). In the case of control due to $z_B(t)$, we obtain

$$\Phi_2(p) = \frac{kp^2}{Tp+1} e_{AB}(p) = \frac{kp^2(\beta p + \gamma)}{(Tp+1)\Delta(p)}. \tag{2.119}$$

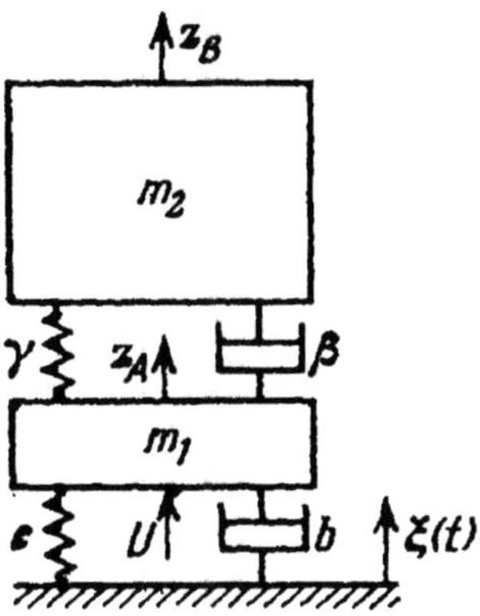

FIGURE 2.27.

For further analysis we adopt the following numerical values for the parameters:

$$\begin{aligned} m_1 &= 300kg,\ m_2 = 1000kg,\ c = 52\cdot 10^4 N/m,\ \gamma = 4\cdot 10^6 N/m, \\ b &= 500Ns/m,\ \beta = 5000\, Ns/m,\ T = 0,3s. \end{aligned}$$

It is easy to see that the chosen values correspond to a system with soft lower and stiff upper stages. The actuator time constant T is relatively large, i.e. at frequencies $\omega \gg T^{-1}$ (approximately at $\omega > 6-8$ Hz) the control is very close to that due to the absolute velocity. Let us now determine the roots of the characteristic equation $\Delta(p) = 0$ which reduces to the following form

$$\begin{aligned} p^4 + \left(\frac{\beta}{m_2} + \frac{b+\beta}{m_1}\right) p^3 + \left(\frac{c+\gamma}{m_1} + \frac{\gamma}{m_2} + \frac{b\beta}{m_1 m_2}\right) p^2 \\ + \left(\frac{c\beta}{m_1 m_2} + \frac{b\gamma}{m_1 m_2}\right) p + \frac{c\gamma}{m_1 m_2} = 0. \end{aligned} \tag{2.120}$$

This equation has two pairs of complex-valued roots. The fact that the roots drastically differ in the absolute value simplifies solving this equation. The result is

$$p_{1,2} \approx -2,08 \pm 20,8i,\ p_{3,4} \approx -11,65 \pm 126i.$$

Transfer functions (2.118) and (2.119) are now cast in the form

$$\Phi_1(p) = \frac{kp^2(\tau^2 p^2 + 2\tau\eta p + 1)}{c(Tp+1)(T_1^2 p^2 + 2\zeta_1 T_1 p + 1)(T_2^2 p^2 + 2\zeta_2 T_2 p + 1)}, \tag{2.121}$$

$$\Phi_2(p) = \frac{kp^2(\tau_1 p + 1)}{(Tp+1)c(T_1^2p^2 + 2\zeta_1 T_1 p + 1)(T_2^2 p^2 + 2\zeta_2 T_2 p + 1)}. \qquad (2.122)$$

Here, see eq. (1.60)

$$T_1 = |p_{1,2}|^{-1} = 0,048\,s, \quad T_2 = |p_{3,4}|^{-1} = 0,008s,$$

$$\zeta_1 = \operatorname{Im} p_1 \cdot T_1 = 0,1, \quad \zeta_2 = \operatorname{Im} p_3 \cdot T_2 = 0,092,$$

$$\tau = 0,016\,s, \ \eta = 0,08, \quad \tau_1 = \beta\gamma^{-1} = 0.00125\,s.$$

Figure 2.28 displays the Bode plot for transfer functions (2.121) and (2.122) for those gains k which ensure that $|\Phi_1(i\omega)| = 1$ and $|\Phi_2(i\omega)| = 1$ at $\omega = 1$. It is clear that the LAC for $\Phi_2(p)$ has no antiresonant "trough at frequency τ_1^{-1} essentially deteriorating control efficiency. It might seem that the control force $U(t)$ due to the upper mass displacement would be preferable for efficient control at frequencies $0 \le \omega \le 100$ s^{-1}. However, this variant of feedback is bounded by the system stability condition. In the next Section it is shown that for stability the closed-loop system having the transfer function

$$k_2(p) = [1 + \Phi_2(p)]^{-1}$$

it is necessary to reduce the gain so that the maximum $|\Phi_2(i\omega)|$ at frequency T_2^{-1} lies below the abscissa axis. This means that the control will be efficient only in a narrow frequency band in the vicinity of T_1^{-1}. Hence, an active force applied to the lower mass turns out to be more efficient.

As mentioned above, while constructing the system transfer function we ignored the fact that the elements of the system are not ideal. This is no doubt admissible if the time constants of the elements, T_s, are chosen so that the minimum value of T_s^{-1} exceeds the crossover frequency of the system. For transfer function $\Phi_1(p)$ this frequency is equal to 3000 s^{-1}.

Let us assume now that a vibration protection facility serves to reduce the force transmitted to the base under a force excitation. Attaching an active element to the lower stage leads to the appearance of an additional action on the base. If the active isolator does not contain inertial elements, this additional force is $-U(t)$. Let us consider an astatic active system forming the following control force

$$U(t) = -kw_1(p)z_A = -kp^{-1}z_A. \qquad (2.123)$$

The designated purpose for this system has already been discussed above. The transfer function of the corresponding closed-loop system

$$\varkappa(p) = [1 + e_A(p)kw_1(p)]^{-1} \qquad (2.124)$$

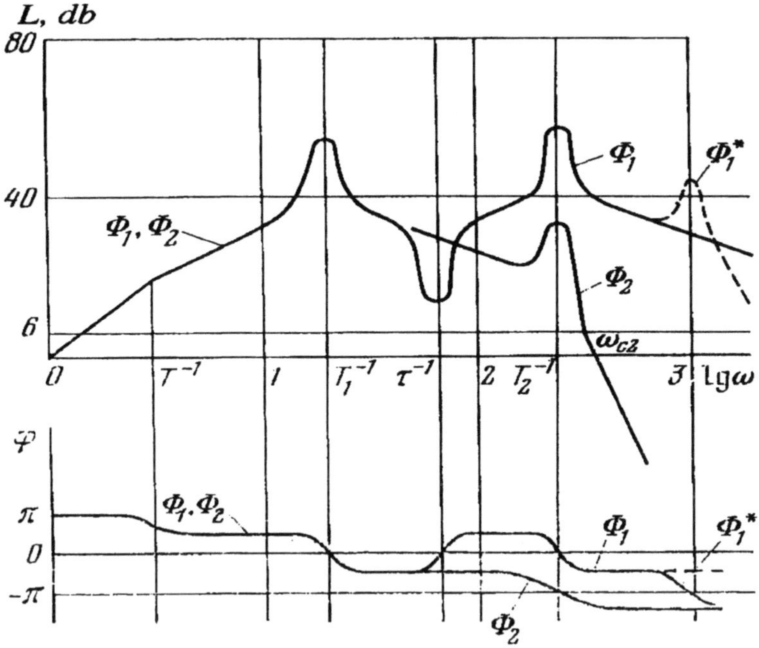

FIGURE 2.28.

characterises the influence of control on the displacement of the lower mass. Action acting on the base is related to this displacement by means of the following expression

$$R = [bp + c + kw_1(p)]\, z_A .$$

Comparing the values of forces with and without control we obtain a transfer function describing the efficiency of protecting the base

$$\varkappa_1(p) = \frac{bp + c + kw_1(p)}{bp + c}\left[1 + e_A(p)kw_1(p)\right]^{-1} = \frac{1 + e'(p)kw_1(p)}{1 + e_A(p)kw_1(p)}, \quad (2.125)$$

where $e'(p) = (bp + c)^{-1}$ is the dynamic compliance of the lower stage on the passive system.

Also in this case, the efficiency analysis can be carried out by means of the LAC. To this end, we introduce transfer function $\Phi_1(p)$ related to $\varkappa_1(p)$ as follows

$$\varkappa_1(p) = \left[1 + \Phi_1(p)\right]^{-1} .$$

Transforming $\Phi_1(p)$ yields

$$\Phi_1(p) = \frac{[e_A(p) - e'(p)]\, kw_1(p)}{1 + e'(p)\, kw_1(p)} = -\frac{\left[m_1 m_2 p^4 + (m_1 + m_2)\, p^2 (\beta p + \gamma)\right] kw_1(p)}{[bp + c + kw_1(p)]\, \Delta(p)}.$$

Inserting $w_1(p) = p^{-1}$ we obtain

$$\Phi_1(p) = -k\frac{p^2[m_1 m_2 p^2 + (m_1 + m_2)(\beta p + \gamma)]}{\Delta(p)(bp^2 + cp + k)}. \tag{2.126}$$

We take the same numerical values for the object parameters as before. As will be shown in the next Section the value of gain k can not exceed the following value

$$k_{\max} = T_1^{-1}\left|e_A^{-1}\left(iT_1^{-1}\right)\right| \approx 7kg/s^3.$$

It is easy to see that such a small value of k has actually no influence on the denominator. Taking this fact into account, we recast $\Phi_1(p)$ in the following form

$$\Phi_1(p) = -k_1 p\frac{\tau^2 p^2 + 2\tau\eta p + 1}{(T_1^2 p^2 + 2\zeta_1 T_1 p + 1)(T_2^2 p^2 + 2\zeta_2 T_2 p + 1)(T'p + 1)}, \tag{2.127}$$

where

$$k_1 = k\frac{m_1 + m_2}{c^2}, \qquad \tau = \sqrt{\frac{\gamma(m_1 + m_2)}{m_1 m_2}} = 0,0076\, s,$$

$$\eta = \frac{\beta(m_1 + m_2)}{2 m_1 m_2 \tau} = 0,083, \quad T' = \frac{b}{c} = 0,00096\, s.$$

Figure 2.29 displays LAC of $\Phi_1(p)$ for $k_1 = 1$ s. The line $L = 6$ db defines the domain of the control efficiency.

Various aspects of practical design of active facilities are closely connected to the problem of efficiency. Investigating efficiency allows one to settle the requirements to sensors and actuators in definite terms. For instance, let us consider the two stage active system depicted in Fig. 2.27 whose open-loop transfer function is given by eq. (2.118). Let us derive the dependence between the base displacement $\xi(t)$ which is a kinematic excitation and the control force $U(t)$ formed by the active facility. From the equation of motion for the system

$$\left.\begin{aligned} m_1 p^2 z_A &= -(bp + c)(z_A - \xi) - (\beta p + \gamma)(z_A - z_B) - \frac{kp^2}{Tp + 1} z_A, \\ m_2 p^2 z_B &= -(\beta p + \gamma)(z_B - z_A), \end{aligned}\right\} \tag{2.128}$$

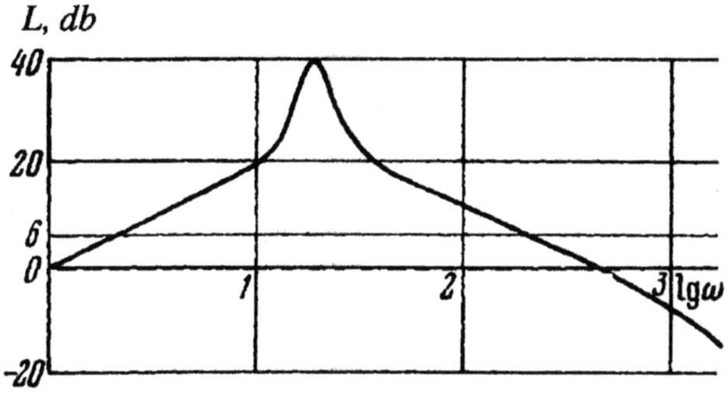

FIGURE 2.29.

we obtain

$$z_A = \frac{(bp+c)\left(m_2p^2+\beta p+\gamma\right)}{\Delta(p)+\dfrac{kp^2}{Tp+1}}\xi.$$

Therefore,

$$U = -\frac{kp^2}{Tp+1}z_{A,} = -\frac{\Phi_1(p)}{1+\Phi_1(p)}(bp+c)\xi. \qquad (2.129)$$

This expression indicates that at those frequencies, for which $|\Phi_1(i\omega)| \gg 1$ and the control becomes highly efficient, one can adopt the following

$$U(t) \approx -(bp+c)\,\xi(t)\,. \qquad (2.130)$$

In other words, a highly efficient system of closed-loop control turns out actually equivalent to a control system due to disturbance, providing a complete compensation of this disturbance. For the chosen parameters, at frequencies at which $\omega b \ll c$ we obtain

$$U_{\max} \approx c\xi_{\max}.$$

If $\xi_{\max} = 10^{-3}$ m, then $U_{\max} = 520$ N which is quite acceptable from the perspective of practical design.

Let us estimate the vibration amplitude of the lower mass. It follows from eq. (2.129) that

$$z_{A_{\max}} = \frac{\sqrt{1+T^2\omega^2}}{k\omega^2}U_{\max}.$$

At frequency $\omega = 10$ s^{-1} and $k = 5\cdot 10^5$ kg which corresponds to the LAC of Fig. 2.28 we obtain

$$z_{A\max} \approx \frac{\sqrt{1+T^2\omega^2}}{k\omega^2}c\xi_{\max} \approx 0,03\xi_{\max}.$$

For $\xi_{\max} = 10^{-3}$ m we get that $z_{A\max} = 3 \cdot 10^{-5}$ m and the acceleration of the lower mass is $3 \cdot 10^{-3}$ m/s^2 (at frequency $\omega = 10\ s^{-1}$). This is the required sensitivity of the sensor which is an accelerator in this example.

2.3 Conditions for stability of active systems

In order to properly function, any active system must be *stable* and possesses a certain *stability margin*, i.e. the system stability must not be affected by occasional deviations in system parameters. In what follows, we remember the basics of the stability theory of linear time-invariant systems, details being adequately covered in the vast literature, see e.g. [35], [24], [99], [20] etc.

Consider a system of linear homogeneous differential equations with constant coefficients which can be written in a vector form

$$\dot{\mathbf{x}} = A\mathbf{x}. \tag{2.131}$$

Here $\mathbf{x}$ is an $n-$ dimensional vector and A is a constant matrix. System (2.131) is said to be *stable* if, all its solutions $\mathbf{x} = \mathbf{x}(t, \mathbf{x}_0, t_0)$ due to initial conditions $\mathbf{x}(t_0, \mathbf{x}_0, t_0) = \mathbf{x}_0$ are bounded for all $t > t_0$. In this definition, $\mathbf{x}(t)$ is termed to be *bounded* if its components are all bounded. If all solutions of eq. (2.131) are stable in the above sense, and in addition to this, satisfy the following condition

$$\lim_{t\to\infty} \mathbf{x}(t) = \mathbf{0}, \tag{2.132}$$

then system (2.131) is called *asymptotically stable.*

If system (2.131) is stable it is easy to prove that all solutions of the inhomogeneous equation

$$\dot{\mathbf{y}} = A\mathbf{y} + \mathbf{f}(t) \tag{2.133}$$

are bounded for any bounded function $\mathbf{f}(t)$. In other words, the presence of function $\mathbf{f}(t)$ does not affect stability.

For system (2.131) to be asymptotically stable, it is necessary and sufficient that the eigenvalues of λ_1, ..., λ_n of matrix A, i.e. the roots of the following characteristic equation

$$\det(A - \lambda I_n) = 0 \tag{2.134}$$

have negative real values. For mere (non-asymptotic) stability it is necessary and sufficient that the real parts of all eigenvalues are non-positive, and additionally, the deficiency number of matrix $A - \lambda I_n$ is equal to multiplicity of eigenvalue λ_k.

Designing vibration protection systems one usually deals with conditions of asymptotic stability. Vibration protection systems which are not asymptotically stable are not encountered in practice.

Determinant (2.134) can be cast as follows

$$a_0\lambda^n + a_1\lambda^{n-1} + \ldots + a_{n-1}\lambda + a_n = 0 \ (a_0 > 0). \tag{2.135}$$

There exist a number of criteria which enable one to judge the stability of system (2.131) in terms of the factors of the characteristic equation (2.135) without solving this equation. In what follows, we use the *Stodola criterion* and the *Hurwitz criterion.*

According to the Stodola criterion, for asymptotic stability it is *necessary* that all coefficients are positive (note that $a_0 > 0$, eq. (2.135)).

The Hurwitz criterion says that a *necessary and sufficient condition* for asymptotic stability is that all the principal minor determinants of the *Hurwitz determinant*

$$H = \begin{vmatrix} a_1 & a_0 & 0 & 0 & \ldots\ldots & 0 \\ a_3 & a_2 & a_1 & a_0 & \ldots\ldots & 0 \\ a_5 & a_4 & a_3 & a_2 & \ldots\ldots & 0 \\ \ldots\ldots & \ldots\ldots & \ldots\ldots & \ldots\ldots & \ldots\ldots & \ldots\ldots \\ a_{2n-1} & a_{2n-2} & a_{2n-3} & a_{2n-4} & \ldots\ldots & a_n \end{vmatrix} \tag{2.136}$$

are positive. The principle of determinant construction is transparent from formula (2.136), furthermore $a_i = 0$ for $i > n$.

Governing equations for control systems are conventionally cast in the form which differs from that given by eq. (2.133). Provided that a system has k inputs and s outputs, these equations are written as follows

$$Q(p)\mathbf{x} = R(p)\mathbf{f}\,(t). \tag{2.137}$$

Here $Q(p)$ and $R(p)$ are respectively $s\times s$ and $s\times k$ polynomial matrices, $\mathbf{f}\,(t)$ is $k-$dimensional vector of input variables and $\mathbf{x}$ is $s-$dimensional vector of output variables. It is evident that system (2.137) can be reexpressed in the form of eq. (2.131), however this is not required for stability analysis. It is easy to show that the characteristic equation in this case is cast in the following form

$$\det Q(\lambda) = 0. \tag{2.138}$$

All stability conditions stated above hold. Provided that the system has only one input and only one output eq. (2.137) is a scalar equation

$$q(p)x = r(p)f\,(t), \tag{2.139}$$

which can be recast as follows

$$x = v(p)f\,(t). \tag{2.140}$$

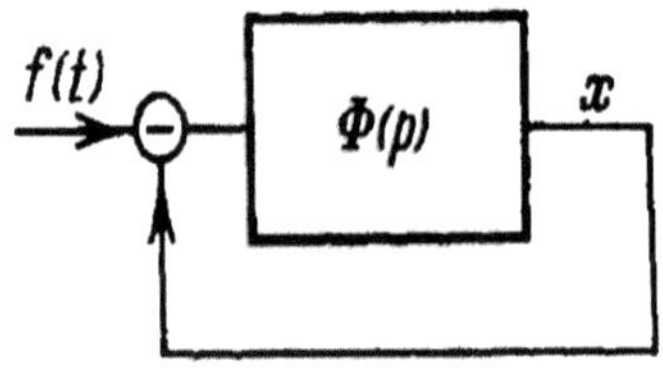

FIGURE 2.30.

Here $q(p)$ and $r(p)$ are polynomials and $v(p) = r(p)/q(p)$ is the system transfer function. The characteristic equation

$$q(\lambda) = 0 \tag{2.141}$$

readily reduces to the form of eq. (2.135). As the roots of eq. (2.141) are the same as the poles of the transfer function, one can state that for asymptotic stability of system (2.140) it is necessary and sufficient that all poles of the transfer function lie in the left half of the complex λ-plane. Therefore, the Hurwitz criterion which is an algebraic one can be used for analysing stability of one-dimensional systems. In practice, however, this criterion, as well as the Rowth criterion, turns out to be convenient only for systems of not very high order. This criterion is also not convenient for synthesis of stable systems since it does now allow one to determine the stability margin. For this reason, another criteria directly related to properties of transfer functions of open-loop system are used in control theory.

Assume that system (2.139) has the block diagram shown in Fig. 2.30 where $\Phi(p)$ denotes the open-loop transfer function of the system. Then,

$$v(p) = \frac{\Phi(p)}{1 + \Phi(p)}. \tag{2.142}$$

Let us also assume that function $\Phi(p)$ has no poles in the right half of the complex plane, however poles on the imaginary axis and the origin may exist. This assumption is valid for any control system with a single feedback, the latter being composed of stable elements. Additional feedbacks applied to the elements are assumed to be absent. In this case the *Nyquist criterion* holds: for asymptotic stability of system (2.139) it is *necessary and sufficient* that the polar plot $\Phi(i\omega)$ of the open-loop system, which is referred to as the Nyquist plot, does not encircle point $-1 + i0$ and do not coincide with this point when the frequency ω grows from $-\infty$ to $+\infty$.

If function $\Phi(p)$ has s poles in the right half plane then a necessary and sufficient condition for asymptotic stability of the closed-loop system is that the Nyquist plot $\Phi\,(i\omega)$ encircles point $-1 + i0$ in a counterclockwise sense as many times as $s/2$ when ω increases from 0 to $+\infty$. In practice, one counts the number of intersections of the polar plot with the abscissa axis on the part $(-\infty, -1)$ rather than number of encirclements. An intersection is treated to be positive when the abscissa axis is crossed from below and

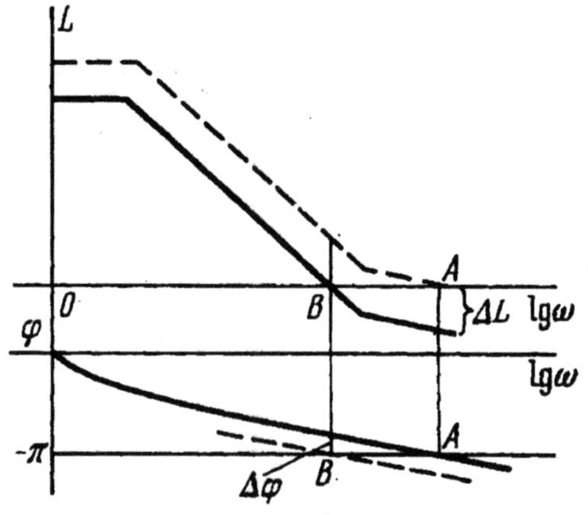

FIGURE 2.31.

negative when from above. For asymptotic stability the algebraic sum of these intersections must be equal to $s/2$. If the origin or end point of polar plot lies on the part $(-\infty, -1)$ of the abscissa axis, this is deemed as a half intersection.

A number of intersections can be easily determined with the help of the Bode plot for the open-loop system. If $\Phi(i\omega)$ is a real-valued number which lies between $-\infty$ and -1, then $\log|\Phi(i\omega)| > 0$ and $\arg\Phi(i\omega) = -\pi$. Thus, each intersection of polar plot $\Phi(i\omega)$ with the abscissa axis is equivalent to the intersection of the LPC with a level $-\pi$ provided that the LAC is positive. For asymptotic stability it is necessary and sufficient that number of these intersections from below exceeds that from above by $s/2$.

While investigating the Bode plot of open-loop systems one can estimate the degree of stability of the closed-loop system. *The gain margin* is the number of decibels by which the gain can be increase before the closed-loop system becomes unstable. *The phase margin* is the difference between the value of phase at the *gain-crossover frequency* (more precisely, at frequency ω_* at which $|\Phi(i\omega_*)| = 1$ and $|\Phi(i\omega_*)| < 1$ when $\omega > \omega_*$) and $-\pi$. Figure 2.31 shows a way of determining gain ΔL and phase $\Delta\varphi$ margins by means of the Bode plot. It is seen that multiplying the gain by $10^{0,05\Delta L}$ we shift the LAC upwards by ΔL db. The condition of the Nyquist criterion is now violated since the LPC intersects the level $-\pi$ at point A for which $L > 0$. A similar violation occurs at point B when the phase lag is increased by $\Delta\varphi$.

Let the coefficients of the characteristic equation of a linear system depend upon α_1, ..., α_k which are parameters of this system. A set of values of these parameters, for which the stability conditions hold, forms a *stability region* in parameter space. For global analysis of system behaviour and design there often requires the need to find the boundaries of the stability

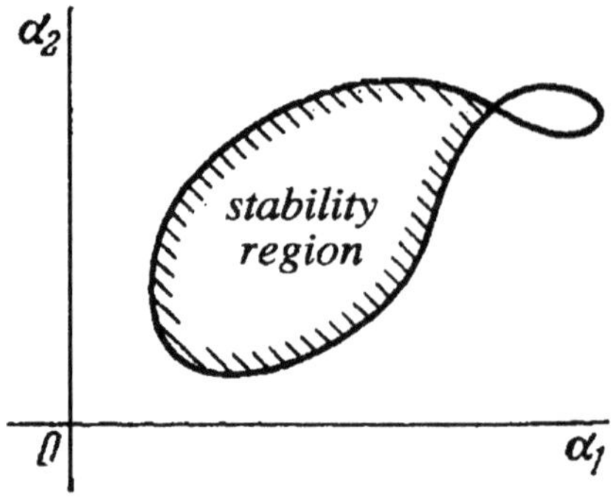

FIGURE 2.32.

region, i.e. the critical values for the system parameters at which the system is no longer stable. There are several approaches to determining the boundaries of the stability region. In simple cases one can apply the Hurwitz criterion. It can be shown that the determinant of $(n-1) \times (n-1)$ principal minor becomes zero on the boundary of the stability region. In complicated cases, the method of subdivision is used, and is briefly explained in what follows.

Assume that the coefficients of the characteristic equation (2.135) and, thus, its roots are continuous functions of parameters. As all roots have negative real values in the region of asymptotic stability, the real value of at least one root must become zero on the region boundary. Either a zero root exists which is possible only if $a_n = 0$ or a pair of pure imaginary roots appear. Let us fix the values of all parameters except, say, α_1 and α_2, and find the boundary of the stability region in the plane $\alpha_1\alpha_2$. To this end, we substitute $\lambda = i\omega$ into eq. (2.135). Equating real and imaginary values of the equation obtained to zero yields

$$\left.\begin{aligned} a_0\omega^n - a_2\omega^{n-2} - a_4\omega^{n-4} - \ldots &= 0, \\ a_1\omega^{n-1} - a_3\omega^{n-3} - a_5\omega^{n-5} - \ldots &= 0. \end{aligned}\right\} \tag{2.143}$$

As coefficients a_i depend only on α_1 and α_2 (for fixed $\alpha_3, \ldots, \alpha_k$) eq. (2.143) can be recast as follows

$$\left.\begin{aligned} f_1(\alpha_1, \alpha_2, \omega) &= 0, \\ f_2(\alpha_1, \alpha_2, \omega) &= 0. \end{aligned}\right\} \tag{2.144}$$

These are equations for a locus in plane $\alpha_1\alpha_2$ which ensures the existence of a pair of purely imaginary roots of the characteristic equation. The boundary of the stability region must be a part of this locus. Plotting curve (2.143) in plane $\alpha_1\alpha_2$, see Fig. 2.32, one can decide which part of the plane separated by this locus corresponds to the stability region. This

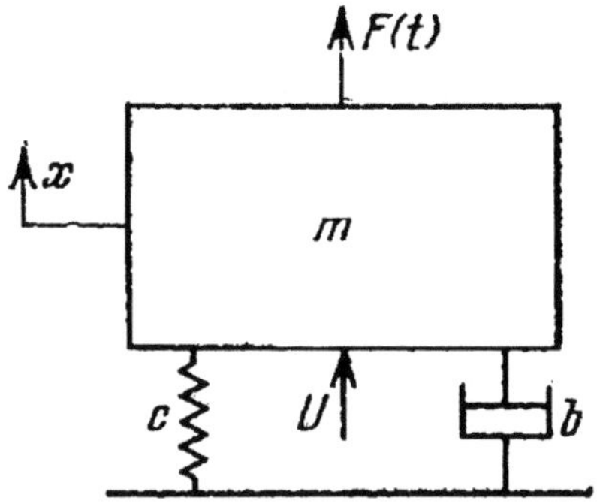

FIGURE 2.33.

can be easily proven for example by the Hurwitz criterion for some chosen points.

Linear passive vibration protection systems are known to be asymptotically stable. Introducing active elements possessing additional energy sources may lead to a system instability. For example, we consider a simple single-degree-of-freedom system which has an active facility forming a control force proportional to the integral of the absolute displacement, Fig. 2.33. With $F(t)$ denoting dynamic action, the governing equation for the mass m is written in the form

$$(mp^2 + bp + c)x = F(t) + U(t). \tag{2.145}$$

Inserting

$$U(t) = -\frac{k}{p}x \tag{2.146}$$

into eq. (2.145) we obtain

$$(mp^2 + bp + c + \frac{k}{p})x = F(t), \tag{2.147}$$

or

$$x(t) = \frac{p}{mp^3 + bp^2 + cp + k}F(t). \tag{2.148}$$

Rewriting the characteristic equation in the form of eq. (2.141)

$$m\lambda^3 + b\lambda^2 + c\lambda + k = 0 \tag{2.149}$$

and applying the Hurwitz criterion we obtain conditions for the asymptotic stability in the following form

$$m > 0,\ b > 0,\ c > 0,\ k > 0,\ mk < bc. \tag{2.150}$$

The latter condition indicates that feedback gain k can not be arbitrarily large since this is bounded by the following value

$$k_{\max} = \frac{bc}{m}. \tag{2.151}$$

The efficiency of control has been shown to increase with the growth of feedback gain k, therefore, condition (2.151) limits the achievable efficiency. This observation is also typical for other, more complex cases which are convenient to study in terms of Bode plots. Variants of active facilities analysed in Section 2.2 are considered below in detail.

By analysing the Bode plot of a system with control proportional to the absolute displacement, see Fig. 2.17, it is easy to find that an increase in feedback gain k does not lead to instability since the LPC does not cross the level of $-\pi$ at all. This result can be easily explained as follows. Control proportional to the absolute displacement can be realised by means of a passive facility, namely, a spring mounting between point A and a fixed point. Clearly, regardless of the spring rigidity (corresponding to feedback gain k) the system remains passive and, thus, stable.

A similar result is also obtained for control due to velocity and accelerations with transfer functions (2.90) and (2.92). However, feedback gain may be increased without bound only in the case in which control involves ideal non-inertial elements, i.e. passive elements. Using real active elements inevitably results in limitations with respect to feedback gain.

Indeed, as shown in Section 2.2, the LPC crosses level $-\pi$ at point A in Fig. 2.17 when one introduces an aperiodic element (2.106) into the feedback which forms a control force proportional to displacement. For stability of the closed-loop system it is necessary and sufficient that the gain-crossover frequency ω_c is smaller than the phase-crossover frequency ω_* corresponding to $\varphi = -\pi$. Hence, feedback gain k can be increased until the gain-crossover frequency ω_c reaches the phase-crossover frequency ω_*.

It is worthwhile noting that in all cases $\omega_* > T^{-1}$. However the value of difference $\omega_* - T^{-1}$ depends on the relationships between T_s, τ_k and T and may be relatively small. Thus, in order to ensure system stability it is expedient to meet the following condition

$$\omega_c < T^{-1}. \tag{2.152}$$

If the feedback transfer function has a more complex appearance than that of eq. (2.106), i.e. its denominator contains a product of several polynomials with time constants T, T' , T'' etc., then $\omega_* > T_{\max}^{-1}$ can be guaranteed. For this reason, it is sufficient to satisfy the following condition

$$\omega_c < T_{\max}^{-1}. \tag{2.153}$$

Let us proceed to the system with control force due to velocity. In the presence of an ideal feedback (2.90) phase φ varies between $-\pi/2$ and $\pi/2$, see Fig. 2.18. In this case this system has a phase margin not less than $\pi/2$ for any feedback gain k. For this reason, introducing an aperiodic element which shifts the phase by $\pi/2$ can not result in instability. Instability becomes feasible if a vibratory element or two aperiodic elements in series are introduced in feedback. This case is shown in Fig. 2.18. In order to ensure stability it is sufficient to satisfy condition (2.153) and, in general, the phase margin turns out to be larger than in the previous case.

Let us consider a system with control force due to acceleration, Fig. 2.19. With an ideal transfer function (2.92) phase φ varies between 0 and π. Thus, intersecting level $\varphi = -\pi$ is possible only by introducing an element with phase lag greater than π. For example, this is the case for the LAC and the LPC shown in Fig. 2.19 with a dash-dot line. Notice that in this case $\omega_* > T_*^{-1}$ where T_* is the second largest time constant. Stability is provided if

$$\omega_c < T_*^{-1}. \tag{2.154}$$

Strict limitations on feedback gain are imposed by stability conditions on systems with control forces due to the integral of displacement. As shown above, in this case instability becomes possible for ideal transfer function (2.100), see condition (2.151). Essentially, introducing a control force due to the integral is equivalent to rotation of the polar plot for dynamic compliance $e_A(i\omega)$ (Fig. 1.8) by angle $\pi/2$ in a clockwise sense. It is easy to see that phase φ is $-\pi$ at the first resonance frequency T_0^{-1} and the stability condition is violated if $\left|\Phi(T_0^{-1})\right| > 1$.

Condition $\left|\Phi(T_0^{-1})\right| < 1$ is equivalent to the following one

$$L(T_0^{-1}) < 0. \tag{2.155}$$

In other words, the first resonant maximum of the LAC must lie below the abscissa axis. This implies that the region of efficient control due to the integral must cover a frequency domain of rather low frequencies, namely $\omega \ll T_0^{-1}$. An admissible location of the LAC is shown in Fig. 2.20 with a dash-dot line. Clearly, there is no need to account for the fact that the feedback elements are not ideal since any change in the LAC due to these elements lie beyond the crossover frequency.

The above-said allows us to draw some conclusions as to what information about the dynamic characteristics of an object it is necessary to possess in order to design an active vibration protection system. It is evident that it is sufficient to have a general idea of dynamic compliance $e_A(i\omega)$ at frequencies lower than crossover frequency ω_c because this part of the frequency characteristic determines whether or not the conditions for efficiency and stability are satisfied. Taking into account eqs. (2.153) and (2.154) we can assert that the region of identification of $e_A(i\omega)$ must be determined by time constants of the feedback elements. The smaller these constants the broader the frequency domain in which the control can be efficient and the frequency characteristic can be identified.

An exception is the case of control due to displacement integration. In this case for synthesis it is sufficient to know the value of the first resonant frequency of the object and the value of $|e_A(i\omega)|$ at this frequency.

It follows from the above-said that control having transfer function (2.94) which compensates for the "trough" in the LAC may lead, in general, to instability because the control is analogous to the control due to displacement with an additional vibratory element. By compensating for the "trough" by means of feedback of type (2.99) we can avoid instability.

Next we consider the question as to what features are typical for synthesis of active systems in the case when the point of control and the point under observation do not coincide. To this purpose, we consider the open-loop polar plots shown in Fig. 2.22.

In the case of displacement control the polar plot has a form schematically depicted in Fig. 2.22a. This plot differs from the polar plot for $e_{AB}(i\omega)$, Fig. 1.9, only in that regard that each radius vector is multiplied by k. Polar plot $\Phi(i\omega)$ crosses the abscissa axis at some frequency ω_*. Consistent with the Nyquist criterion it is necessary that the following inequality

$$\Phi(i\omega_*) = ke_{AB}(i\omega_*) > -1 \tag{2.156}$$

holds, otherwise the system is unstable. Remember that the value of ω_* lies between those adjacent eigenfrequencies of the system λ_{r-1} and λ_r for which the coefficients of the normal modes at points A and B satisfy the following conditions

$$l_{As}l_{Bs} > 0, \; s = 1, ..., r-1; \; l_{Ar}l_{Br} < 0. \tag{2.157}$$

This permits us to explain the violation of stability conditions from a physical perspective. In the case of free vibration with eigenfrequencies $\lambda_s \leq \lambda_{r-1}$ the phases of vibration at points A and B ensure that a rigid active feedback is negative for $k > 0$. At free vibration with frequency λ_r the phase of displacement $z_B(t)$ increases by π, and, in actuality, the feedback becomes positive and supports free vibrations. Clearly, an instability is inevitable for a sufficiently large gain.

An analogous situation takes place in the case of control due to velocity, $w(p) = k_1 p$, see Fig. 2.22b. In this case the polar plot crosses the negative part of the abscissa axis at a frequency which is approximately equal to λ_r which is the first eigenfrequency for which the coefficients of the normal modes l_{Ar} and l_{Br} have opposite signs. In the case of control due to acceleration, see Fig. 2.22c, polar plot $\Phi(i\omega)$ crosses the negative part of the abscissa axis at that frequency ω_{**} at which polar plot $e_{AB}(i\omega)$ returns from the upper half plane to the lower one. Frequency ω_{**} lies between those eigenfrequencies λ_{r-1} and λ_r for which

$$l_{A,r-1}l_{B,r-1} < 0, \; l_{Ar}l_{Br} > 0. \tag{2.158}$$

This is indicative of a more broad potential inherent in control due to acceleration. In principle, this type of control can suppress vibrations with any frequencies under ω_{**} whereas the frequency domain of control due to displacement is bounded by frequency ω_*.

In the case of control due to integration of the displacement, Fig. 2.22d, the relative position of the control point and the point under observation do not actually influence the stability condition because the latter is determined by the first system eigenfrequency corresponding to rigid-body vibration of the object.

One of the most efficient approaches to the synthesis of stable active systems is using compensators in feedback circuits. The idea behind a compensator is to provide such a change in the LPC of the open-loop system which prevents the crossing level $\varphi = -\pi$. The elements providing phase leads, i.e. first and second order elements (2.83), satisfy these conditions.

Let, for instance, the LPC of a open-loop system cross level $\varphi = -\pi$ at some frequency ω_*. Let us introduce a first order element with the following transfer function

$$w(p) = \tau' p + 1, \tag{2.159}$$

where τ' is chosen so that $\tau' > \omega_*^{-1}$. Then, at all frequencies higher than $(\tau')^{-1}$, and therefore, at frequency ω_* phase φ increases. The increase is about $\pi/2$ if $\tau' \gg \omega_*^{-1}$. Using a second-order element with the following transfer function

$$w(p) = \tau'^2 p^2 + 2\tau'\eta' p + 1 \tag{2.160}$$

one can increase the phase by π however introducing element (2.160) results in a drop of the LAC at frequency $\omega = (\tau')^{-1}$ which is analogous to an antiresonant drop considered above.

Compensators can provide stability to astatic systems with control due to displacement integration if the transfer function is taken in the following form

$$w(p) = \frac{\tau' p + 1}{p}. \tag{2.161}$$

In this case, the value of $(\tau')^{-1}$ must be essentially smaller than the first eigenfrequency of the system. Notice that realisation of first and second-order elements with large time constants is associated with certain technical difficulties since it requires the design of mechanical facilities to provide high damping and low rigidity.

Let us investigate the stability of active facilities whose efficiency was studied in Section 2.2. To begin with, we consider the system depicted in

Fig. 2.23. Figure 2.24 indicates that the LPC of the corresponding open-loop system varies in the interval $(-\pi/2, \pi/2)$. Hence, the closed-loop system remains stable for any value of gain k. It is evident that this assertion is valid inasmuch as transfer function (2.110) may be treated as being adequate for the system under consideration. In reality, however, the crossover frequency increases with the growth of gain k, which enlarges the frequency domain under which the system can be completely modelled. This may result in the necessity to account for elastic vibrations of the object to be protected and inertial properties of feedback elements with small time constants. Associated changes in the system structure may lead to violation of the stability conditions for sufficiently large gain.

Let us now consider a system with control force due to displacement integration, Fig. 2.25. As one sees from Fig. 2.26 the system is unstable for the chosen value of gain k $(k = 1)$. Indeed, the LPC crosses level $\varphi = \pi$ at frequency $\omega = T_0^{-1}$, at which $L(\omega) \approx 0$. In the system under consideration there will occur vibrations of the protected object with frequency T_0^{-1} which is frequency of free vibration of the object mounted on passive isolators. The amplitude of this vibration will increase until the system nonlinearities do not bound it. In other words, self-sustained vibration supported by the active feedback will occur in the system.

One can indicate several ways of removing instability. Firstly, the gain can be decreased to such a value at which $L\left(T_0^{-1}\right) < 0$. In this case it is necessary to shift the LAC downwards by 3 db which corresponds to $k = 0{,}7$. Clearly, the control efficiency will be reduced, too. Secondly, one can increase the damping in the passive isolator by increasing ζ_0 up to 0,14-0,15. Thirdly, one can introduce an additional compensator into the feedback circuit which provides an increase in phase φ at $\omega < T_0^{-1}$. For instance, a lead element with transfer function $w(p) = \tau' p + 1$ $(\tau' > T)$ may be used.

Let us investigate the stability of the active vibration protection systems shown in Fig. 2.27. If the control force is due to displacement of the lower mass, then, as the Bode plot for the open-loop system shows, see Fig. 2.28, the closed-loop system is stable for any value of gain because the LPC for $\Phi_1(p)$ does not reach level $\varphi = -\pi$. However, increasing the gain results in increasing crossover frequency ω_c. This may lead to the need to account for the time constants of feedback elements which were neglected while constructing transfer function $\Phi_1(p)$.

Let for example the largest from these time constants T_* be 0,001 s and the corresponding element be a vibratory one.

Then, the transfer function of the open-loop system should be written in the form

$$\Phi_1^*(p) = \frac{kp^2(m_2p^2 + \beta p + \gamma)}{\Delta(p)(Tp+1)(T_*^2p^2 + 2\zeta_* T_* p + 1)}, \tag{2.162}$$

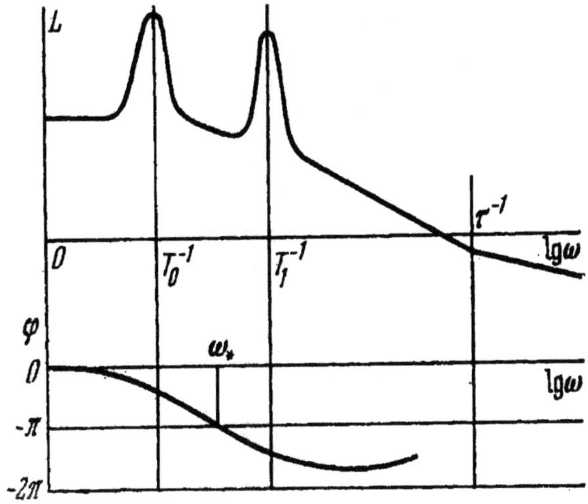

FIGURE 2.34.

where ζ_* is damping factor which we will assume to be $\zeta_* = 0,1$. Changes in the Bode plot caused by the vibratory element introduced are shown in Fig. 2.28 with dash-dot lines. The LPC crosses level $\varphi = -\pi$ at frequency $\omega = T_*^{-1}$ at which $L\left(T_*^{-1}\right) > 0$, i.e. the system is unstable. The instability is due to violation of condition (2.154). To remove the cause of this, it is necessary either to decrease the gain or introduce a compensator having the following transfer function

$$w(p) = {\tau'}^2 p^2 + 2\tau'\eta' p + 1, \tag{2.163}$$

where τ' is chosen so that $T_2 > \tau' > T_*$. Introducing this correction provides an increase in phase φ by π at frequency $\omega > (\tau')^{-1}$ which removes the instability.

Figure 2.34 displays the Bode plot for a two-mass system having control due to displacement of the upper mass. The corresponding transfer function is given by

$$\Phi(p) = \frac{k}{c+\gamma} \cdot \frac{\tau p + 1}{(T_0^2 p^2 + 2\zeta_0 T_0 p + 1)(T_1^2 p^2 + 2\zeta_1 T_1 p + 1)}, \tag{2.164}$$

where $T_0 > T_1 > \tau$. Phase φ crosses level $-\pi$ on the part $T^{-1} < \omega < T_1^{-1}$. If the corresponding value of the phase crossover frequency ω_* is lower than the gain crossover frequency the systems becomes unstable. Thus, provided that the control force acting on the lower mass is proportional to the displacement of the upper mass, then increasing gain causes instability.

Consider now the system shown in Fig. 2.35. Here the object to be protected is a rigid body whose position is defined by coordinates x_C, y_C and

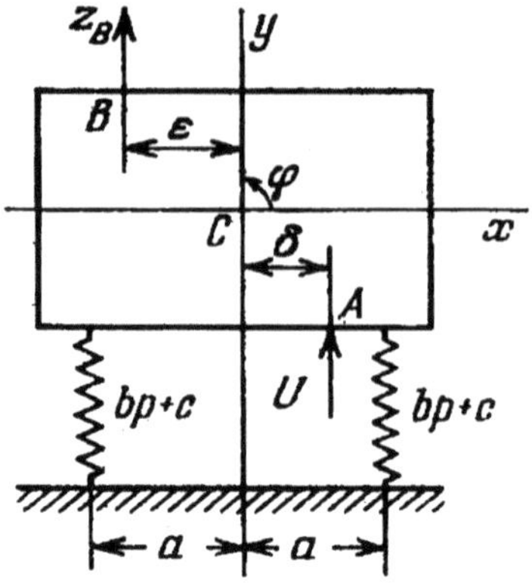

FIGURE 2.35.

φ. Passive isolators of rigidity c and damping factors b are attached to the object at a distance a from the object axis. The object mass and the inertia radius about axis C are m and ρ, respectively. In this case

$$z_A = y_C + \delta\varphi, \; z_B = y_C - \varepsilon\varphi. \tag{2.165}$$

Constructing an expression for $e_{AB}(p)$ we obtain, after some manipulation, the following equation for dynamic compliance

$$\begin{aligned} e_{AB}(p) &= \frac{m(\rho^2 - \varepsilon\delta)p^2 + 2(a^2 - \varepsilon\delta)(bp + c)}{[mp^2 + 2(bp + c)][m\rho^2 p^2 + 2a^2(bp + c)]} \\ &= e_{AB}(0)\frac{\tau^2 p^2 + 2\tau\eta p + 1}{(T_0^2 p^2 + 2\zeta_0 T_0 p + 1)(T_1^2 p^2 + 2\zeta_1 T_1 p + 1)}, \end{aligned} \tag{2.166}$$

where

$$\begin{aligned} \tau^2 &= \frac{m(\rho^2 - \varepsilon\delta)}{2c(a^2 - \varepsilon\delta)}, \; T_0^2 = \frac{m}{2c}, \; T_1^2 = \frac{m\rho^2}{2ca^2}, \; \eta = \frac{b}{2c\tau}, \\ \zeta_0 &= \frac{b}{2cT_0}, \; \zeta_1 = \frac{ba^2}{2cT_1\rho^2}, \; e_{AB}(0) = \frac{a^2 - \varepsilon\delta}{2ca^2}. \end{aligned} \tag{2.167}$$

If values of ε and δ are chosen so that

$$1 < \frac{\rho^2 - \varepsilon\delta}{a^2 - \varepsilon\delta} < \frac{\rho^2}{a^2}, \tag{2.168}$$

or

$$\frac{\rho^2}{a^2} < \frac{\rho^2 - \varepsilon\delta}{a^2 - \varepsilon\delta} < 1, \tag{2.169}$$

then values of T_0, τ and T_1 satisfy the following condition

$$T_0^{-1} < \tau^{-1} < T_1^{-1}. \tag{2.170}$$

If additionally $a^2 - \varepsilon\delta > 0$ and $\rho^2 - \varepsilon\delta > 0$, then expression (2.166) has the same structure as that of dynamic compliance at a point. This coincidence is the most favourable from the perspective of satisfying the stability condition. It is easy to show that conditions (2.168) and (2.169) are satisfied if and only if $\varepsilon\delta < 0$, i.e. if points A and B lie on the same side of the object axis.

Assume that by increasing the feedback gain we achieve the boundary of the system stability region. In this case, due to the Nyquist criterion the polar plot $\Phi(i\omega)$ must cross point $-1 + i0$. On the other hand, the gain crossover frequency must coincide with the phase crossover frequency when the boundary of the stability region is achieved. Hence, the gain crossover frequency satisfies the following condition

$$\Phi(i\omega_c) = -1. \tag{2.171}$$

A differential equation corresponding to the closed-loop system is written in the form (2.5). From this it follows that the roots of the characteristic equation must in general coincide with the roots of the following equation

$$1 + \Phi(\lambda) = 0. \tag{2.172}$$

Comparing eqs. (2.171) and (2.172) it is easy to see that on the boundary of the stability region one of the roots of the characteristic equation is $i\omega_c$. This means that harmonic "free" vibrations of frequency ω_c are observed when the boundary of the stability region is achieved. After the boundary has been crossed these vibrations increase in time and, in the presence of nonlinear constraints, become self-sustained vibrations of approximately the same frequency.

Thus, we can state that the frequency of self-sustained vibration occurring in active systems outside the stability region (however in the vicinity of the boundary) is approximately equal to the crossover frequency for that gain which brings the system to the stability boundary.

The following conclusions can be drawn from Sections 2.2 and 2.3:

1. The choice of a place for providing information about a vibrational field (the so called point under observation) and a place where the control force formed by feedback is applied (the point of control) is of crucial importance for synthesis of active vibration protection systems. Choosing these places, it is necessary to account for values of coefficients of normal modes in both points. If the control objective is to reduce vibration at a certain point C, the above-said does not necessarily imply that the sensor must measure vibration at this point. It may turn out that a maximum control efficiency is achieved when another point under observation, e.g. the point of control, is chosen.

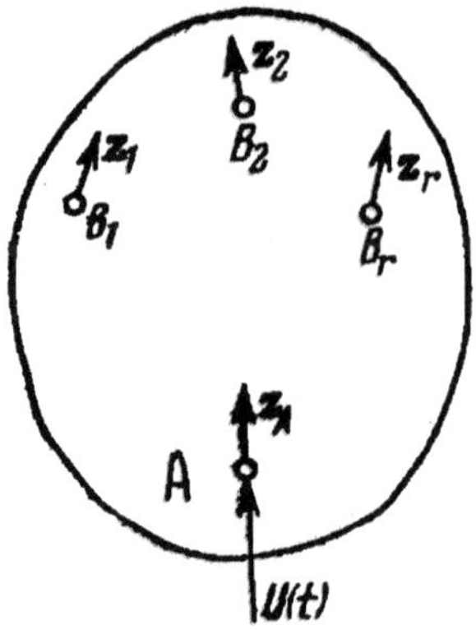

FIGURE 2.36.

2. Let an efficient controller in the frequency band $0 < \omega < \omega_{\max}$ be sought. The synthesis of an appropriate active system demands knowledge about the frequency characteristic $e_{AB}(i\omega)$ of the object in the frequency band up to the fundamental frequency which must be higher than $\omega_{\max}$. For synthesis it is also necessary to take into account the values of the time constants of the feedback elements which exceed $\omega_{\max}^{-1}$.

3. A highly efficient closed-loop control system forms a control force which nearly completely compensates for the disturbing dynamic excitations. The feasibility of the practical realisation of such a system depends primarily on the sensor sensitivity.

2.4 Systems with several measurement points

We consider the question as to what properties an active system may possess if several sensors are placed at various points on the object. Let A be the point at which a control force acts and $B_1, B_2, \ldots, B_r$ be the points under observation. One of the latter points may coincide with point A, see Fig. 2.36.

Assume that the control force due to the displacement of points B_s is feedback in accordance with the following law

$$U = -k \sum_{\rho=1}^{r} \chi_\rho z_\rho, \tag{2.173}$$

where z_ρ denotes the displacement of point B_ρ. On the other hand,

$$z_\rho = e_{AB_\rho}(p) U(t) + \zeta_p \; (\rho = 1, \ldots, r), \tag{2.174}$$

where ζ_p is the displacement of point B_ρ if the control force $U(t)$ is absent. Let us choose the following quantity

$$z_B = \sum_{\rho=1}^{r} \chi_\rho z_\rho \tag{2.175}$$

as the output variable characterising the vibration field of the object. Multiplying eq. (2.174) by χ_ρ and adding the equations obtained yields

$$z_B = -k \sum_{\rho=1}^{r} \chi_\rho e_{AB_\rho}(p) z_B + \zeta_B. \tag{2.176}$$

Here

$$\zeta_B = \sum_{\rho=1}^{r} \chi_\rho \zeta_\rho.$$

Comparing eqs. (2.176) and (1.92) it is easy to notice that the system under consideration is equivalent to a system with a single observed point B and the following dynamic compliance operator

$$e_{AB}(p) = \sum_{\rho=1}^{r} \chi_\rho e_{AB_\rho}(p) \tag{2.177}$$

together with an ideal amplifier in the feedback circuit.

Two objectives can be pursued by introducing control force (2.173). Firstly, it may turn out that function (2.175) under a certain choice of weighting coefficients χ_ρ describes the level of vibration in the system better than the displacement of one of the observed points. Next, the control efficiency can be increased in such a way. In the latter case there remains the problem of a rational choice of coefficients χ_ρ.

Let us study expression (2.177) in detail. Inserting eq. (1.53) into (2.177) yields

$$e_{AB}(p) = \det{}^{-1}(I_n p^2 + Bp + K) \sum_{\rho=1}^{r} \chi_\rho \sum_{s,m=1}^{n} W_{sm}(p) l_{As} l_{Bm}^{(\rho)}, \tag{2.178}$$

where $l_{Bm}^{(\rho)}$ denotes coefficients of the $m-th$ normal mode at point B_ρ. It is evident that the resonant frequencies of characteristic (2.178) are close to the eigenfrequencies of the object. For this function to retain the properties inherent in the point dynamic compliance, it is necessary and sufficient that its resonant and antiresonant frequencies alternate. In this case the Bode plot of the open-loop system will have the form shown in Fig. 2.17 which implies that the system remains stable for any gain (provided that the inertial properties of the feedback elements are neglected).

To establish the conditions under which resonant and antiresonant frequencies alternate we consider next the system without dissipation. Substituting the expression for dynamic compliance (1.49) into eq. (2.177) we obtain

$$e_{AB}(p) = \left[\prod_{r=1}^{n}(k_r^2+p^2)\right]^{-1} \sum_{\rho=1}^{r}\chi_\rho \sum_{s=1}^{n} l_{As} l_{Bs}^{(\rho)} \prod_{m=1}^{n}{}_{s} (k_m^2+p^2). \qquad (2.179)$$

Antiresonant frequencies $\varkappa_l$ are the roots of the following equation

$$\begin{aligned} V(i\varkappa) &= \sum_{\rho=1}^{r}\chi_\rho \sum_{s=1}^{n} l_{As} l_{Bs}^{(\rho)} \prod_{m=1}^{n}{}_{s} (k_m^2-\varkappa^2) \\ &= \sum_{s=1}^{n} l_{As} \sum_{\rho=1}^{r}\chi_\rho l_{Bs}^{(\rho)} \prod_{m=1}^{n}{}_{s} (k_m^2-\varkappa^2) = 0. \qquad (2.180) \end{aligned}$$

A sufficient condition for the roots of this equation to alternate with the eigenfrequencies is that real-valued quantities $V(ik_\alpha)$ and $V(ik_{\alpha+1})$, where k_α and $k_{\alpha+1}$ are two adjacent eigenfrequencies, have opposite signs, i.e. function $V(i\omega)$ changes its sign within each interval $(k_\alpha, k_{\alpha+1})$, $(\alpha = 1, ..., n-1)$. But

$$V(ik_\alpha) = l_{A\alpha} \sum_{\rho=1}^{r}\chi_\rho l_{B\alpha}^{(\rho)} \prod_{m=1}^{n}{}_{\alpha} \left(k_m^2 - k_\alpha^2\right),$$

$$V(ik_{\alpha+1}) = l_{A,\alpha+1} \sum_{\rho=1}^{r}\chi_\rho l_{B,\alpha+1}^{(\rho)} \prod_{m=1}^{n}{}_{\alpha+1} \left(k_m^2 - k_{\alpha+1}^2\right).$$

It is easy to see that the following products

$$\prod_{m=1}^{n}{}_{\alpha} \left(k_m^2 - k_\alpha^2\right), \quad \prod_{m=1}^{n}{}_{\alpha+1} \left(k_m^2 - k_{\alpha+1}^2\right)$$

have opposite signs. Hence, the alternation in question requires that the following coefficients

$$l_{A\alpha} \sum_{\rho=1}^{r}\chi_\rho l_{B\alpha}^{(\rho)}, \quad l_{A,\alpha+1} \sum_{\rho=1}^{r}\chi_\rho l_{B,\alpha+1}^{(\rho)}$$

have the same sign. It is known that a sign of the normal modes can be chosen arbitrarily since these modes are determined up to an arbitrary constant. Taking $l_{As} > 0$, $(s = 1, ..., n)$ the condition for alternate frequencies takes the form

$$\sum_{\rho=1}^{r} \chi_\rho l_{Bs}^{(\rho)} > 0 \ (s = 1, ..., n). \tag{2.181}$$

These conditions remain valid also for systems with weak dissipation (up to second order of smallness). It is clear that they determine an admissible region of coefficients χ_ρ up to a common factor which can be deemed as a multiplier of gain k.

As an example we consider the two-mass system shown in Fig. 1.12. Let one of the observed points be coincident with point A at which the control force is applied whereas the second point be point B. Eigenfrequencies k_1 and k_2 of the system are the roots of the frequency equation

$$k^4 + \left(\frac{\gamma}{m_2} + \frac{c+\gamma}{m_1}\right) k^2 + \frac{c\gamma}{m_1 m_2} = 0. \tag{2.182}$$

Coefficients l_{As} and l_{Bs} are defined as the amplitudes of forced vibrations at frequencies k_1 and k_2, respectively, the dissipation being assumed to be absent. As one of the coefficients can be chosen arbitrarily, we take

$$l_{A1} = l_{A2} = 1. \tag{2.183}$$

Then,

$$l_{B1} = \frac{e_{AB}^0 (ik_1)}{e_A^0 (ik_1)}, \ l_{B2} = \frac{e_{AB}^0 (ik_2)}{e_A^0 (ik_2)},$$

where e_{AB}^0 and e_A^0 are determined from eq. (1.85) for $\beta = b = 0$. Therefore

$$l_{B1} = \frac{\gamma}{\gamma - m_2 k_1^2}, \ l_{B2} = \frac{\gamma}{\gamma - m_2 k_2^2}. \tag{2.184}$$

Let us proceed now to conditions (2.181). Taking $\chi_1 = 1$ and assuming that B_1 coincides with A, and B_2 coincides with B, we obtain

$$1 + \chi_2 \frac{\gamma}{\gamma - m_2 k_1^2} > 0, \ 1 + \chi_2 \frac{\gamma}{\gamma - m_2 k_2^2} > 0. \tag{2.185}$$

Taking into account that $m_2 k_1^2 \leq \gamma \leq m_2 k_2^2$ we find the admissible region for parameter χ_2

$$\frac{\gamma - m_2 k_1^2}{\gamma} < \chi_2 < \frac{m_2 k_2^2 - \gamma}{\gamma}. \tag{2.186}$$

Let us construct expression (2.177) by inserting eq. (1.85) into it

$$e_{AB}(i\omega) = \frac{1}{\Delta(i\omega)} [\gamma - m_2\omega^2 + \beta i\omega + \chi_2(\beta i\omega + \gamma)]. \tag{2.187}$$

By virtue of eq. (2.182) it is easy to see that antiresonance occurs at frequencies k_1 and k_2 when χ_2 takes respectively a minimum and maximum value prescribed by condition (2.186). For non-extreme values of χ_2 antiresonance occurs at frequencies between k_1 and k_2. It is not important that the antiresonant frequency varies. Important is that antiresonant "trough" and resonant peak disappear when antiresonant frequency approaches resonant one. This provides us with an opportunity for increasing the control efficiency. Indeed, let

$$\chi_2 = -\frac{\gamma - m_2 k_1^2}{\gamma}. \tag{2.188}$$

From eq. (2.187) we obtain

$$\begin{aligned} e_{AB}(ik_1) &= \frac{1}{\Delta(ik_1)}\left[\gamma - m_2 k_1^2 + \beta i k_1 - \frac{\gamma - m_2 k_1^2}{\gamma}(\beta i k_1 + \gamma)\right] \quad (2.189) \\ &= \frac{\beta m_2 k_1^2}{\gamma[(b+\beta)(\gamma - m_2 k_1^2) + \beta(c + \gamma - m_1 k_1^2) + \beta(b+\beta) i k_1]}. \end{aligned}$$

If b and β are values of the same order of smallness, the dynamic compliance at frequency k_1 is real-valued (up to second order of smallness) and its absolute value is no longer large. Consequently, the antiresonant trough disappears. It is evident that the efficiency of the system increases and the stability conditions are determined only by the inertial properties of the feedback elements. An analogous effect can be obtained at frequency k_2 if we take

$$\chi_2 = -\frac{\gamma - m_2 k_2^2}{\gamma}. \tag{2.190}$$

Let us return to the general case and consider expression (2.179). If coefficients χ_ρ are chosen so that the following relationship

$$\sum_{\rho=1}^{r} \chi_\rho l_{Bj}^{(\rho)} = 0 \tag{2.191}$$

holds, then the $j-th$ component in eq. (2.179) vanishes. At $\omega = k_j$ the dynamic compliance is no longer unbounded since all terms in the numerator contain a multiplier $(k_j^2 + p^2)$ which is cancelled with the corresponding multiplier in the denominator. This implies that under condition (2.191) the system becomes non-observable in the principal coordinate q_j (or weakly observable if dissipation is taken into account). Clearly, one resonance and one antiresonance disappear.

As there are $r-1$ independent parameters χ_ρ they can be chosen so that the following condition

$$\sum_{\rho=1}^{r} \chi_\rho l_{Bs}^{(\rho)} = 0 \ (s = s_1, s_2, \dots, s_{r-1}) \tag{2.192}$$

is satisfied.

Such a choice of values of χ_ρ turns out to be expedient in many cases. Indeed, let an active facility be aimed at the control of low frequency vibrations of the object. If only one point is observed, then an increase in the gain may lead to instability and excitation of vibration at one of the eigenfrequencies of the object. By satisfying condition (2.192) for $s = 1, 2, \ldots, r-1$ the system may become non-observable in the first $r-1$ eigenfrequencies which excludes the very possibility of exciting these vibrations. The gain may be increased up to such a level at which the crossover frequency approaches the $r-th$ eigenfrequency (provided that the time constants of the feedback elements are sufficiently small).

In practice, the coefficients of the normal modes $l_{Bs}^{(\rho)}$ are always determined with a certain inaccuracy. For this reason, it is recommended that values of χ_ρ are chosen so that conditions (2.181) are satisfied for all possible deviations in the coefficients of the normal modes.

Assume now that the control force $U(t)$ is formed due to the velocities of points B_ρ, i.e.

$$U = -k \sum_{\rho=1}^{r} \psi_\rho \dot{z}_\rho. \tag{2.193}$$

The transfer function of the open-loop system is as follows

$$\Phi(p) = k \sum_{\rho=1}^{r} \psi_\rho p e_{AB_\rho}(p). \tag{2.194}$$

Let us show that under the following conditions

$$\sum_{\rho=1}^{r} \psi_\rho l_{Bs}^{(\rho)} > 0 \; (s = 1, \ldots, n) \tag{2.195}$$

the closed-loop system is stable regardless of the value of gain k. Indeed, if conditions (2.195) are satisfied, then, as shown above, the frequency characteristic

$$e(i\omega) = \sum_{\rho=1}^{r} \psi_\rho e_{AB_\rho}(i\omega)$$

possesses properties inherent of a point dynamic compliance. In particular, this means that

$$0 \geq \arg e(i\omega) > -\pi,$$

and therefore

$$\frac{\pi}{2} \geq \arg \Phi(i\omega) > -\frac{\pi}{2}.$$

Thus, the LPC of the open-loop system does not cross level $\varphi = -\pi$ at any value of k which implies that the closed-loop system is stable.

Introducing control due to velocity is related to a rather intensive dissipation of vibrational energy. Let us find the conditions under which the control force (2.193) possesses the properties of a dissipative force. To this end, we determine the power of this force which is

$$\begin{aligned} N &= U(t)\,\dot{z}_A(t) = -k\sum_{\rho=1}^{r}\psi_\rho \dot{z}_\rho \dot{z}_A \\ &= -k\sum_{\rho=1}^{r}\psi_\rho \sum_{s,m=1}^{n} l_{Bm}^{(\rho)} l_{As}\dot{q}_m\dot{q}_s = -k\sum_{s,m=1}^{n}\alpha_{ms}\dot{q}_m\dot{q}_s, \end{aligned} \tag{2.196}$$

where

$$\alpha_{ms} = \alpha_{sm} = \frac{1}{2}\sum_{\rho=1}^{r}\psi_\rho (l_{Bm}^{(\rho)} l_{As} + l_{Bs}^{(\rho)} l_{Am}). \tag{2.197}$$

For force $U(t)$ to be dissipative, the power must be negative for any vibration of the object. For this purpose, the following quadratic form

$$\sum_{m,s=1}^{n}\alpha_{ms}\dot{q}_m\dot{q}_s$$

must be positive definite, i.e. it must satisfy Sylvester' s criterion. This is not always possible. For example, assuming for the two-mass system under consideration that $B_1 = A$ and $B_2 = B$ we obtain

$$\begin{aligned} l_{B1}^{(1)} &= l_{A1} = 1, & l_{B2}^{(1)} &= l_{A2} = 1, \\ l_{B1}^{(2)} = l_{B1} &= \frac{\gamma}{\gamma - m_2 k_1^2}, & l_{B2}^{(2)} = l_{B2} &= \frac{\gamma}{\gamma - m_2 k_2^2}. \end{aligned}$$

Inserting these values into eq. (2.197) we get

$$\begin{aligned} \alpha_{11} &= \psi_1 + \psi_2\frac{\gamma}{\gamma - m_2 k_1^2}, \\ \alpha_{12} &= \psi_1 + \frac{1}{2}\psi_2\left(\frac{\gamma}{\gamma - m_2 k_1^2} + \frac{\gamma}{\gamma - m_2 k_2^2}\right), \\ \alpha_{22} &= \psi_1 + \psi_2\frac{\gamma}{\gamma - m_2 k_2^2}. \end{aligned}$$

The matrix $\{\alpha_{ms}\}$ satisfies Sylvester's criterion if

$$\alpha_{11} > 0,\ \alpha_{11}\alpha_{22} - \alpha_{12}^2 > 0.$$

However

$$\alpha_{11}\alpha_{22} - \alpha_{12}^2$$

$$= \psi_1^2 + \psi_1\psi_2\left(\frac{\gamma}{\gamma - m_2k_1^2} + \frac{\gamma}{\gamma - m_2k_2^2}\right) + \psi_2^2\frac{\gamma^2}{(\gamma - m_2k_1^2)(\gamma - m_2k_2^2)}$$

$$= \psi_1^2 - \psi_1\psi_2\left(\frac{\gamma}{\gamma - m_2k_1^2} + \frac{\gamma}{\gamma - m_2k_2^2}\right) - \frac{1}{4}\psi_2^2\left(\frac{\gamma}{\gamma - m_2k_1^2} + \frac{\gamma}{\gamma - m_2k_2^2}\right)$$

$$= \frac{\psi_2^2}{4}\left[\frac{2\gamma^2}{(\gamma - m_2k_1^2)(\gamma - m_2k_2^2)} - \frac{\gamma^2}{(\gamma - m_2k_1^2)^2} - \frac{\gamma^2}{(\gamma - m_2k_2^2)^2}\right] < 0.$$

Thus, for any values of ψ_1 and ψ_2 ($\psi_2 \neq 0$) control force $U(t)$ does not possesses the properties of a dissipative force.

2.5 Transient processes in active systems and protection from impacts

As mentioned above, motion of any time-invariant linear system can be described by equations of the type (2.133). General solutions of this equation subjected to the initial condition

$$t = 0, \; \mathbf{y} = \mathbf{y}_0 \tag{2.198}$$

may be cast in the following form

$$\mathbf{y}(t) = H(t)\,\mathbf{y}_0 + \int_0^t H(t-\tau)\mathbf{f}(\tau)\,d\tau. \tag{2.199}$$

Here $H(t)$ is a solution of the following matrix equation

$$\dot{X} = AX, \tag{2.200}$$

satisfying the initial condition

$$H(0) = I_n.$$

The first component in eq. (2.199)

$$\mathbf{y}_t(t) = H(t)\,\mathbf{y}_0 \tag{2.201}$$

is conventionally referred to as a *transient process*. It is easy to see that a transient process is a motion of the system under initial conditions (2.198) and $\mathbf{f}(t) = \mathbf{0}$. If the system is asymptotically stable, then the transient process decays as time increases. In this case, for sufficiently large values of t the motion of the system is slightly different from a *steady-state motion*

$$\mathbf{y}_*(t) = \int_{-\infty}^{t} H(t-\tau)\mathbf{f}(\tau)\,d\tau. \tag{2.202}$$

A necessity of investigating transient processes in vibration protection systems arises while analysing non-stationary regimes. If the system is subjected to a stationary dynamic action, e.g. a harmonic one, then to judge the performance of the system it is necessary not only to evaluate its efficiency in the steady-state regime but also to estimate the maximum values of the output quantities in the transient regime. Though the transient regime is of a short duration, an excess beyond admissible values of the output quantities may be dangerous for the body to be protected. These arguments become especially essential when the external excitations are short-time processes with large amplitudes of the input quantities. Such an action is ordinarily called *impact.* Problems related to protection from impact excitations are considered in what follows.

Let us consider a transient process in a one-dimensional control system in detail. Let the governing equation be given by eq. (2.139) in which

$$\begin{aligned} q(p) &= a_0 p^n + a_1 p^{n-1} + \dots + a_n, \\ r(p) &= b_0 p^m + b_1 p^{m-1} + \dots + b_m \end{aligned} \tag{2.203}$$

and $m < n$. Assume that characteristic equation (2.141) has roots λ_1, ..., λ_n, no multiple or zero roots being among them. Then, as shown for example in [106], the general solution of eq. (2.139) is given by

$$x(t) = C_1 e^{\lambda_1 t} + \dots + C_n e^{\lambda_n t} + \int_0^t h(t-\tau) f(\tau)\,d\tau. \tag{2.204}$$

Here $h(t)$ is the *impulse response* of system (2.139) for which the following expression is valid

$$h(t) = \sum_{k=1}^{n} \frac{r(\lambda_k)}{q'(\lambda_k)} e^{\lambda_k t}, \tag{2.205}$$

where $q' = dq/dp$. It can be shown that the impulse response satisfies the conditions

$$h(0) = \dot{h}(0) = \dots = h^{(n-2)}(0) = 0, \ \ h^{(n-1)}(0) = 1 \tag{2.206}$$

and is equal to the inverse Fourier transform of the system frequency characteristic

$$h(t) = \frac{1}{2\pi} \int_{-\infty}^{\infty} \frac{r(i\omega)}{q'(i\omega)} e^{i\omega t}\,d\omega. \tag{2.207}$$

Integration constants $C_1, ..., C_k$ do not depend on the particular form of function $f(t)$ and are expressed in terms of the initial conditions

$$x(0) = x_0, \ \dot{x}(0) = \dot{x}_0, \ ..., \ x^{(n-1)}(0) = x_0^{(n-1)} \tag{2.208}$$

as follows, [82]

$$C_k = \frac{1}{\lambda_k q'(\lambda_k)} \sum_{s=0}^{n-1} x_0^{(s)} \Delta_s(\lambda_k) \ (k = 1, ..., n). \tag{2.209}$$

Here $\Delta_s(\lambda_k)$ are polynomials satisfying the following recurrent relationships

$$\left. \begin{array}{l} \Delta_0(\lambda_k) = q(\lambda_k) - a_n, \\ \Delta_s(\lambda_k) = \lambda_k^{-1}[\Delta_{s-1}(\lambda_k) - a_{n-s}\lambda_k] \\ (s = 1, ..., n-1). \end{array} \right\} \tag{2.210}$$

Expression (2.210) can be rewritten in another form if we conditionally adopt the following abbreviation $x_0^{(s)} = p^s x_0$. After simple transformations we obtain

$$C_k = \frac{1}{q'(\lambda_k)} \frac{q(p)}{p - \lambda_k}(x_0). \tag{2.211}$$

It is clear that $q(\lambda_k) = 0$, thus, the expressions in (2.210) can also be simplified, to give

$$\Delta_0(\lambda_k) = -a_n, \ \Delta_1(\lambda_k) = \frac{1}{\lambda_k}(-a_n - a_{n-1}\lambda_k), ... \tag{2.212}$$

The above-said indicates that the transient processes in the system may be very different, depending upon the initial conditions. In order to judge the transient processes it is common practice to examine motion of the system subject to a certain "standard" action and trivial initial conditions. A unit step function

$$f(t) = \eta(t) = \begin{cases} 0 & \text{if} \quad t < 0, \\ 1 & \text{if} \quad t \geq 0. \end{cases} \tag{2.213}$$

is often chosen as an external action. Substituting eq. (2.213) into eq. (2.204) under trivial initial conditions we obtain

$$x(t) = \int_0^t h(t-\tau)\eta(\tau)\,d\tau = \int_0^t h(t-\tau)\,d\tau = \int_0^t h(\xi)d\xi = g(t). \tag{2.214}$$

Function $g(t)$ is termed the *step response* of the system. As $t \to \infty$ this function tends to a certain finite value which can be obtained by substituting eq. (2.205) into (2.214), to get

$$g(\infty) = \int_0^\infty \sum_{k=1}^{n} \frac{r(\lambda_k)}{q'(\lambda_k)} e^{\lambda_k t} dt = -\sum_{k=1}^{n} \frac{r(\lambda_k)}{q'(\lambda_k)\lambda_k}. \tag{2.215}$$

On the other hand, it is clear that $g(\infty)$ must be a time-independent solution of the following equation

$$q(p)\, x = r(p)\, \eta(t).$$

One easily obtains that

$$g(\infty) = \frac{r(0)}{q(0)} = \frac{b_m}{a_n}. \tag{2.216}$$

If all $\operatorname{Re}\lambda_k < 0$, then it follows from eq. (2.205) that

$$\begin{aligned} \dot{g}(\infty) &= h(\infty) = 0, \\ \ddot{g}(\infty) &= \dot{h}(\infty) = 0, ..., g^{(n)}(\infty) = h^{(n-1)}(\infty) = 0. \end{aligned}$$

For this reason, motion (2.214) may be considered as a transition of the system into a new equilibrium position, defined as follows

$$x = g(\infty), \ \dot{x} = \ddot{x} = ... = x^{(n-1)} = 0. \tag{2.217}$$

It can be shown that motion (2.214) may be viewed as a transient process, i.e. function $g(t)$ is one of solutions of the homogeneous equation

$$q(p)\, x = 0.$$

The initial conditions corresponding to this solution are in general nontrivial and depend on a particular form of operator $r(p)$. A way of defining these initial condition is explained in [7].

Quantitative measures of performance of transient processes are *maximum overshoot* and *settling time.* Maximum overshoot is a non-dimensional quantity defined as

$$\gamma = \frac{x_{\max} - x(\infty)}{x(\infty)}, \tag{2.218}$$

which is ordinarily expressed as a percentage. For an astatic system $r(0) = b_m = 0$, thus $x(\infty) = 0$ and expression (2.218) makes no sense. In this case one introduces the *maximum displacement* $x_{\max}$ rather than the maximum overshoot. The maximum overshoot and maximum displacement are important characteristics of transient processes in vibration protection systems as they determine the required clearance of the buffers. If F_0 is the maximum possible change in the static load, then in order to avoid impacts with buffers in the transient regime these buffers must have an required distance

$$\Delta > F_0 g_{\max}$$

from the original equilibrium position, where $g_{\max}$ denotes the maximum deviation due to action (2.213).

The settling time of a transient process t_s is defined as the time required for the step response to decrease to and remain within some bounds of the final value. In other words, the settling time satisfies the following condition

$$|x(t) - x(\infty)| \leq \alpha x(\infty) \text{ for } t \geq t_s, \tag{2.219}$$

where a commonly used value for α is 0,01-0,05. For an astatic system condition (2.219) is replaced by the following one

$$|x(t)| < \varepsilon \text{ for } t \geq t_s. \tag{2.220}$$

The transient function describes the transient processes in the system due to a change in the static load or constant component of the base acceleration sufficiently well. In order to describe the behaviour of the system subjected to short-duration shock excitation, one uses another standard excitation, namely Dirac's delta-function

$$f(t) = \delta(t). \tag{2.221}$$

Using the known property of the delta-function and taking into account that all $C_k = 0$ in eq. (2.204) under trivial initial conditions we obtain the system reaction on action (2.221)

$$x(t) = \int_0^t h(t-\tau)\,\delta(\tau)\,d\tau = h(t). \tag{2.222}$$

Thus, the law of motion coincides with the step response. The most important measure of quantitative quality of process (2.222) is the value of $x_{\max}$.

Along with the above-mentioned, there are other estimates of quality of transient processes, e.g. integral estimates. The most wide-spread measure is the integral of the square of the output quantity

$$J = \int_0^\infty x^2(t)\,dt. \tag{2.223}$$

As mentioned in Section 1.4 by comparing values of J for two systems a rather objective judgement on the efficiency of the introduced control can be performed.

Using Parseval's formula yields

$$J = \frac{1}{\pi} \int_0^\infty |X(i\omega)|^2 \, d\omega, \tag{2.224}$$

where $X(i\omega)$ is Fourier transform of process $x(t)$. Since the Fourier transform of the step function is the system frequency characteristic, the integral estimate (2.222) takes the following form

$$J = \frac{1}{\pi} \int_0^\infty \left| \frac{r(i\omega)}{q(i\omega)} \right|^2 d\omega. \tag{2.225}$$

We obtain by analogy that

$$J_1 = \int_0^\infty g^2(t) \, dt = \frac{1}{\pi} \int_0^\infty \left| \frac{r(i\omega)}{q(i\omega) i\omega} \right|^2 d\omega. \tag{2.226}$$

Evaluating integrals (2.225) and (2.226) for rational transfer function can be easily carried out, see e.g. [106].

To establish the relationships between the quality measures, like $x_{\max}$, maximum overshoot and settling time, and frequency characteristics of the system is a challenging problem. For this purpose one uses expressions relating functions $h(t)$ and $g(t)$ with real and imaginary parts of the frequency characteristic. If

$$\frac{r(i\omega)}{q(i\omega)} = P(\omega) + iQ(\omega),$$

then

$$h(t) = \frac{2}{\pi} \int_0^\infty P(\omega) \cos \omega t d\omega = -\frac{2}{\pi} \int_0^\infty Q(\omega) \sin \omega t d\omega, \tag{2.227}$$

$$g(t) = \frac{2}{\pi} \int_0^\infty \frac{P(\omega) \sin \omega t}{\omega} d\omega = P(0) + \frac{2}{\pi} \int_0^\infty \frac{Q(\omega) \cos \omega t}{\omega} d\omega. \tag{2.228}$$

These relations are used for approximate plots of transient processes. In the literature, see e.g. [7] and [124], dependences of the quality measures for transient processes on properties of function $P(\omega)$ are studied in detail. In particular, it is shown that increasing the value of ω_* which is the least root of function $P(\omega)$, see Fig. 2.37, leads to a decrease in the settling time of the transient process whereas decreasing $P_{\max}(\omega)$ results in a decrease in the maximum overshoot.

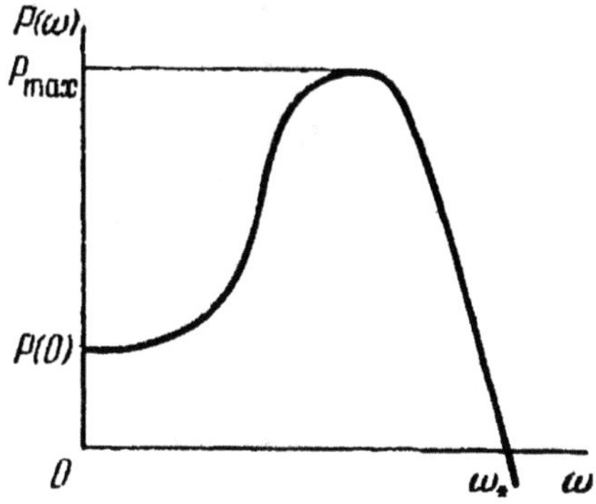

FIGURE 2.37.

It is much more difficult to establish relationships between transient processes and frequency characteristics of the open-loop system. The common strategy is utilising special alignment charts to determine $P(\omega)$ in terms of amplitude and phase characteristics of the open-loop system and then to use eq. (2.227) to find $h(t)$. It is worthwhile noting that the settling time decreases as crossover frequency increases, suggesting that the settling time may be assumed to be approximately equal to the inverse of the crossover frequency.

We proceed now to an investigation of transient processes in active one-dimensional vibration protection systems. Consider a system schematically depicted in Fig. 2.38 where control force $U(t)$ is due to the displacement of the point at which the control force acts

$$U(t) = -w_z(p) z_A(t), \tag{2.229}$$

$w_z(p)$ being the feedback transfer function. To investigate the quality of transient processes we apply a standard disturbance $F(t)$ of the type (2.213) or (2.221). Then

$$z_A(t) = e_A(p)[U(t) + F(t)]. \tag{2.230}$$

By virtue of eqs. (2.229) and (2.230) we find

$$z_A(t) = \frac{e_A(p)}{1 + e_A(p) w_z(p)} F(t). \tag{2.231}$$

Thus, the transfer function due to disturbance is given by

$$\frac{r(p)}{q(p)} = \frac{e_A(p)}{1 + e_A(p) w_z(p)}. \tag{2.232}$$

Let us point out some general properties of the step response $g(t)$. To this aim, we substitute eq. (2.205) into (2.214), to get

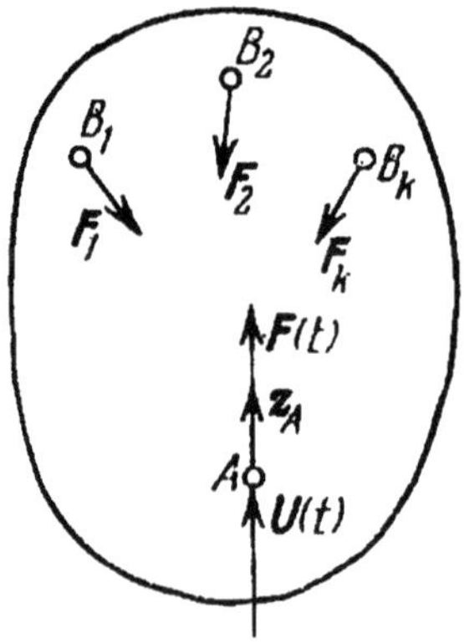

FIGURE 2.38.

$$g(t) = g(\infty) + \sum_{k=1}^{n} \frac{r(\lambda_k)}{q'(\lambda_k)\lambda_k} e^{\lambda_k t}, \tag{2.233}$$

where, as shown above

$$g(\infty) = \frac{r(0)}{q(0)} = \frac{e_A(0)}{1 + e_A(0) w_z(0)}. \tag{2.234}$$

Considering (2.215) we obtain

$$g(t) = \sum_{k=1}^{n} \frac{r(\lambda_k)}{q'(\lambda_k)\lambda_k} (e^{\lambda_k t} - 1). \tag{2.235}$$

As $\left|e^{\lambda_k t}\right| \leq 1$ for $\operatorname{Re}\lambda_k \leq 0$, we have

$$\left|e^{\lambda_k t} - 1\right| \leq 2,$$

and therefore

$$|g(t)|_{\max} \leq 2 \sum_{k=1}^{n} \left| \frac{r(\lambda_k)}{q'(\lambda_k)\lambda_k} \right| = 2 \sum_{k=1}^{n} |s_k|, \tag{2.236}$$

where

$$s_k = \frac{r(\lambda_k)}{q'(\lambda_k)\lambda_k}. \tag{2.237}$$

Consider an open-loop system, i.e. $w_z(p) = 0$, for which

$$\frac{r(p)}{q(p)} = e_A(p) = e_A(0) \frac{\pi_a(p)}{\pi_r(p)}, \tag{2.238}$$

where

$$\pi_a(p) = \prod_{k=1}^{n-1} \left(\tau_k^2 p^2 + 2\tau_k \eta_k p + 1\right),$$
$$\pi_r(p) = \prod_{s=1}^{n} \left(T_s^2 p^2 + 2T_s \zeta_s p + 1\right).$$

Characteristic equation

$$q(p) = \pi_r(p) = \prod_{s=1}^{n} (T_s^2 p^2 + 2T_s \zeta_s p + 1) = 0 \tag{2.239}$$

has $2n$ complex conjugate roots

$$\lambda_k = -T_k^{-1}(\zeta_k \pm i) \ \ (k = 1, ..., n).$$

Expression (2.235) can be cast in the form

$$g(t) = 2\sum_{k=1}^{n} \{e^{-T_k^{-1}\zeta_k} |s_k| \cos(T_k^{-1} t + \varphi_k) - \operatorname{Re} s_k\}, \tag{2.240}$$

where $\varphi_k = \arg s_k$, and s_k are complex conjugate coefficients as they correspond to complex conjugate roots λ_k.

Let us proceed to evaluation of coefficients s_k. We take the derivative of $\pi_r(p)$ with respect to p and substitute $p = \lambda_k$ in the result. With the absence of multiple roots assumed, all terms in the expression for the derivative vanish except the $k-th$ term. Thus,

$$\pi_r'(\lambda_k) = \prod_{s=1}^{n}{}_{k} \left(1 + 2T_s \zeta_s \lambda_k + T_s^2 \lambda_k^2\right) 2\left(T_k \zeta_k + T_k^2 \lambda_k\right).$$

As a first approximation for evaluating s_k we adopt that all $\zeta_s = 0$. Then

$$\pi_r'(\lambda_k) = \prod_{s=1}^{n}{}_{k} \left(1 - T_s^2 T_k^{-2}\right)(\pm 2T_k i).$$

For all $\eta_l = 0$ we obtain by analogy that

$$\pi_a(\lambda_k) = \prod_{l=1}^{n-1} \left(\tau_l^2 \lambda_k^2 + 2\tau_l \eta_l \lambda_k + 1\right) \approx \prod_{l=1}^{n-1} \left(1 - \tau_l^2 T_k^{-2}\right).$$

Hence

$$\frac{r(\lambda_k)}{q'(\lambda_k)} = e_A(0)\,\frac{\pi_a(\lambda_k)}{\pi_r'(\lambda_k)} \approx e_A(0)\,\frac{\prod_{l=1}^{n-1}(1 - \tau_l^2 T_k^{-2})}{\prod_{s=1}^{n}{}_{k}(1 - T_s^2 T_k^{-2})}\left(\mp\frac{1}{2}T_k^{-1} i\right), \tag{2.241}$$

$$s_k = \frac{r(\lambda_k)}{q'(\lambda_k)} \approx \frac{1}{2} e_A(0) \frac{T_k^{-1} i \prod_{l=1}^{n-1} (1 - \tau_l^2 T_k^{-2})}{-T_k^{-1}(\zeta_k \pm i) \prod_{s=1}^{n}{}_k (1 - T_s^2 T_k^{-2})}$$

$$\approx -\frac{1}{2} e_A(0) \frac{\prod_{l=1}^{n-1} (1 - \tau_l^2 T_k^{-2})}{\prod_{s=1}^{n}{}_k (1 - T_s^2 T_k^{-2})}. \tag{2.242}$$

Expression (2.242) can be simplified if we take into account that values of τ_l and T_s alternate and assume that these values are well spaced. Then

$$\begin{aligned} a)\quad & \text{for } l \geq k \quad \tau_l^2 T_k^{-2} \ll 1, \\ & \text{for } l > k \quad T_l^2 T_k^{-2} \ll 1, \\ b)\quad & \text{for } l < k \quad \tau_l^2 T_k^{-2} \gg 1,\ T_l^2 T_k^{-2} \gg 1. \end{aligned}$$

It follows from eq. (2.242) that

$$s_k \approx -\frac{1}{2} e_A(0) \prod_{l=1}^{k-1} \frac{\tau_l^2}{T_s^2}. \tag{2.243}$$

Thus, under the above assumptions coefficients s_k are real-valued. Moreover, as $\tau_l < T_l$ they decrease with the growth of k. Coefficients s_k are negative since the products in the numerator and in the denominator in eq. (2.242) have an equal number of negative factors, namely $k-1$. Equation (2.243) confirms negativity of s_k.

Inserting eq. (2.242) into eq. (2.240) we obtain the following approximate expression for $g(t)$

$$g(t) \approx e_A(0) \sum_{k=1}^{n} \frac{\prod_{l=1}^{n-1} (1 - \tau_l^2 T_k^{-2})}{\prod_{s=1}^{n}{}_k (1 - T_s^2 T_k^{-2})} \left(1 - e^{-T_k^{-1} \zeta_k t} \cos T_k^{-1} t\right). \tag{2.244}$$

This equation turns out to be sufficient for practical design. In some cases one can use another, less accurate approximation which is obtained by substituting eq. (2.243) into (2.240)

$$g(t) \approx e_A(0) \sum_{k=1}^{n} \prod_{l=1}^{k-1} \frac{\tau_l^2}{T_s^2} \left(1 - e^{-T_k^{-1} \zeta_k t} \cos T_k^{-1} t\right). \tag{2.245}$$

As all coefficients $s_k < 0$, estimate (2.236) simplifies, to give

$$|g(t)|_{\max} \leq 2 \sum_{k=1}^{n} s_k = 2g(\infty). \tag{2.246}$$

Thus, the maximum overshoot does not exceed 100%. In reality, this value of overshoot can not be achieved since this overshoot is possible only if all $\cos T_k^{-1} t = -1$ at the same instant of time t_* , and additionally all $\zeta_k = 0$.

A more accurate estimate is obtained by accounting for values of ζ_k. Presuming that $t_* = \pi T_1$ ($\cos T_k^{-1} t$ can not be equal -1 if $t < t_*$) yields

$$|g(t)|_{\max} \leq g(\infty)\{1 + \exp[-T_1 \pi (T_k^{-1} \zeta_k)_{\min}]\}. \tag{2.247}$$

Expression (2.244) enables us to estimate the settling time, too. Condition (2.219) is met if

$$e^{-T_k^{-1} \zeta_k t} < \alpha \; (k = 1, ..., n). \tag{2.248}$$

This yields an upper bound for t_s

$$t_s \leq \left(\frac{T_k}{\zeta_k} \right)_{\max} \ln \frac{1}{\alpha}. \tag{2.249}$$

Using expression (2.241) we can define the impulse response of the system which is the response to a delta-function. Substituting eq. (2.241) into (2.205) we obtain after some simple transformation that

$$h(t) = e_A(0) \sum_{k=1}^{n} \frac{\prod_{l=1}^{n-1} (1 - \tau_l^2 T_k^{-2})}{\prod_{s=1, s \neq k}^{n} (1 - T_s^2 T_k^{-2})} T_k^{-1} e^{-T_k^{-1} \zeta_k t} \sin T_k^{-1} t. \tag{2.250}$$

The latter equation enables the following estimate for $|h(t)|_{\max}$

$$|h(t)|_{\max} \leq e_A(0) \sum_{k=1}^{n} \frac{\prod_{l=1}^{n-1} (1 - \tau_l^2 T_k^{-2})}{\prod_{s=1, s \neq k}^{n} (1 - T_s^2 T_k^{-2})} T_k^{-1}. \tag{2.251}$$

Let us estimate the error due to using expression (2.244). The Laplace transform of process $g(t)$ is as follows

$$L\{g(t)\} = e_A(p) L\{\eta(t)\} = \frac{e_A(0)}{p} \frac{\prod_{k=1}^{n-1} (\tau_k^2 p^2 + 2\tau_k \eta_k p + 1)}{\prod_{s=1}^{n} (T_s^2 p^2 + 2T_s \zeta_s p + 1)}.$$

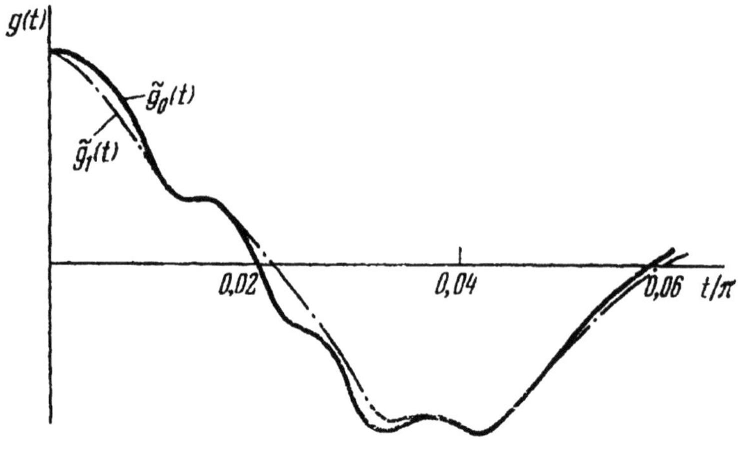

FIGURE 2.39.

Assuming that $\eta_k^2 \ll 1$ and $\zeta_s^2 \ll 1$ for all s and k and recovering the original yields

$$g(t) = e_A(0)\sum_{k=1}^{n}\left\{\frac{\prod_{l=1}^{n-1}(1-\tau_l^2T_k^{-2})}{\prod_{s=1}^{n}{}_k\,(1-T_s^2T_k^{-2})}\left(1-e^{-T_k^{-1}\zeta_k t}\cos T_k^{-1}t\right)\right. \quad (2.252)$$

$$+\prod_{l=1}^{n-1}\left[\frac{-\zeta_k(1-\tau_l^2T_k^{-2})}{\prod_{s=1}^{n}{}_k\,(1-T_s^2T_k^{-2})}+2\frac{\tau_l}{T_k}\frac{\eta_l-\tau_lT_k^{-1}\zeta_k}{\prod_{s=1}^{n}{}_k\,(1-T_s^2T_k^{-2})}+\right.$$

$$\left.\left.\frac{(1-\tau_l^2T_k^{-2})}{\prod_{s=1}^{n}{}_k\,(1-T_s^2T_k^{-2})}\prod_{s=1}^{n}\frac{T_s}{T_k}2\frac{\zeta_s-T_sT_k^{-1}\zeta_k}{(1-T_s^2T_k^{-2})}\right]e^{-\zeta_kT_k^{-1}t}\sin T_k^{-1}t\right\}.$$

Comparing this expression with eq. (2.244) it is easy to see that the error is due to the second term in brace brackets.

Figure 2.39 displays two plots of $\tilde{g}(t) = g(t) - g(\infty)$. Here $\tilde{g}_0(t)$ is due to eq. (2.244) and $\tilde{g}_1(t)$ is due to eq. (2.252) for the two-mass system of Fig. 1.12. The following numerical values of parameters were taken:

$$T_1 = 0,04,\quad T_2 = 0,01,\quad \tau_1 = 0,02,$$

$$\zeta_1 = 0,2,\quad \zeta_2 = 0,1,\quad \eta_1 = 0,1.$$

The plot indicates that the error caused by using eq. (2.244) is not significant in this case.

Next, we consider the question as to how feedback introduced can affect the transient process of the system. To begin with, we notice that in accordance with eq. (2.234) the feedback changes the value of $g(\infty)$. If $g^0(\infty)$ denotes the value of the static deflection before a feedback has been introduced, then

$$\frac{g(\infty)}{g^0(\infty)} = \frac{1}{1 + e_A(0)w_z(0)}. \tag{2.253}$$

Let us study the influence of control proportional to displacement. In this case, the feedback transfer function is given by eq. (2.84). Inserting this equation into (2.232) and considering (2.238) we have

$$\frac{r(p)}{q(p)} = \frac{e_A(0)\pi_a(p)}{\pi_r(p) + ke_A(0)\pi_a(p)}. \tag{2.254}$$

As mentioned above, this is the transfer function of a certain passive system which is obtained from the original system by an additional spring of stiffness k. Thus, the form of the transient process must be given by eq. (2.244). It is clear that antiresonant frequencies in systems (2.238) and (2.254) coincide. Let us investigate the change in resonant frequencies. To this end, we consider the system without dissipation. The LAC for transfer function $ke_A(p)$ is plotted in Fig. 2.40 where τ_l^{-1} and T_s^{-1} designate antiresonant and resonant frequencies of the open-loop system. Resonant frequencies T_s' of the closed-loop system are the same as the roots of the denominator in eq. (2.254). The corresponding equation can be cast in the form

$$\frac{ke_A(0)\pi_a[i(T_s')^{-1}]}{\pi_r[i(T_s')^{-1}]} = ke_A[i(T_s')^{-1}] = -1. \tag{2.255}$$

Therefore, the resonant frequencies T_s' are given by the points of intersection of the LAC with the abscissa axis (the phase is equal to $-\pi$ at these points), see Fig. 2.40. This Figure shows that at frequencies lower than the crossover frequency ω_c the resonant frequencies T_s' of the closed-loop system are rather close to the antiresonant frequencies τ_l of the open-loop system while at frequencies higher than ω_c the resonant frequencies of the closed-loop system and open-loop system are actually coincident. Let us remember that the crossover frequency ω_c is that frequency at which the asymptotic LAC crosses the abscissa axis.

This implies that for the closed-loop system

$$\left.\begin{aligned} \tau_l' = \tau_l, \quad T_s' = \tau_s \quad &\text{for} \quad T_s' > \omega_c^{-1}, \\ T_s' = T_s \quad &\text{for} \quad T_s' < \omega_c^{-1}. \end{aligned}\right\} \tag{2.256}$$

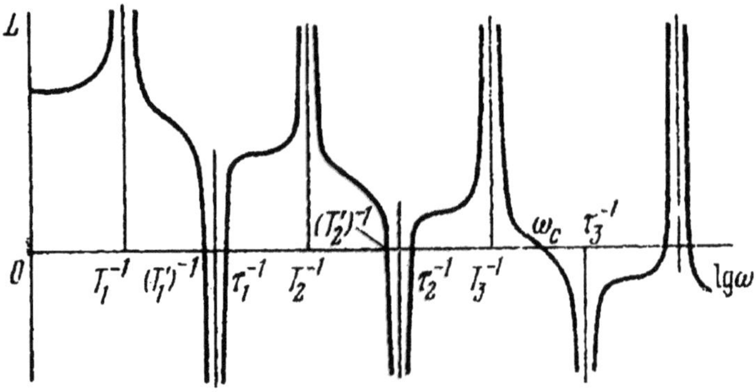

FIGURE 2.40.

Returning to eq. (2.244) we note that the factors in the numerators are close to zero for all k for which $T_k' = \tau_k$.

Thus, introducing a displacement feedback leads to reducing amplitudes of those "harmonics" of the transient process whose frequencies are lower that the crossover frequency, while the amplitudes of high frequency components are not significantly affected. Note in passing that the crossover frequency ω_c is one of the solutions of eq. (2.255), i.e. this is one of the resonant frequencies of the closed-loop system.

The above-said is valid also for systems with weak dissipation. In this case, the roots of characteristic equation (2.255) are complex-valued, and relations (2.256) will be valid for their real parts.

Suppressing the low frequency components results in a decrease in the settling time. The latter is ordinarily determined by a component with the lowest frequency which is close to the crossover frequency of the closed-loop system. An estimate for t_s can be obtained with the help of eq. (2.249) if one replaces T_k by ω_c^{-1}. The corresponding value of ζ may be obtained from the characteristic equation

$$e_A(\zeta + i\omega_c) = -\frac{1}{k}.$$

We consider now a more general case. Let control point A not coincide with the point under observation B. Then transfer function of the system with disturbance feedback can be represented in the form analogous to eq. (2.232)

$$\frac{r(p)}{q(p)} = \frac{e_{AB}(p)}{1 + e_{AB}(p)w_z(p)} = \frac{v(p)\pi_{ab}(p)}{u(p)\pi_{ab}(p) + v(p)\pi_r(p)}. \tag{2.257}$$

Here $u(p)$ and $v(p)$ are respectively the numerator and the denominator of transfer function $w_z(p)$ while $\pi_{ab}(p)$ and $\pi_r(p)$ are those of $e_{AB}(p)$. Let us show that the transient process cannot contain an essential, slowly decaying harmonic with frequency much lower than the system crossover frequency.

Indeed, such a component must be described by a complex-valued root $p = -\mu \pm i\nu$ of the denominator of transfer function (2.257), μ being a small value and $\nu < \omega_c$. It is clear that in this case

$$u(-\mu \pm i\nu)\,\pi_{ab}(-\mu \pm i\nu) = -v(-\mu \pm i\nu)\,\pi_r(-\mu \pm i\nu),$$

and, therefore,

$$|M(-\mu \pm i\nu)| = |N(-\mu \pm i\nu)|, \tag{2.258}$$

where

$$u(p)\,\pi_{ab}(p) = M(p), \quad v(p)\,\pi_r(p) = N(p).$$

On the other hand, $M(p)$ and $N(p)$ are the numerator and the denominator of the transfer function $\Phi(p)$ of the open-loop system. In general, at frequency $\nu \ll \omega_c$ one has

$$|M(i\nu)| \gg |N(i\nu)|. \tag{2.259}$$

This condition contradicts condition (2.258) because polynomials $M(p)$ and $N(p)$ are continuous functions of the argument and μ is small.

Condition (2.259) can be violated at frequencies ν of antiresonant troughs of the LAC, if $\nu < \omega_c$. At these frequencies $|M(i\nu)/N(i\nu)|$ may be close to unity, i.e. condition (2.258) can be satisfied. In this case $|\pi_{ab}(-\mu \pm i\nu)|$ is small which results in the following factor being small

$$\begin{aligned} s_k(-\mu \pm i\nu) &= \frac{r(-\mu \pm i\nu)}{q(-\mu \pm i\nu)(\mu \pm i\nu)} = \\ &= \frac{-v(-\mu \pm i\nu)\,\pi_{ab}(-\mu \pm i\nu)}{u(-\mu \pm i\nu)\,\pi_{ab}(-\mu \pm i\nu) + v(-\mu \pm i\nu)\,\pi_r(-\mu \pm i\nu)}. \end{aligned}$$

On the other hand, at frequencies ω much higher than the crossover frequency of the open-loop system we have

$$|M(i\omega)| \ll |N(i\omega)|.$$

In this case, for any root of the characteristic equation of the closed-loop system $p = -\mu \pm i\omega$ and small μ we obtain

$$\frac{r(-\mu \pm i\omega)}{q(-\mu \pm i\omega)} \approx \frac{\pi_{ab}(-\mu \pm i\omega)}{\pi_r(-\mu \pm i\omega)} = e_{AB}(-\mu \pm i\omega).$$

It follows from the latter equation that the feedback affects neither the root value nor the corresponding value of factor s_k.

Hence, in the general case introducing feedback leads to suppression of the slowly decaying components of the transient process with frequencies under the crossover frequency and no suppression of components with frequencies essentially higher than the crossover frequency.

As an example we study the transient process of the two-mass system of Fig. 2.27 in the case of control due to displacement of the lower mass, in which the transfer function of the open-loop system is determined by eq. (2.121).

The transient process of a passive system is obtained by means of eq. (2.244). For the numerical values of parameters adopted in Section 2.2 we get

$$\begin{aligned} g(t) &= \frac{z_A(t)}{F_0} = 10^{-6}\left[1,75\left(1-e^{-2,08t}\cos 20,8t\right)\right. \\ &\quad \left. +0,163\left(1-e^{-11,65}\cos 126t\right)\right] m/N. \end{aligned}$$

Here F_0 is value of the applied force in Newtons and $z_A(t)$ is the displacement of the lower mass in meters. For the closed-loop system with a transfer function due to disturbance we obtain

$$\begin{aligned} \frac{r(p)}{q(p)} &= \frac{e_{AB}(p)}{1+\Phi_1(p)} = (Tp+1)(\tau^2p^2+2\eta\tau p+1) \\ &\quad \times\left[c(Tp+1)(T_1^2p^2+2\zeta_1T_1p+1)\right. \\ &\quad \left.\times(T_2^2p^2+2\zeta_2T_2p+1)+kp^2(\tau^2p^2+2\eta\tau p+1)\right]^{-1}. \end{aligned} \tag{2.260}$$

Let us take $k = 5,2\cdot 10^5$ kg which corresponds to the LAC plotted in Fig. 2.28. Substituting numerical values of the parameters and defining roots of the denominator in eq. (2.260) yields

$$\begin{aligned} \frac{r(p)}{q(p)} &= \frac{(0,3p+1)}{5,2\cdot 10^5\left(1,73\cdot 10^{-4}p+1\right)} \\ &\quad \times\frac{\left(0,016^2p^2+2\cdot 0,08\cdot 0,016p+1\right)}{\left(0,99^2p^2+0,31p+1\right)\left(0,016^2p^2+0,00256p+1\right)}. \end{aligned}$$

It is easy to see that the numerator and the denominator contain identical polynomials. Cancelling these, we obtain

$$\frac{r(p)}{q(p)} = \frac{0,3p+1}{5,2\cdot 10^5(1,73\cdot 10^{-4}p+1)(0.99^2p^2+0,31p+1)}. \tag{2.261}$$

The time constant in the cancelled polynomial $0,016^2p^2+2\cdot 0,08\cdot 0,016p+1$ corresponds to an antiresonant frequency of the object. For large gains in the feedback circuit this antiresonant frequency becomes one of the system eigenfrequencies, however $|r(i\omega)|$ is small at this frequency.

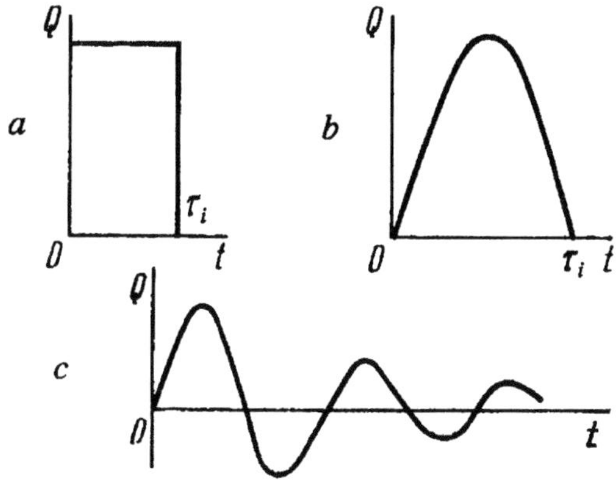

FIGURE 2.41.

We obtain the transient process corresponding to transfer function (2.261) by virtue of eq. (2.233), to get

$$\begin{aligned} g(t) &= 10^{-6}\left(1,93+0,048e^{-5800t}+0,192e^{-0,15t}\cos 0,99t\right.\\ &\left.-0,086e^{-0,155t}\sin 0,99t\right) m/N. \end{aligned}$$

Hence, this transient process contains an aperiodic component which decays very quickly and a harmonic component with a frequency lower than the frequencies of the region of efficiency of active feedback.

Let us proceed to some problems which are relevant to protection from shock excitation. Let us remember that one commonly refers to an action of high intensity and short-duration as an *impact.* Some idealised forms of impact force $Q(t)$ are displayed in Fig. 2.41. Impact can be also viewed as a kinematic action provided that some points of the system undergo a jump in velocity within a short time.

If the impact duration τ_i is much smaller than the smallest period of free vibration of linear system, then with a good degree of accuracy we can treat this impact as being instantaneous and let

$$Q(t) = Q_0\delta(t). \tag{2.262}$$

The reaction of the system is determined by integrating the equation of motion under zero initial conditions.

Consider the system shown in Fig. 2.42. An impact is assumed to act at point C, and the objective of the protection system is to reduce the displacement of point A where a control force $U(t)$ is applied. Without a control force, displacement ξ of point A relates to force $Q(t)$ by the dynamic compliance operator $e_{AC}(p)$, i.e.

$$\xi(t) = e_{AC}(p)Q(t). \tag{2.263}$$

FIGURE 2.42.

Introducing active feedback

$$U(t) = -w_x(p)x(t) \tag{2.264}$$

results in the following

$$x(t) = \xi(t) + e_A(p)U(t),$$

which, after substituting eq. (2.259) into (2.260) takes the following form

$$x(t) = e_{AC}(p)[1 + e_A(p)w_x(p)]^{-1}Q(t). \tag{2.265}$$

Thus, in order to determine $x(t)$ it is sufficient to obtain the impulse response $h(t)$ for the following transfer function

$$w(p) = e_{AC}(p)[1 + e_A(p)w_x(p)]^{-1}. \tag{2.266}$$

Here

$$x(t) = \int_0^t h(t-\tau)Q(\tau)\,d\tau. \tag{2.267}$$

Let τ_i denote the impact duration, then $Q(t) \equiv 0$ for $t > \tau_i$ and

$$x(t) = \int_0^{\tau_i} h(t-\tau)Q(\tau)\,d\tau \quad (t > \tau_i). \tag{2.268}$$

In some practical cases, the form of impact action is unknown, and only the maximum of $|Q(t)|$ termed the *impact magnitude* and impact duration are known. In these cases one can estimate the maximum value of $|x(t)|$ by considering the problem of accumulation of disturbances in linear systems, [82]

$$|x(t)| \leq Q_{\max} \int_0^t |h(\tau)| \, d\tau \ (t \leq \tau_i). \tag{2.269}$$

The motion is especially simple to find in the case of a rectangular form of impact, Fig. 2.41a, where

$$Q(t) = [\eta(t) - \eta(t - \tau_i)] Q_0, \tag{2.270}$$

$\eta(t)$ being the unit step function (2.213). Inserting eq. (2.270) into (2.267) yields

$$x(t) = [g(t) - g(t - \tau_i)] Q_0, \tag{2.271}$$

where

$$g(t) = \int_0^t h(t - \tau) \, d\tau.$$

Hence, in order to find $x(t)$ it is sufficient to determine the transient function for transfer function (2.266).

Comparing eqs. (2.263) and (2.265) we find that, similar to the stationary case, the efficiency of the active facility is characterised by the following operator

$$k(p) = [1 + e_A(p) w_x(p)]^{-1}. \tag{2.272}$$

The properties of this operator were analysed in detail in Section 2.2. In particular, it was shown that the region of control efficiency is completely determined by condition (2.75). Provided that the entire spectrum of the process $\xi(t)$ lies in this region, then the active facility unconditionally reduces the displacement caused by the impact action. Conversely if this is not the case, analysis of the efficiency becomes more complicated.

So far, we have assumed that the system subjected to an impact remains within the region of linearity of passive and active elements. In reality, this condition is met only for relatively weak impacts. In the majority of cases, relative displacements are so considerable that stops or buffers are used to meet the dimensional constraints. The isolators' characteristics then become nonlinear, and investigation of the system dynamics is complicated.

2.6 Work and power in active systems

Energy methods can be applied to the investigation of vibration protection systems. For example, isolator efficiency can be evaluated by the amount of energy transmitted to the protected object per second.

Let us consider the one-dimensional system depicted in Fig. 1.14 and determine the power of control force $U(t)$. Taking into account eq. (1.91) we obtain

$$N(t) = U(t)\,\dot{z}_A(t) = -w_z(p)z_B(t)\,\dot{z}_A(t)\,. \qquad (2.273)$$

In the harmonic regime

$$z_A = z_{A0}\cos\omega t,\;\; z_B = z_{B0}\cos(\omega t + \varphi). \qquad (2.274)$$

Inserting eq. (2.274) into (2.273) yields

$$\begin{aligned} N(t) &= -\left|w_z(i\omega)\right| z_{B0}\cos[\omega t + \varphi + \arg w_z(i\omega)][-z_{A0}\omega\sin\omega t] \\ &= \frac{1}{2}\left|w_z(i\omega)\right| z_{A0}z_{B0}\omega\,\{\sin[2\omega t + \varphi + \arg w_z(i\omega)] \\ &\quad - \sin[\varphi + \arg w_z(i\omega)]\} \end{aligned} \qquad (2.275)$$

The average power is evidently equal to the constant component in the latter expression, which yields

$$\begin{aligned} N_{av} &= -\frac{1}{2}\left|w_z(i\omega)\right| z_{A0}\cdot z_{B0}\omega\sin[\varphi + \arg w_z(i\omega)] \\ &= -\frac{1}{2}\{\operatorname{Re} w_z(i\omega)\sin\varphi + \operatorname{Im} w_z(i\omega)\cos\varphi\}z_{A0}z_{B0}\omega. \end{aligned} \qquad (2.276)$$

If points A and B coincide, then

$$z_{A0} = z_{B0},\;\; \varphi = 0,$$

and

$$N_{av} = -\frac{1}{2}\operatorname{Im} w_z(i\omega)z_{A0}^2\omega. \qquad (2.277)$$

This formula indicates that the energy transmitted by a control facility to the object is proportional to the square of the amplitude of vibration of the point at which the control force is applied. This formula also shows that the average power becomes zero if $\operatorname{Im} w_z(i\omega) = 0$. In particular, this occurs in the case of control due to displacement and acceleration provided that the feedback transfer function is assumed to be ideal. If

$$\operatorname{Im} w_z(i\omega) > 0, \qquad (2.278)$$

then $N_{av} < 0$, i.e. the control force takes energy away from the object. If

$$\operatorname{Im} w_z(i\omega) < 0 \qquad (2.279)$$

the control force supplies additional energy to the object. For the case of control due to integral of displacement this is observed at all frequencies.

Energetic characteristics can be used for estimating the efficiency of vibration protection facilities. For example, let us consider the base isolator shown in Fig. 1.17. In the case of a rigid attachment, the power of force $-R$ acting on the base is given by

$$N(t) = -R(t)\,\dot{\eta}(t)\,. \tag{2.280}$$

Taking into account eq. (1.134) yields

$$N(t) = e_0(p)R(t)\,\dot{R}(t)\,. \tag{2.281}$$

In the case of harmonic vibration

$$R = R_0 \cos\omega t$$

we obtain by analogy with eq. (2.277) that

$$N_{av} = \frac{1}{2}\,\mathrm{Im}\, e_0(i\omega)R_0^2\omega. \tag{2.282}$$

When the base is isolated, Fig. 1.17c, we obtain

$$N'(t) = Q(t)\,\dot{z}_c(t)\,,$$

or considering eq. (1.137)

$$N'(t) = Q(t)\,e_0(p)\,\dot{Q}(t)\,, \tag{2.283}$$

In the harmonic regime

$$N'(t) = \frac{1}{2}\,\mathrm{Im}\, e_0(i\omega)Q_0^2\omega. \tag{2.284}$$

Because of relationship (1.139)

$$Q_0 = |k_Q(i\omega)|\,R_0,$$

and, hence,

$$\frac{N'(t)}{N(t)} = \frac{Q_0^2}{R_0^2} = |k_Q(i\omega)|^2\,. \tag{2.285}$$

Therefore, the relationship between the powers is the same as between squares of the force amplitudes. For this reason, efficiency estimates due to power and amplitude are equivalent.

In some cases, utilising an energy method becomes difficult. For example, if the base is assumed to be rigid, then regardless of the presence of isolators the average transmitted power will be zero which implies that it is not feasible to evaluate the efficiency of isolators. It is also difficult to evaluate the efficiency of dynamic vibration absorbers since their action on the object is not explicitly related to reducing the vibration energy.

An energy approach is of crucial importance while investigating active systems, and the power of the active facility which determines energy consumption for purposes of control is an important characteristic. This latter characteristic becomes especially essential in those cases when large control forces or displacements are required, and the feasibility of an active system is limited by its admissible power.

Next we define the average power of an active facility in some special cases.

1. To obtain the power of a one-dimensional dynamic vibration absorber with the following force

$$U(t) = -w_z(p)\, z_A(t),$$

one can use eq. (2.277). As a simple example, we consider a passive vibration absorber which consists of mass m attached to the object via a damper with resistance factor b and a spring of rigidity c. It is easy to show that in this case

$$w_z(p) = \frac{mp^2(bp+c)}{mp^2+bp+c}.$$

Hence,

$$\operatorname{Im} w_z(i\omega) = -\operatorname{Im}\frac{m\omega^2(bi\omega+c)}{c-m\omega^2+bi\omega} = -\frac{bm^2\omega^5}{(c-m\omega^2)^2+b^2\omega^2}.$$

Substituting the latter expression into eq. (2.277) yields

$$N_{av} = -\frac{1}{2}\frac{bm^2\omega^6 z_0^2}{(c-m\omega^2)^2+b^2\omega^2} < 0.$$

For $c = m\omega^2$, i.e. at the tuned frequency of the dynamic vibration absorber, we obtain

$$N_{av} = -\frac{1}{2}\frac{m^2\omega^4}{b}z_0^2 = -\frac{c^2}{2b}z_0^2.$$

It should be borne in mind that reducing the resistance factor decreases the vibration amplitude z_0 at the tuned frequency, i.e. absorbed power tends to zero.

2. Consider now one-dimensional isolator having an active facility. Let $U(t)$ be the force applied by the active facility to the object and $U_0(t)$ be the force acting on the base. We denote the base motion and displacement of the object point on which force $U(t)$ acts as $\xi(t)$ and $x(t)$, respectively,

then the total power of the forces produced by the active facility is given by

$$N(t) = U(t)\,\dot{x}(t) + U_0(t)\,\dot{\xi}(t)\,. \qquad (2.286)$$

If the base is at rest, then $\dot{\xi}(t) = 0$ and

$$N(t) = U(t)\,\dot{x}(t)\,. \qquad (2.287)$$

Provided that the active facility does not contain inertial elements, i.e. $U_0(t) = U(t)$, then

$$N(t) = U(t)\left[\dot{x}(t) - \dot{\xi}(t)\right] = U(t)\,\dot{y}(t)\,, \qquad (2.288)$$

where $y(t)$ is the relative displacement of the point of attachment of the facility to the object. The forthcoming analysis will be limited to two particular cases.

If in the system with an active isolator the control force is due to absolute displacement, i.e.

$$U(t) = -w_x(p)\,x(t)\,,$$

then, as follows from eq. (2.287), we obtain by analogy with eq. (2.277) that

$$N_{av} = -\frac{1}{2}\operatorname{Im} w_x(i\omega)\omega x_0^2. \qquad (2.289)$$

Provided that the object is isolated eq. (2.288) yields

$$N(t) = -w_x(p)x(t)\,\dot{y}(t)\,. \qquad (2.290)$$

For a harmonic process

$$x = x_0\cos\omega t,\;\; y = y_0\cos(\omega t + \varphi)$$

we obtain the following expression for the average power

$$N_{av} = -\left[\operatorname{Im} w_x(i\omega)\cos\varphi - \operatorname{Re} w_x(i\omega)\sin\varphi\right]\omega x_0 y_0. \qquad (2.291)$$

In the case of control due to relative displacement

$$U(t) = -w_y(p)\,y(t)$$

we obtain by means of eq. (2.288) that

$$N(t) = -w_y(p)\,y(t)\,\dot{y}(t)\,,$$

from which we have

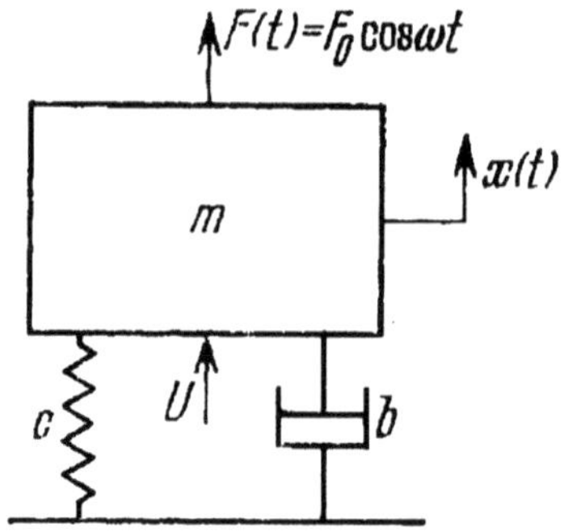

FIGURE 2.43.

$$N_{av} = -\frac{1}{2} \operatorname{Im} w_y(i\omega)\omega y_0^2. \tag{2.292}$$

As an example we consider now the system depicted in Fig. 2.43 assuming that an active facility forms control due to acceleration

$$U(t) = -\frac{k}{T_1 p + 1} p^2 x(t). \tag{2.293}$$

Let us determine the dependence of the average power of the active facility on amplitude and frequency of dynamic excitation $F(t)$. In this case,

$$w_x(i\omega) = -\frac{k\omega^2}{T_1 i\omega + 1}, \quad \operatorname{Im} w_x(i\omega) = \frac{k\omega^3 T_1}{1 + T_1^2\omega^2}.$$

The governing equation for the object

$$\left[mp^2 + bp + c + w_x(p)\right] x = F(t)$$

under the assumption $F(t) = F_0 \cos\omega t$ yields

$$\begin{aligned} x_0 &= \frac{F_0}{\left|c - m\omega^2 + bi\omega + w_x(i\omega)\right|} \\ &= F_0 \left[\left(c - m\omega^2 - \frac{k\omega^2}{1 + T_1^2\omega^2}\right)^2 + \left(b\omega + \frac{kT_1\omega^3}{1 + T_1^2\omega^2}\right)^2\right]^{-1/2}. \end{aligned}$$

Hence,

$$N_{av} = -\frac{1}{2}\frac{kT_1\omega^4}{1 + T_1^2\omega^2} \frac{F_0^2}{\left(c - m\omega^2 - \frac{k\omega^2}{1 + T_1^2\omega^2}\right)^2 + \left(b\omega + \frac{kT_1\omega^3}{1 + T_1^2\omega^2}\right)^2}.$$

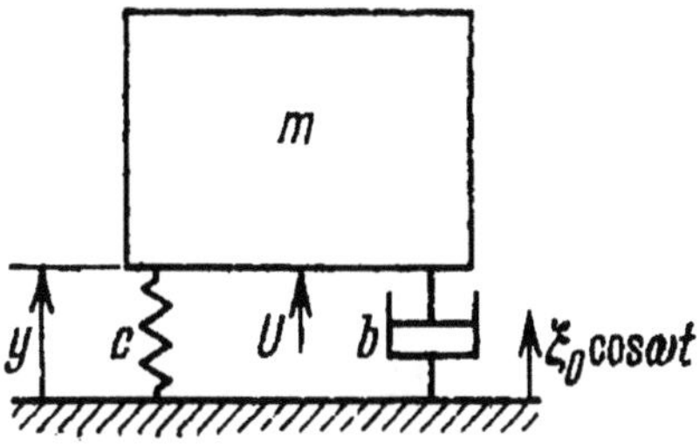

FIGURE 2.44.

Clearly, $N_{av} < 0$ at any frequency ω.

Let the system shown in Fig. 2.44 have the following control force

$$U(t) = -\frac{k}{p} y(t). \tag{2.294}$$

In this case

$$\omega_y(p) = \frac{k}{p}, \ \mathrm{Im}\, w_y(i\omega) = -\frac{k}{\omega}.$$

As

$$\begin{aligned} y_0 &= \frac{m\omega^2 \xi_0}{|c - m\omega^2 + bi\omega + \omega_y(i\omega)|} = \\ &= m\omega^2 \xi_0 \left[\left(c - m\omega^2\right)^2 + \left(b\omega - \frac{k}{\omega}\right)^2 \right]^{-1/2}, \end{aligned}$$

eq. (2.292) yields

$$N_{av} = \frac{1}{2} k m^2 \omega^4 \xi_0^2 \left[\left(c - m\omega^2\right)^2 + \left(b\omega - \frac{k}{\omega}\right)^2 \right]^{-1}.$$

In this case $N_{av} > 0$. This implies that, in the case of control due to integration of displacement, the active facility transmits energy to the isolated body.

Up to now, it has been assumed that the dynamic excitations are harmonic functions of time. The relationships obtained can be easily generalised to the cases of polyharmonic and random excitations. As an example, we determine an expression for the average power of a one-dimensional system under polyharmonic excitation assuming that

$$N(t) = -w_z(p) z_A(t)\, \dot{z}_A(t). \tag{2.295}$$

Let

$$\left.\begin{aligned} z_A &= \sum_{k=1}^{N} a_k \cos(\omega_k t + \psi_k), \\ \dot{z}_A &= \sum_{k=1}^{N} -\omega_k a_k \sin(\omega_k t + \psi_k). \end{aligned}\right\} \tag{2.296}$$

Inserting these expressions into (2.295) we obtain

$$\begin{aligned} N(t) &= -\sum_{k=1}^{N} |w_z(i\omega_k)|\, a_k \cos\left[\omega_k t + \psi_k + \arg w_z(i\omega_k)\right] \\ &\quad \times \sum_{k=1}^{N} -\omega_k a_k \sin(\omega_k t + \psi_k). \end{aligned}$$

The constant component of this expression is the average power given by

$$N_{av} = -\frac{1}{2}\sum_{k=1}^{N} \operatorname{Im} w_z(i\omega_k)\omega_k a_k^2. \tag{2.297}$$

If $z_A(t)$ is a stationary random process having spectral density $S_z(\omega)$, then defining the correlation function of $w_z(p)z_A(t)$ and $\dot{z}_A(t)$ we obtain, using known formulae [128] that

$$K(\tau) = \frac{1}{2\pi}\int_{-\infty}^{\infty} e^{i\omega\tau} w_z(-i\omega)\, i\omega S_z(\omega)\, d\omega.$$

The power averaged over realisations of process $z_A(t)$ is equal to $K(0)$, i.e.

$$\begin{aligned} N_{av} &= \frac{1}{2\pi}\int_{-\infty}^{\infty} w_z(-i\omega)\, i\omega S_z(\omega)\, d\omega = \\ &= \frac{1}{2\pi}\int_{-\infty}^{\infty} i\omega\left[\operatorname{Re} w_z(i\omega) - i\operatorname{Im} w_z(i\omega)\right] S_z(\omega)\, d\omega. \end{aligned}$$

Taking into account that $S_z(\omega) = S_z(-\omega)$ yields

$$N_{av} = \frac{1}{2\pi}\int_{-\infty}^{\infty} \operatorname{Im} w_z(i\omega)\, \omega S_z(\omega) d\omega. \tag{2.298}$$

If $z_A(t)$ is an ergodic random process, then expression (2.298) is also the time–averaged power for "nearly all" realisations of a random process.

The power computed by means of the above equations is the average useful power of the actuator of the active facility. It is clear that the parameters of the facility may be chosen in accordance with this power. One must keep in mind that an active facility having low mean power must often be capable of providing considerable instantaneous power. This must especially be taken into account in those cases for which the average power of the active facility is negative. However the instantaneous power turns out to be positive on some portions of the system motion. An active facility is said to function in the regime of *energy recuperation* if $N_{av} < 0$. If $N(t) \leq 0$ at any instant in time we refer to this regime as the *damping regime*. For a system to function in the damping regime, it is necessary that the sign of $U(t)$ be opposite to the sign of velocity. For harmonic motions in systems for which the power is defined by an equation of the type (2.275) this is equivalent to the following condition

$$\arg w_z(i\omega) = -\frac{\pi}{2}. \tag{2.299}$$

Since $w_z(i\omega)$ is a continuous function of its argument this condition can be met in general only at a few fixed frequencies.

The recuperation regime, as follows from the above examples, may be realised over the entire frequency domain. This regime occurs under the following condition

$$\operatorname{Im} w_z(i\omega) < 0. \tag{2.300}$$

We show now that by introducing a compensator into the feedback circuit one can obtain an astatic system in which the recuperation regime is observed at sufficiently high frequencies. To this aim, we introduce a vibratory element into the system with a control force due to integration of the relative displacement

$$U(t) = -\frac{k}{p(T^2p^2 + 2T\zeta p + 1)} y(t) = w_y(p) y(t). \tag{2.301}$$

In this case

$$\begin{aligned} w_y(i\omega) &= -\frac{k}{i\omega(1 - T^2\omega^2 + 2T\zeta i\omega)}, \\ \operatorname{Im} w_y(i\omega) &= -\frac{k(1 - T^2\omega^2)}{\omega[(1 - T^2\omega^2)^2 + 4T^2\zeta^2\omega^2]}. \end{aligned} \tag{2.302}$$

Therefore, at $\omega > T^{-1}$ condition (2.300) holds. It is easy to prove that the system with control force (2.301) is astatic.

It is necessary to notice that an efficient vibration protection system considerably reduces the vibration amplitude of the point at which the control force is applied. Thus, the power which is ordinarily proportional to the square of the amplitude turns out to be negligible in a steady-state regime. For this reason, in the majority of cases the power is determined rather by the necessity to control the transient process than by the condition to control the stationary regime.

Let us demonstrate a way to define the power of an active facility under motion of the system subjected to a unit force. Consistent with (2.295) we have

$$N(t) = -w_x(p)g(t)\,\dot{g}(t) = -w_z(t)g(t)\,h(t), \tag{2.303}$$

since $z_A(t) = g(t)$. Inserting $g(t)$ and $h(t)$ from eqs. (2.235) and (2.207) yields

$$N(t) = -\sum_{k=1}^{n}\sum_{s=1}^{n}\frac{r(\lambda_k)\,r(\lambda_s)}{q'(\lambda_k)\,q'(\lambda_s)\,\lambda_k}[w_z(\lambda_k)e^{\lambda_k t} - w_z(0)]w_z(\lambda_s)e^{\lambda_s t}. \tag{2.304}$$

The total amount of energy of the transient process is thus given by

$$A = \int_0^\infty N(t)\,dt = \sum_{k=1}^{n}\frac{r(\lambda_k)}{\lambda_k q'(\lambda_k)}\sum_{s=1}^{n}\frac{r(\lambda_s)\,w_z(\lambda_s)}{q'(\lambda_s)}\left[\frac{w_z(\lambda_k)}{\lambda_k+\lambda_s} - \frac{w_z(0)}{\lambda_s}\right]. \tag{2.305}$$

2.7 Multidimensional systems

Vibration protection systems forming several control forces U_1, U_2, ..., U_N acting on the object at points A_1, A_2, ..., A_N are called *multidimensional* systems, Fig. 2.45. Consistent with the concept of Section 1.1 any passive system with several isolators is multidimensional because the isolators' reaction may be viewed as control forces. However, in what follows we study systems several active elements, as a rule.

It is known that in passive systems the generalisation to multidimensional systems leads to considerable complication, mostly because of complicated dynamical modelling the object. In active systems, there arise additional difficulties related to the presence of feedbacks.

Consider the N–dimensional system shown in Fig 2.45. Let the vector of control forces $\mathbf{U}(t)$ be related to the vector $\mathbf{z}(t)$ of measured parameters of the vibration field as follows

$$\mathbf{U}(t) = -W_{Uz}(p)\mathbf{z}(t), \tag{2.306}$$

where $W_{Uz}(p)$ stands for the matrix of feedback transfer functions. In turn, vector $\mathbf{z}$ due to the arrangement of Fig. 2.2 can be expressed in terms of

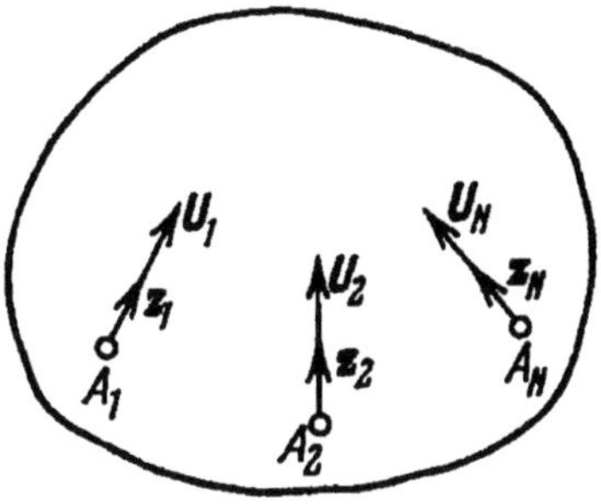

FIGURE 2.45.

the vector of control forces $\mathbf{U}$ and the vectors of dynamic actions $\mathbf{F}(t)$ and $\boldsymbol{\xi}(t)$

$$\mathbf{z}(t) = W_{0U}(p)\mathbf{U} + W_{0F}(p)\mathbf{F} + W_{0\xi}(p)\boldsymbol{\xi}, \tag{2.307}$$

where W_{0U}, W_{0F} and $W_{0\xi}$ denote transfer function matrices of the object. Inserting eq. (2.306) into (2.307) yields

$$\mathbf{z}(t) = [I_N + W_{0U}(p)W_{Uz}(p)]^{-1}\{W_{0F}(p)\ \mathbf{F} + W_{0\xi}(p)\boldsymbol{\xi}\}. \tag{2.308}$$

By virtue of eq. (2.307) the expression in braces represents a vector of output parameters for $\mathbf{U}(t) = 0$, or the disturbance of the object output in accordance with the terminology of automatic control. Designating this vector as $\boldsymbol{\zeta}(t)$ we obtain

$$\mathbf{z}(t) = [I_N + W_{0U}(p)W_{Uz}(p)]^{-1}\boldsymbol{\zeta}(t) = K(p)\boldsymbol{\zeta}(t). \tag{2.309}$$

Operator $K(p)$ relating the vectors of the output parameters in the system with and without control is a matrix defining the control efficiency which can be described by a matrix norm. If, for instance, all dynamic actions and thus all output parameters obey a harmonic law then condition (1.131) may be taken as an efficiency criterion. It also follows from eq. (2.308) that the characteristic equation of the system can be recast in the following form

$$\det[I_N + W_{0U}(\lambda)W_{Uz}(\lambda)] = 0. \tag{2.310}$$

If vector $\mathbf{z}$ is the vector of displacements of points $A_1, A_2, \ldots, A_N$ at which control forces are applied, then matrix $W_{0U}(p)$ coincides with the dynamic compliance matrix of the object at these points, i.e.

$$W_{0U}(p) = E_A(p) = \{e_{rk}(p)\}_1^N.$$

Assume that the control forces $U_s(t)$ are linear in the displacement of points A_m

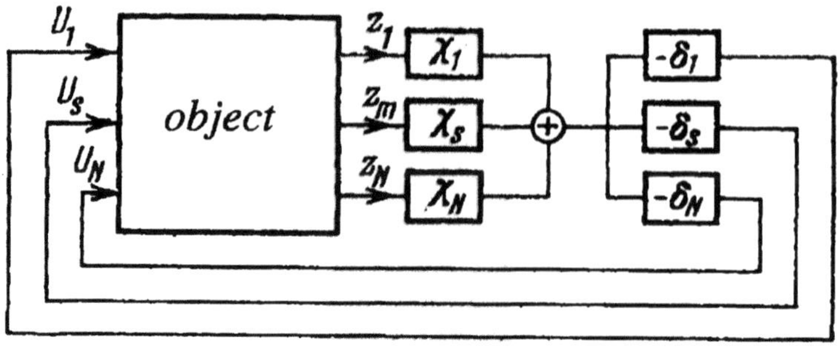

FIGURE 2.46.

$$U_s(t) = -\sum_{m=1}^{N} \gamma_{sm} z_m. \tag{2.311}$$

If matrix

$$\Gamma = \{\gamma_{sm}\}_1^N \tag{2.312}$$

is diagonal, i.e.

$$\Gamma = \operatorname{diag}\{\gamma_1, \gamma_2, ..., \gamma_N\},$$

and its entries γ_s are positive, then control (2.311) is equivalent to introducing additional elastic elements between points A_m and fixed points. Clearly, the passive system obtained in such a manner is stable for any value of γ_s. If matrix (2.312) is not diagonal, i.e. if inputs and outputs are linked by cross-feedbacks, the system may become unstable. Vibrations due to the vibration modes of the original system studied in Sections 2.3 and 2.4 may be observed in such a system. In the general case, stability analysis reduces to determining the roots of the characteristic equation (2.310) with the further requirement that the real parts of these roots are negative.

Consider another special case. Let

$$\gamma_{sm} = \delta_s \chi_m, \tag{2.313}$$

which implies that the system has the block diagram shown in Fig. 2.46. First, the following variable

$$z_* = \sum_{m=1}^{N} \chi_m z_m, \tag{2.314}$$

is formed in terms of the displacements of points A_m and then the control forces proportional to the above signal are formed as follows

$$U_s = -\delta_s z_*. \tag{2.315}$$

Since

$$z_m = \sum_{r=1}^{N} e_{rm}(p) U_r,$$

we obtain from eqs. (2.314) and (2.315) that

$$z_* = -\sum_{m=1}^{N}\sum_{r=1}^{N} \chi_m \delta_r e_{rm}(p) z_*. \tag{2.316}$$

Hence, the characteristic equation for the closed-loop system takes the following form

$$1 + \sum_{m=1}^{N}\sum_{r=1}^{N} \chi_m \delta_r e_{rm}(\lambda) = 0. \tag{2.317}$$

By analogy with Section 2.4 it is easy to show that the system under consideration is stable if the following function

$$e(i\omega) = \sum_{m=1}^{N}\sum_{r=1}^{N} \chi_m \delta_r e_{rm}(i\omega)$$

possesses the same structure as the dynamic compliance at a point, i.e. if its resonant and antiresonant values alternate. Taking into account that all coefficients δ_s are positive we can write sufficient conditions for stability in a form analogous to (2.181)

$$\sum_{m=1}^{N} \chi_m l_{As}^{(m)} > 0 \ (s = 1, ..., n), \tag{2.318}$$

where $l_{As}^{(m)}$ is the coefficient of the $s-th$ normal form at point A_m. Choosing coefficients χ_m so that the following condition

$$\sum_{m=1}^{N} \chi_m l_{As}^{(m)} = 0 \ (s = s_1, s_2, ..., s_m) \tag{2.319}$$

is met, one can ensure non-observability with respect to the corresponding principal coordinates and remove the very possibility of exciting vibration with these normal modes. Similar to one-dimensional systems, this structure of feedbacks enables increasing total feedback gain and thus increases the system performance.

All of the above remains valid in the case when the points under observation do not coincide with the points of action of the control forces. However, it is worthwhile noting that the methods considered assume that the modal coefficients are prescribed at the points under observation. For actual systems, this is possible only in tests implying excitation of resonant vibrations in the system and measuring vibration amplitudes at the points of interest. Clearly, one can expect only approximate coefficients $l_{As}^{(m)}$.

Another method of forming control which is also based on given coefficients of normal modes is worthy of attention.

Let $B_1, ..., B_M$ and $A_1, ..., A_N$ be the points under observation and the points where the control forces are applied, respectively. Let us introduce matrices of coefficients of normal modes at these points as follows

$$\begin{aligned} L_A &= \left\{l_{As}^{(m)}\right\}, \ L_B = \left\{l_{Br}^{(u)}\right\}, \\ (s &= 1, ..., n; \ m = 1, ..., N; \ r = 1, ..., n; \ u = 1, ..., M). \end{aligned}$$

According to eqs. (2.311) and (2.312) we obtain

$$\mathbf{U} = -\Gamma \mathbf{z}_B = -\Gamma L_B \mathbf{q}, \tag{2.320}$$

where $\mathbf{U}$ is the vector of control forces, $\mathbf{z}_B$ is the vector of displacements of points B_u, $\mathbf{q}$ is the vector of principal coordinates, and Γ is an $N \times M$ matrix of feedback gains γ_{mu}. Let us determine the generalised forces corresponding to control forces. From eq. (1.38) we obtain

$$\mathbf{Q} = L_A^T \mathbf{U} = -L_A^T \Gamma L_B \mathbf{q}. \tag{2.321}$$

Assume that coefficients γ_{mu} are chosen so that the matrix

$$D = L_A^T \Gamma L_B \tag{2.322}$$

is diagonal with positive entries. In other words, coefficients γ_{mu} satisfy the following conditions

$$\sum_{m=1}^{N} \sum_{u=1}^{M} l_{As}^{(m)} l_{Br}^{(u)} \gamma_{mu} = d_s \delta_{sr} \ (s, r = 1, ..., n), \tag{2.323}$$

where $d_s > 0$ and δ_{sr} denotes Kronecker's symbol. Then each generalised coordinate depends only on the corresponding principal coordinate

$$Q_s = -d_s q_s \tag{2.324}$$

and, therefore, principal coordinates in the closed-loop system are related only via dissipative forces which implies a weak relationship. Such an "autonomy" of control with respect to each principal coordinate is expedient in many cases. It is evident that with resonant vibrations excited "autonomous" control behaves like control due to displacement in a one-dimensional system, i.e. this will suppress vibrations due to resonant principal coordinate and hardly affect vibrations due to another principal coordinate. Coefficients γ_{mu} can be defined from condition (2.323) only if

$MN \geq n^2$, i.e. if the number of equations does not exceed the number of unknowns. For this reason, "autonomy" of control can be achieved only for sufficiently large numbers of observed and controlled points. When condition (2.323) holds, "autonomy" of control remains if all feedback circuits contain the same arbitrary transfer function, i.e. if

$$\mathbf{U} = -w(p)\Gamma z_B. \tag{2.325}$$

For $w(p) = kp$ and $w(p) = k_1 p^2$ we obtain a system with "autonomous" control due to generalised velocities $\dot{q}_s$ and generalised accelerations $\ddot{q}_s$, respectively. These systems are similar to one-dimensional systems with control due to velocity and accelerations, see Sections 2.2 and 2.3.

Analysis of efficiency and stability for the case of "autonomous" control does not cause any difficulty as it reduces to the investigation of one-dimensional systems. All that is needed for this investigation is the transfer function matrix of the object in principal coordinates. With the dissipation in the object ignored, this matrix is given by eq. (1.46). When dissipation is small, one can neglect terms of order higher than the first order of smallness and adopt the following

$$\begin{aligned} W(p) &= (I_n p^2 + Bp + K)^{-1} \qquad (2.326) \\ &\approx \operatorname{diag}\{(p^2 + b_{11}p + k_1^2)^{-1}, ..., (p^2 + b_{nn}p + k_n^2)^{-1}\}, \end{aligned}$$

i.e. account only for diagonal elements of matrix B.

Consider another particular case of a system with autonomous control. Assume that the isolated object may be treated as a rigid body mounted on N isolators, Fig. 2.47. The total reaction of the $s-th$ isolator which also includes the reaction of passive elements is denoted as U_s. Let

$$U_s = -\gamma_s w(p) z_s, \tag{2.327}$$

where γ_s denotes some factors and z_s is the displacement of the point of attachment of the $s-th$ isolator to the object. For example, characteristics of the type (2.327) are typical for isolators of the same design solution.

Next we replace active isolators by elastic elements of the following rigidity

$$c_s = \gamma_s c, \tag{2.328}$$

where c is a certain constant parameter. We consider free vibration of the object mounted on these linear elastic elements. Let $\boldsymbol{\theta}$ designate the vector of principal coordinates of this conservative system which has six degrees of freedom. In order to apply Lagrange's equation we need the kinetic energy

$$T = \frac{1}{2}\sum_{k=1}^{6} \dot{\theta}_k^2.$$

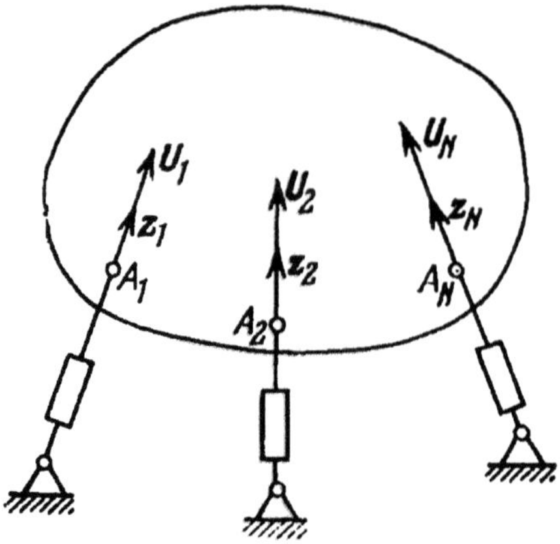

FIGURE 2.47.

One sees that

$$\ddot{\theta}_k = \Theta_k \ (k = 1, ..., 6), \tag{2.329}$$

where Θ_k are generalised forces. It is known that for conservative systems

$$\Theta_k = -\lambda_k^2 \theta_k, \tag{2.330}$$

where λ_k $(k = 1, ..., 6)$ are the eigenfrequencies.

Let

$$z_s = \sum_{k=1}^{6} h_{sk}\theta_k.$$

Then from the following relation

$$\sum_{k=1}^{6} \Theta\delta\dot{\theta}_k = \sum_{s=1}^{N} U_s \delta z_s = \sum_{s=1}^{N} U_s \sum_{k=1}^{6} h_{sk}\delta\theta_k = \sum_{k=1}^{6} \delta\theta_k \sum_{s=1}^{N} U_s h_{sk}$$

we obtain

$$\begin{aligned} \Theta_k &= \sum_{s=1}^{N} h_{sk} U_s = -\sum_{s=1}^{N} h_{sk} c \gamma_s \sum_{l=1}^{6} h_{sl}\theta_l = \\ &= -c \sum_{l=1}^{6} \theta_l \sum_{s=1}^{N} h_{sk}\gamma_s h_{sl} \ (k = 1, ..., 6). \end{aligned} \tag{2.331}$$

Comparing eqs. (2.330) and (2.331) we find that

$$c\sum_{s=1}^{N} h_{sk}\gamma_s h_{sl} = \lambda_k^2 \delta_{kl}, \tag{2.332}$$

δ_{kl} denoting Kronecker's symbol.

Let us now return to the system with active isolators. The equations of motion in this case can be cast in the form of eq. (2.329). In order to obtain generalised force Θ_k it is necessary to substitute $U_s = -\gamma_s w(p) z_s$ into eq. (2.331) which yields

$$\Theta_k = -w(p)\sum_{l=1}^{6}\theta_l \sum_{s=1}^{N} h_{sk} c\gamma_s h_{sl} \ (k = 1, ..., 6). \tag{2.333}$$

Taking into account eq. (2.332) we obtain

$$\Theta_k = -\frac{w(p)}{c}\lambda_k^2 \theta_k. \tag{2.334}$$

Therefore, "autonomy" of control with respect to all principal coordinates is ensured for any function $w(p)$. In this case, the governing equations are given by

$$\left[p^2 + \frac{w(p)}{c}\lambda_k^2\right]\theta_k = 0 \ (k = 1, ..., 6). \tag{2.335}$$

Thus, the characteristic equation for the closed-loop system is split into six uncoupled equations

$$p^2 + \frac{w(p)}{c}\lambda_k^2 = 0 \ (k = 1, ..., 6). \tag{2.336}$$

It is evident that ratios λ_k^2/c do not depend upon the choice of c. It is also clear that when constructing eq. (2.336) there is no need to determine the main coordinates. It is sufficient to find the eigenfrequencies of the obtained conservative system for an arbitrary value of c.

Assume now that we have carried out the analysis of a one-dimensional system consisting of a mass m_s on the $s-th$ isolator of the system under consideration as a preliminary for further investigation. The equation of motion for such a system is written in the form

$$\left[\frac{m_s}{\gamma_s}p^2 + w(p)\right]x_s = 0. \tag{2.337}$$

Let us presume that this system is stable under the following conditions

$$\alpha_1 \leq \frac{m_s}{\gamma_s} \leq \alpha_2.$$

Then, the multidimensional system will be stable if inequalities

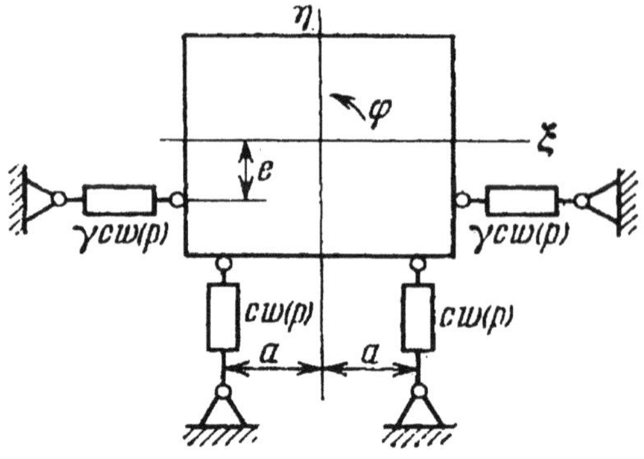

FIGURE 2.48.

$$\alpha_1 \leq \frac{m_s}{\lambda_s^2} \leq \alpha_2 \tag{2.338}$$

hold for $s = 1, ..., 6$.

Clearly, eq. (2.337) can be used not only for stability analysis but also for estimating the performance of transient processes.

As an example, we find the stability conditions for the system shown in Fig. 2.48 provided that

$$w(p) = -c - bp - \frac{k}{p(Tp+1)}. \tag{2.339}$$

The latter equation implies that each isolator contains passive elements with rigidity c and resistance factor b and an active element forming the control force due to integration of displacement. Let us replace the active isolators by vertical elastic elements with stiffness c and horizontal elastic elements with stiffness γc and consider this conservative system. The eigenfrequencies of the system are as follows

$$\lambda_1^2 = \frac{2c}{m}, \ \lambda_{2,3}^2 = \lambda_1^2 \left[1 + \frac{a^2}{\rho^2} + \gamma \frac{e^2}{\rho^2} \pm \sqrt{\left(1 - \frac{a^2}{\rho^2} - \gamma \frac{e^2}{\rho^2}\right)^2 + 4\gamma \frac{e^2}{\rho^2}} \right]. \tag{2.340}$$

Here ρ is the radius of inertia about the centre of mass. In accordance with eq. (2.336) the characteristic equation of the system is reduced to the following form

$$p^2 + \lambda_k^2 \left[1 + \frac{b}{c} p + \frac{k}{cp(Tp+1)} \right] = 0 \ (k = 1, ..., 6). \tag{2.341}$$

Applying Hurwitz's criterion yields the following constraint on the value of gain of active feedback

$$k < \frac{\lambda_k^2 b(1 + \lambda_k^2 T^2 + \lambda_k^2 bT)}{(c + bT\lambda_k^2)^2} c^2. \tag{2.342}$$

For $T = 0$ we obtain

$$k < b\lambda_k^2. \tag{2.343}$$

Therefore, an admissible value of feedback gain is determined by the lowest eigenfrequency of the system.

3

Nonlinear passive single-degree-of-freedom systems

3.1 Methods of analysis for nonlinear system

The dynamic characteristics of the passive elements of vibration protection systems and their active feedbacks were assumed to be linear in Chapter 2. Frequently, while analysing actual engineering systems one has to use nonlinear dynamic models. The main reasons for the existence of nonlinearities in vibration protection systems are listed below.

1. The passive elastic elements can be considered as being linear only under small deformations. When the deformations are large, one observes nonlinearity of the overall elastic characteristics caused either by material properties (e.g. rubber), constructive peculiarities of the element (e.g. presence of some initial compressive force, use of a conical spring etc.), or the presence of stops. Figure 3.1 displays schematically the most typical nonlinear elastic restoring characteristics. The characteristic shown in Fig. 3.1a corresponds to a linear vibration isolator with high stiffness buffers. The governing equation is given by

$$\left.\begin{aligned} f(x) &= cx && \text{if } -\Delta_1 \le x \le \Delta, \\ f(x) &= c\Delta + c_1(x-\Delta) && \text{if } x > \Delta, \\ f(x) &= -c\Delta_1 + c_1(x-\Delta_1) && \text{if } x < \Delta_1. \end{aligned}\right\} \tag{3.1}$$

Using the Heaviside step function

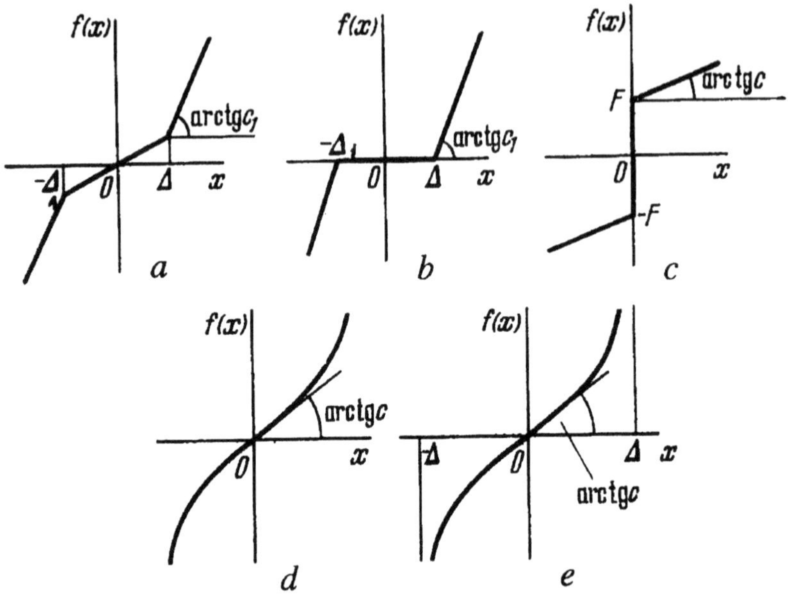

FIGURE 3.1.

$$\eta(x) = \begin{cases} 1 \text{ if } x > 0, \\ 0 \text{ if } x \leq 0, \end{cases} \tag{3.2}$$

we can represent eq. (3.1) in the form of a single equation

$$\begin{aligned} f(x) &= c_1 x + (c_1 - c)\,\Delta_1 + (c - c_1)\,(x + \Delta_1)\eta(x + \Delta_1) \\ &\quad + (c_1 - c)\,(x - \Delta)\eta(x - \Delta). \end{aligned} \tag{3.3}$$

Letting $c = 0$ yields a system representing "clearance", see Fig. 3.1b. Figure 3.1c shows characteristic of an elastic element with some initial compressive force F, i.e.

$$f\,(x) = cx + F \text{ sign} x. \tag{3.4}$$

Some smooth nonlinear restoring forces are displayed in Figs. 3.1d and 3.1e. The symmetric characteristic of Fig. 3.1d can be approximated by a linear-plus-cubic function of the displacement

$$f\,(x) = cx + \gamma x^3. \tag{3.5}$$

Figure 3.1e shows the characteristic of a nonlinear elastic element with rigid stops. This type of the nonlinear restoring force may be conveniently approximated as follows

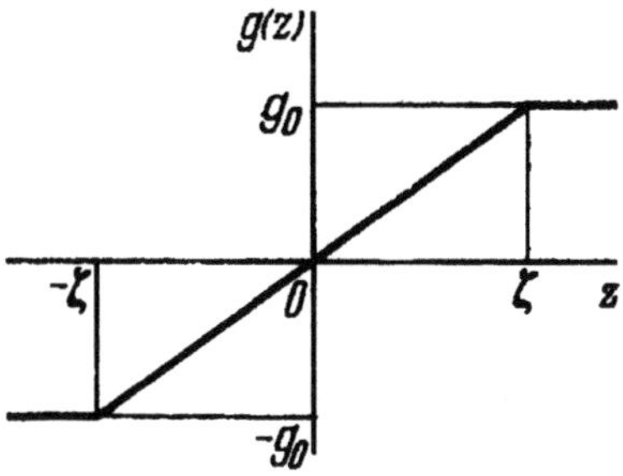

FIGURE 3.2.

$$f(x) = \frac{2c\Delta}{\pi} \tan\frac{\pi x}{2\Delta}. \tag{3.6}$$

2. Nonlinear characteristics may also contain some dissipative elements. Elements utilising interface friction are frequently used. To a first approximation, they obey Coulomb's law

$$h(\dot{x}) = h_0 \mathrm{sign}\dot{x}. \tag{3.7}$$

In some cases, the internal friction in the material of the elastic element should be taken into account. The modelling of the internal friction is studied in detail in [96] and [95].

3. Last but not least, nonlinear elements can be used in active systems of vibration protection. As mentioned in Section 3.1 the sensors, e.g. the relative displacement transducers, may have nonlinear characteristics. Typically, the nonlinearity of the sensor has a constrained output

$$\left.\begin{array}{ll} g(z) = k_z z & \text{if } |z| \leq \zeta, \\ g(z) = k_z \zeta \mathrm{sign} z & \text{if } |z| > \zeta, \end{array}\right\} \tag{3.8}$$

see Fig. 3.2. The nonlinear elements may be introduced into the compensators in order to provide the system with some special properties. For example, the elements restraining the maximum displacement of the object under impact can be introduced in the case of the control due to relative displacement, [104].

The actuators are essentially nonlinear elements. As mentioned above, the flow rate through a control valve depends nonlinearly upon the position of the control valve, see Fig. 2.11. As follows from eq. (2.58) a pneumatic facility exhibits nonlinear properties. The nonlinearity of electromagnetic systems is caused by the nonlinear force-current relationship, cf. (2.65).

We turn now to the methods of analysis which will be applied in this book. In nonlinear systems, harmonic excitations usually excite periodic vibrations, the first harmonic being dominant. In order to investigate these vibrations, use will be made of the method of harmonic linearisation. We recall the basic concept of the method.

Let there exist a nonlinear element with the following characteristic

$$f = f(x, \dot{x}), \tag{3.9}$$

where x stands for the input while $f(x, \dot{x})$ denotes some nonlinear function.

In accordance with the method of harmonic linearisation we assume that the function $x(t)$ is approximately the sum of a harmonic process of frequency ω and a constant component

$$x \approx a_0 + a\cos\omega t, \ \dot{x} \approx -a\omega\sin\omega t. \tag{3.10}$$

Substituting (3.10) into eq. (3.9) yields a periodic function whose Fourier's series is as follows

$$f(a_0 + a\cos\omega t, -a\omega\sin\omega t) = f_0 + f_{1c}\cos\omega t + f_{1s}\sin\omega t + \ldots \tag{3.11}$$

The Fourier coefficients are calculated by the formulae

$$\left.\begin{aligned}
f_0 &= \frac{1}{2\pi}\int_0^{2\pi} f(a_0 + a\cos\psi, -a\omega\sin\psi)d\psi = f_0(a_0, a),\\
f_{1c} &= \frac{1}{\pi}\int_0^{2\pi} f(a_0 + a\cos\psi, -a\omega\sin\psi)\cos\psi d\psi = f_{1c}(a_0, a),\\
f_{1s} &= \frac{1}{\pi}\int_0^{2\pi} f(a_0 + a\cos\psi, -a\omega\sin\psi)\sin\psi d\psi = f_{1s}(a_0, a).
\end{aligned}\right\} \tag{3.12}$$

Applying the method of harmonic linearisation implies that the higher harmonics of the process (3.11) can be ignored as having little influence on the parameters of approximation (3.10). Substantiation for neglecting higher harmonics caused by the filtering properties of the linear parts of the system has been discussed in detail in [109].

Clearly, the nonlinear function (3.9) in the equations of motion may be replaced by any other function $f^*(x, \dot{x})$ provided that the latter has the same three first terms in the Fourier series and eq. (3.10) holds. Let us take a linear function

$$f^*(x, \dot{x}) = qx + r\dot{x} + s. \tag{3.13}$$

Substituting (3.10) into (3.13) yields

$$f^*(a_0 + a\cos\omega t, -a\omega\sin\omega t) = qa\cos\omega t - ra\omega\sin\omega t + qa_0 + s. \tag{3.14}$$

Comparing eqs. (3.14) and (3.11) we conclude that the right hand sides of these equations are identical if the factors q, r, and s are defined as follows

$$q = \frac{f_{1c}(a_0, a)}{a}, \quad r = -\frac{f_{1s}(a_0, a)}{a\omega}, \quad s = f_0(a_0, a) - qa_0. \tag{3.15}$$

Thus, under the above assumptions, the nonlinear function (3.9) can be replaced by the linear function (3.13) with factors q, r, and s given by eq. (3.15). This replacement is called the *harmonic linearisation* of nonlinear functions while factors (3.15) are referred to as the *harmonic linearisation factors.* Notice that these factors depend on the parameters a and a_0 of the process $x(t)$. For this reason, the linearisation factors are determined only after an approximate solution of the linearised equation of motion has been found.

Let $x^0(t)$ denote the harmonic component in $x(t)$, i.e.

$$x = a_0 + x^0, \quad \dot{x} = \dot{x}^0,$$

then, eq. (3.13) can be written down in the form

$$f^*(x, \dot{x}) = qx^0 + r\dot{x}^0 + f_0. \tag{3.16}$$

One of the linearisation factors coincides with the constant term in the Fourier series which makes the linearisation more convenient.

Let us demonstrate the method of harmonic linearisation for the following system

$$M(p)x + N(p)f(x, \dot{x}) = R(p)(A\cos\omega t), \tag{3.17}$$

where M, N and R are polynomials in the differentiation operator p.

An approximate periodic solution of eq. (3.17) is sought in the form of eq. (3.10). Linearising the nonlinear function yields

$$M(p)x + N(p)[(q + rp)x^0 + f_0] = R(p)(A\cos\omega t). \tag{3.18}$$

Taking into account that

$$M(p)x = M(p)(a_0 + x^0) = M(0)a_0 + M(p)x^0,$$

we order and equate the constant terms and harmonics in the left and right hand sides of eq. (3.18). The result is

$$\left.\begin{aligned} M(0)\,a_0 + N(0)\,f_0(a_0, a) &= 0, \\ \{M(p) + N(p)[q(a_0, a) + r(a_0, a)p]\}x^0 &= R(p)(A\cos\omega t). \end{aligned}\right\} \tag{3.19}$$

The first equation yields $a_0(a)$ which, being substituted into $q(a_0, a)$ and $r(a_0, a)$, results in the amplitude-dependent functions q and r. Treating the second equation in (3.19) as linear we obtain the following equation for the amplitude a of the harmonic of frequency ω

$$a = \left| \frac{R(i\omega)}{M(i\omega) + N(i\omega)[q\,(a) + r\,(a)\, i\omega]} \right| A. \tag{3.20}$$

Having determined the amplitude a from this equation we can find the phase-angle between the dynamic excitation and $x\,(t)$ from eq. (3.19)

$$\varphi = \arg \frac{R(i\omega)}{M(i\omega) + N(i\omega)[q\,(a) + r\,(a)\, i\omega]}. \tag{3.21}$$

There exist a number of cases in which the dependence of the linearisation factors on the parameters of the approximate solution due to eqs. (3.12) and (3.15) is simplified. So, if the nonlinear function can be represented in the form

$$f(x, \dot{x}) = g(x) + h(\dot{x}),$$

then

$$q = \frac{1}{\pi a} \int_0^{2\pi} g(a_0 + a\cos\psi)\cos\psi d\psi, \tag{3.22}$$

$$r = \frac{1}{\pi a \omega} \int_0^{2\pi} h(a\omega \sin\psi)\sin\psi d\psi, \tag{3.23}$$

$$f_0 = \frac{1}{2\pi} \int_0^{2\pi} g(a_0 + a\cos\psi) d\psi. \tag{3.24}$$

The formulae for some frequently encountered nonlinearities are given in [102].

Another interpretation may be given to the method of harmonic linearisation. Let the intent of the linearisation be the replacement of the nonlinear function (3.9) by a linear function (3.16). We form the functional

$$J = \frac{\omega}{2\pi} \int_0^{2\pi/\omega} [f(x, \dot{x}) - qx_0 - r\dot{x}_0 - f_0]^2 dt \tag{3.25}$$

and determine the values of parameters q, r and f_0 which render the integral a minimum under the assumption of harmonic motion (3.10). It is easy to prove that the minimum of this functional is achieved by those values of q, r and f_0 which are the harmonic linearisation factors for the function $f(x, \dot{x})$, see [65]. In other words, these factors minimize the mean square of the difference between the nonlinear and linear functions of the harmonic process $x(t)$.

Let us assume that the system undergoes the polyharmonic vibration

$$x(t) = a_0 + \sum_{i=1}^{N} a_i \cos(\omega_i t + \psi_i) = a_0 + x^0(t). \tag{3.26}$$

For example, such a vibration can be excited by several independent, periodic excitations. Let us determine the factors q, r and f_0 which minimize the functional

$$\Phi(q, r, f_0) = \lim_{T\to\infty} \frac{1}{T} \int_0^T \{f[x(t), \dot{x}(t)] - qx^0(t) - r\dot{x}^0(t) - f_0\}^2 dt, \tag{3.27}$$

where $x(t)$ is due to eq. (3.26). Some simple manipulations result in

$$\left.\begin{aligned} q &= \frac{1}{\sigma_x^2} \lim_{T\to\infty} \frac{1}{T} \int_0^T f(x,\dot{x}) x^0 dt, \\ r &= \frac{1}{\sigma_v^2} \lim_{T\to\infty} \frac{1}{T} \int_0^T f(x,\dot{x}) \dot{x}^0 dt, \\ f_0 &= \lim_{T\to\infty} \frac{1}{T} \int_0^T f(x,\dot{x}) dt, \end{aligned}\right\} \tag{3.28}$$

where

$$\sigma_x^2 = \lim_{T\to\infty} \frac{1}{T} \int_0^T (x^0)^2 dt, \quad \sigma_v^2 = \lim_{T\to\infty} \frac{1}{T} \int_0^T (\dot{x}^0)^2 dt. \tag{3.29}$$

It is apparent that q, r and f_0 due to eq. (3.28) are functions of $a_0, a_1, ..., a_N$, $\psi_1, ..., \psi_N$. Inserting the linear function

$$f^*(x, \dot{x}) = qx^0 + r\dot{x}^0 + f_0$$

into the equation of motion and solving these equations one can express the solution parameters in terms of the linearisation factors. This allows one to obtain a closed system of transcendental equations for the solution parameters and the linearisation factors q, r and f_0.

In practice, this modification of the method of harmonic linearisation proves to be impedient due to a large number of unknowns to be determined and difficulties associated with estimating of the integrals (3.28). To simplify the method, it is worthwhile reducing the number of parameters of the polyharmonic process to those which are essential for the linearisation factors. For this purpose, *the method of linearisation based upon the distribution function* was developed in [74]. The method is studied in detail in [65], here our consideration is restricted to the case in which the nonlinear function has a single independent variable

$$f = f(x). \tag{3.30}$$

The basic idea of the method relies on the distribution function $W(u)$ for the polyharmonic process $x(t)$ which implies the relative duration of the time intervals in which

$$x(t) \leq u. \tag{3.31}$$

The derivative $w(u) = dW/du$ is called the *distribution density*. If x_2 and x_1 are the maximum and minimum of $x(t)$, respectively, then according to the definition of $W(u)$

$$W(u) \equiv 0 \ \text{if}\, u \leq x_1, \ W(u) \equiv 1 \ \text{if}\, u \geq x_2,$$

and, therefore,

$$\int_{x_1}^{x_2} w(u)\, du = W(x_2) - W(x_1) = 1. \tag{3.32}$$

The following equations hold for $W(u)$ and $w(u)$

$$W(u) = \lim_{T\to\infty} \frac{1}{T} \int_0^T \eta[u - x(t)]dt, \tag{3.33}$$

$$w(u) = \lim_{T\to\infty} \frac{1}{T} \int_0^T \delta[u - x(t)]dt. \tag{3.34}$$

It is easy to see that $W(u)$ is the probability of the "inequality (3.31) holds true" event when t is randomly taken from the interval $(0, \infty)$. This explains the application of the terminology of probability theory and allows us to introduce the moments of the distribution function by analogy. The time average value of $[x^0(t)]^k$ is referred to as the *$k-th$ central moment*

$$M_k^0\{x(t)\} = \lim_{T\to\infty} \frac{1}{T} \int_0^T [x^0(t)]^k dt. \tag{3.35}$$

The second central moment is called the *variance*

$$M_2^0\{x(t)\} = \sigma_x^2. \tag{3.36}$$

Using the probability density one can replace time averaging over an infinite interval by averaging u over the interval (x_1, x_2). Indeed, by virtue of the known properties of Dirac's delta-function one obtains for an arbitrary piecewise continuous function $\varphi(x)$

$$\lim_{T\to\infty}\frac{1}{T}\int_0^T \varphi[x(t)]dt = \int_{x_1}^{x_2}\varphi(u)\,du \lim_{T\to\infty}\frac{1}{T}\int_0^T \delta[u-x(t)]dt = \int_{x_1}^{x_2}\varphi(u)\,w(u)\,du. \tag{3.37}$$

Specifically,

$$M_k^0\{x(t)\} = \int_{x_1}^{x_2}(u-a_0)^k w(u)\,du. \tag{3.38}$$

We now proceed to the linearisation of function (3.30). Let us find the linear function

$$f^*(x) = qx^0 + f_0 \tag{3.39}$$

that minimizes the functional

$$\Phi(q, f_0) = \lim_{T\to\infty}\frac{1}{T}\int_0^T \{f[x(t) - qx^0(t) - f_0]\}^2 dt. \tag{3.40}$$

Applying eq. (3.37) yields

$$\Phi(q, f_0) = \int_{x_1}^{x_2}[f(u) - q(u-a_0) - f_0]^2 w(u)\,du. \tag{3.41}$$

The conditions for a minimum of integral (3.41) yield

$$q = \frac{1}{\sigma_x^2}\int_{x_1}^{x_2} f(u)(u-a_0)\,w(u)\,du, \tag{3.42}$$

$$f_0 = \int_{x_1}^{x_2} f(u)\,w(u)\,du. \tag{3.43}$$

Hence, the linearisation factors obtained from the condition of minimising the mean square deviation between $f(x)$ from $f^*(x)$ may be determined for a given distribution density of the sought-for solution. Generally speaking, the distribution density of the process (3.26) depends upon all a_i and ψ_i. But, as shown in [65] this distribution density can be represented as a function of a_0 and the central moments of the process. The linearisation factors turn out to be essentially dependent only on the few first moments. With this in view, the linearisation due to the distribution function implies that one chooses a family of the functions

$$w(u, a_0, \sigma_x^2, M_3^0, M_4^0, ...), \tag{3.44}$$

and assumes that the distribution density belongs to this family. Substituting (3.44) into eqs. (3.42) and (3.43) one can express q and f_0 in terms of the moments of the sought-for solution. On the other hand, by solving the linearised equations one can express the amplitudes a_i and the phases ψ_i and thus the moments in terms of the linearisation factors. This means that a closed system of equations for the unknown parameters is obtained.

The approximate solution obtained is known to be relatively insensitive to a particular distribution function. The recommended choices of the distribution function are collected in [65], [74] and [75]. In the majority of these cases, it is sufficient to retain only a_0 and the even moments σ_x^2, M_4^0 and M_6^0 in eq. (3.44). The odd moments may be ignored as they are either small or trivial.

We consider now an example illustrating the above method. To this end, we solve the following equation

$$M(p)x + N(p)x^3 = A_0 + \sum_{i=1}^{N} A_i \cos(\omega_i t + \psi_i). \tag{3.45}$$

An approximate polyharmonic solution is sought in the form

$$x = a_0 + \sum_{i=1}^{N} a_i \cos(\omega_i t + \psi_i + \theta_i) = a_0 + x^0(t). \tag{3.46}$$

Let us linearise the nonlinear function $f(x) = x^3$. Its linearisation factors, due to eqs. (3.42) and (3.43), are given by

$$\begin{aligned} q &= \frac{1}{\sigma_x^2} \int_{x_1}^{x_2} u^3 (u - a_0) w(u) du \\ &= \frac{1}{\sigma_x^2} \int_{x_1}^{x_2} (u - a_0)[(u - a_0)^3 + 3a_0(u - a_0)^2 + 3a_0^2(u - a_0) + a_0^3] w(u) du \\ &= \frac{1}{\sigma_x^2} (M_4^0 + 3a_0 M_3^0 + 3a_0^2 \sigma_x^2), \end{aligned}$$

$$f_0 = \int_{x_1}^{x_2} u^3 w(u) du = M_3^0 + 3a_0\sigma_x^2 + a_0^3.$$

With frequencies ω_i assumed to be incommensurable, $M_3^0 = 0$, see [65], the latter equation can be simplified to

$$q = \frac{M_4^0}{\sigma_x^2} + 3a_0^2, \ f_0 = a_0(3\sigma_x^2 + a_0^2). \tag{3.47}$$

We form now the linearised equation

$$M(p)(x^0 + a_0) + N(p)(qx^0 + f_0) = A_0 + \sum_{i=1}^{N} A_i \cos(\omega_i t + \psi_i) \tag{3.48}$$

and separate the constant terms and the time-dependent centered terms, to obtain

$$\left. \begin{aligned} M(0)\, a_0 + N(0)\, f_0 &= A_0, \\ [M(p) + qN(p)]x^0 &= \sum_{i=1}^{N} A_i \cos(\omega_i t + \psi_i). \end{aligned} \right\} \tag{3.49}$$

Expressing a_0 from the first equation and inserting it into q in the second equation allows us to consider q as a known functions of σ_x^2 and M_4^0

$$q = \varepsilon + 3a_0^2\sigma_x^2, \ \varepsilon = \frac{M_4^0}{\sigma_x^2}. \tag{3.50}$$

Prescribing the value of ε ($3/2 < \varepsilon < 3$ for any polyharmonic process with independent phases), q depends only on σ_x^2

$$q = q(\sigma_x^2).$$

Solving the second equation in (3.49) yields

$$x^0 = \sum_{i=1}^{N} a_i \cos(\omega_i t + \psi_i + \theta_i), \tag{3.51}$$

where

$$a_i = \frac{A_i}{|M(i\omega) + qN(i\omega)|}, \ \theta_i = -\arg\{M(i\omega) + qN(i\omega)\}. \tag{3.52}$$

The following formula holds for the polyharmonic process

$$\sigma_x^2 = \sum_{i=1}^{N} \frac{a_i^2}{2}. \tag{3.53}$$

Substituting a_i from eq. (3.52) into this expression results in the following equation for σ_x^2

$$\sigma_x^2 = \sum_{i=1}^{N} \frac{A_i^2}{\left|M(i\omega) + q(\sigma_x^2)N(i\omega)\right|^2}. \tag{3.54}$$

Solving this equation for σ_x^2 and inserting the solution into the equation for q yields the amplitude a_i. The following equation holds for the process with incommensurable frequencies, see [65]

$$M_4^0 = \frac{9}{4}\left(\sigma_x^4 - \frac{1}{12}\sum_{i=1}^{N} a_i^4\right),$$

which allows one to find the numerical value of ε and to prove whether this value coincides with the prescribed value.

Assume now that the dynamic excitation implies a stationary random process $F(t)$ having a given mathematic expectation m_F and a given spectral density $S_F(\omega)$. The *method of statistical linearisation*, cf. [55], [97] and [108], is used to determine the motion of the nonlinear system. In accordance with this method, a steady-state solution of the nonlinear stochastic equation is sought in the form

$$x(t) = m_x + x^0(t),$$

with m_x and $x^0(t)$ denoting the mathematic expectation and the centered time-dependent part of the solution, respectively. The nonlinear function $f(x)$ is replaced by the linear function

$$f^*(x) = f_0 + qx^0, \tag{3.55}$$

whose factors f_0 and qx^0 are determined by minimising the variance of the following random function

$$\Delta f = f[x(t)] - f^*[x(t)]. \tag{3.56}$$

As $x(t)$ is a stationary function its variance is independent of time and can be found as follows

$$\left\langle (\Delta f)^2 \right\rangle = \int_{-\infty}^{\infty} [f(u) - f_0 - q(u - m_x)]^2 w_x(u)du. \tag{3.57}$$

Here $\langle\rangle$ means averaging over the realisation of the random process and $w_x(u)$ stands for the probability density of the random process $x(t)$. By analogy with the linearisation due to the distribution function we assume that $w_x(u)$ belongs to a certain family of functions depending upon the first few moments. When the external excitation is normal it is then common practice to assume that $x(t)$ is approximately Gaussian, i.e. the distribution density of $x(t)$ is taken in the form

$$w_x(u) = \frac{1}{\sqrt{2\pi}\sigma_x} \exp\left[-\frac{(u-m_x)^2}{2\sigma_x^2}\right]. \tag{3.58}$$

Minimisation of functional (3.57) leads to the known expressions for the linearisation factors (3.42) and (3.43) with the distribution density of the polyharmonic process $w(u)$ being replaced by the probability density of the random process $w_x(t)$. If one substitutes a particular probability density, e.g. (3.58), into eqs. (3.42) and (3.43) the linearisation factors will be functions of the moments m_x and σ_x^2 of the solution.

By means of the linearised equation one can determine the dependence of the mathematic expectation and the variance of the solution on the linearisation factors. As above, this results in a system of transcendental equations for the solution parameters.

Let us solve the following equation

$$M(p)x + N(p)f(x) = F(t) = m_F + F^0(t), \tag{3.59}$$

provided that $F(t)$ is a normal stationary random process with the mathematic expectation m_F and the spectral density $S_F(\omega)$. We linearise $f(x)$, i.e. we replace it by the linear function (3.55), to have

$$[M(p) + qN(p)]x^0 + M(0)\,m_x + N(0)\,f_0 = m_F + F^0(t), \tag{3.60}$$

where

$$q = q\left(m_x, \sigma_x^2\right), \quad f_0 = f_0\left(m_x, \sigma_x^2\right)$$

are due to eqs. (3.42) and (3.43). Separating the constant terms and the time-dependent random component in eq. (3.60) yields

$$\left.\begin{aligned} &M(0)\,m_x + N(0)\,f_0(m_x, \sigma_x^2) = m_F, \\ &[M(p) + qN(p)]x^0 = F^0(t). \end{aligned}\right\} \tag{3.61}$$

The first equation allows us to express m_x in terms of σ_x^2 and substitute the result into the second equation in which $q = q\left[m_x\left(\sigma_x^2\right), \sigma_x^2\right] = q\left(\sigma_x^2\right)$. The second equation gives

$$\sigma_x^2 = \frac{1}{2\pi}\int_{-\infty}^{\infty} S_x(\omega)d\omega = \frac{1}{2\pi}\int_{-\infty}^{\infty} \left|M(i\omega) + q(\sigma_x^2)N(i\omega)\right|^{-2} S_F(\omega)d\omega. \tag{3.62}$$

The variance of the solution is determined from this equation.

A number of modifications of the method of equivalent linearisation have been considered. The solutions obtained do not claim to be of high accuracy and can be viewed as a first approximation. Nevertheless, the method of equivalent linearisation turns out to be effective in the cases in which there is a need to estimate the system behaviour and expose the resonance processes. This method is especially appropriate to the analysis of vibration protection systems which involve estimating, often roughly, the vibration reduction due to the control introduced. There exist methods which permit estimation of the accuracy of the solution obtained by the methods of equivalent linearisation. Accurate estimates have been obtained for the method of harmonic linearisation, see [109], some estimates of the accuracy of the method of linearisation due to the distribution function and the method of statistical linearisation can be found in [66], [67], [69], [71] and [108].

Let us next consider the method of evaluating the linearisation factors for the piecewise linear functions, i.e. the functions whose graph is a broken line. Some examples of the piecewise linear functions are displayed in Figs 3.1a, b and c, see also [90].

Any piecewise linear function $f(x)$ can be represented in the form of linear combination of the following three functions:

a) a linear function

$$f_1(x) = cx + e; \tag{3.63}$$

b) the unit step function at $x = b_k$

$$f_2(x) = \eta(x - b_k), \tag{3.64}$$

where $\eta()$ is the Heaviside unit function (2.213) and

c) the unit ramp function at $x = b_k$

$$f_3(x) = (x - b_k)\eta(x - b_k). \tag{3.65}$$

Let some piecewise linear function $f(x)$ and its derivative $f'(x)$ have jumps h_k and γ_k at $x = b_k$ $(k = 1, ..., s)$, respectively. Let us assume that the most left piece of the broken line $f(x)$ obeys eq. (3.63). In this case, function $f(x)$ can be written down in the form

$$f(x) = cx + e + \sum_{k=1}^{s}[h_k + \gamma_k(x - b_k)]\eta(x - b_k). \tag{3.66}$$

For example, this equation renders eq. (3.3) for the function (3.1).

The linearisation factors are some linear functional of the function to be linearised. Thus, provided that $f(x)$ can be represented in the form (3.66), the linearisation factors are as follows

$$q = c + \sum_{k=1}^{s} (h_k q'_k + \gamma_k q''_k), \tag{3.67}$$

$$f_0 = ca_0 + e + \sum_{k=1}^{s} (h_k f'_{0k} + \gamma_k f''_{0k}), \tag{3.68}$$

where q'_k, q''_k, f'_{0k} and f''_{0k} are the linearisation factors of functions (3.64) and (3.65). In order to obtain these factors we substitute eqs. (3.64) and (3.65) into eqs. (3.22) and (3.24), to get

$$q'_k = \frac{1}{\pi a} \int_0^{2\pi} \eta(a_0 - b_k + a\cos\psi) \cos\psi d\psi, \tag{3.69}$$

$$f'_{0k} = \frac{1}{2\pi} \int_0^{2\pi} \eta(a_0 - b_k + a\cos\psi) d\psi, \tag{3.70}$$

$$q''_k = \frac{1}{\pi a} \int_0^{2\pi} \eta(a_0 - b_k + a\cos\psi)(a_0 - b_k + a\cos\psi) \cos\psi d\psi, \tag{3.71}$$

$$f''_{0k} = \frac{1}{2\pi} \int_0^{2\pi} \eta(a_0 - b_k + a\cos\psi)(a_0 - b_k + a\cos\psi) d\psi. \tag{3.72}$$

The closed form expressions for the linearisation factors depend on the following parameter

$$\alpha_k = \frac{b_k - a_0}{a}.$$

Taking into account that

$$\eta(a_0 - b_k + a\cos\psi) = \eta(\cos\psi - \alpha_k) = \begin{cases} 0 \text{ if } \alpha_k \geq 1 \\ 1 \text{ if } \alpha_k \leq -1, \end{cases}$$

$(a > 0)$ and evaluating the integral yields the formulae collected in Table 3.1.

Analogous evaluations can be carried out for the linearisation factors using the distribution function and for the factors using statistical linearisation. The Gaussian distribution law (3.58) yields the linearisation factors collected in Table 3.2, where

$$\Phi(u) = \frac{1}{\sqrt{2\pi}} \int_0^u \exp(-\frac{t^2}{2}) dt \tag{3.73}$$

	$\alpha_k \geq 1$	$\|\alpha_k\| < 1$	$\alpha_k \leq -1$
q'_k	0	$\frac{2}{\pi a}\sqrt{1-\alpha_k^2}$	0
f'_{0k}	0	$\frac{1}{\pi}\arccos\alpha_k$	1
q''_k	0	$\frac{1}{\pi}\left(\arccos\alpha_k - \alpha_k\sqrt{1-\alpha_k^2}\right)$	1
f''_{0k}	0	$\frac{a}{\pi}\left(\sqrt{1-\alpha_k^2} - \alpha_k\arccos\alpha_k\right)$	$-a\alpha_k$

TABLE 3.1.

q'_k	$\frac{1}{\sqrt{2\pi}\sigma_x}\exp(-\beta_k^2/2)$
f'_{0k}	$\frac{1}{2} - \Phi(\beta_k)$
q''_k	$\frac{1}{2} - \Phi(\beta_k)$
f''_{0k}	$\sigma_x\left[\frac{1}{\sqrt{2\pi}}\exp(-\beta_k^2/2) - \beta_k/2 + \beta_k\Phi(\beta_k)\right]$

TABLE 3.2.

and

$$\beta_k = \frac{b_k - m_x}{\sigma_x}.$$

3.2 Forced vibration under harmonic excitation and linear damping

Consider a nonlinear single-degree-of-freedom system of vibration protection. The equation of motion for such a system may be cast as follows

$$m\ddot{u} + U(u, \dot{u}) = Q(t). \tag{3.74}$$

Here m denotes the mass of the object to be isolated, u denotes the displacement of the elastic vibration isolator relative to the equilibrium point, U denotes the force applied to the vibration isolator in the direction u. $Q(t)$ stands for the vibrational action which implies either an active force or the inertia force acting on the object in its relative motion. In the latter case this is simply the product of the object mass and the base acceleration with a minus sign.

In the present Chapter we investigate some particular cases of the vibration protection systems governed by eq. (3.74).

To begin with, we assume that the dissipation force is proportional to the relative velocity in the vibration isolator $\dot{u}$ which means that the force acting on the object to be protected can be represented as follows

$$U(u, \dot{u}) = U_e(u) + b\dot{u}, \tag{3.75}$$

where $U_e(u)$ stands for the elastic restoring force and b is a constant factor.

Let us also assume that the vibrational action is a single frequency harmonic function of time, i.e.

$$Q(t) = F\cos\omega t. \tag{3.76}$$

Such an action occurs at tests involving harmonic shakers, in rotating machinery due to unbalanced mass etc.

Inserting eqs. (3.75) and (3.76) into (3.74) yields

$$m\ddot{u} + b\dot{u} + U_e(u) = F\cos\omega t. \tag{3.77}$$

The solution of this equation is sought by means of the harmonic linearisation method in the form

$$u = a_0 + u^0(t) = a_0 + a\cos(\omega t + \varphi). \tag{3.78}$$

To this end, we linearise the nonlinear elastic force

$$U_e \approx U_0 (a_0, a) + c_d (a_0, a) u^0, \tag{3.79}$$

where

$$U_0 = \frac{1}{2\pi} \int_0^{2\pi} U_e (a_0 + a \cos\psi) \, d\psi, \tag{3.80}$$

$$c_d = \frac{1}{\pi a} \int_0^{2\pi} U_e (a_0 + a \cos\psi) \cos\psi d\psi. \tag{3.81}$$

Substituting eq. (3.79) into (3.77) we obtain the linearised equation

$$m\ddot{u}^0 + b\dot{u}^0 + U_0 + c_d u^0 = F \cos\omega t. \tag{3.82}$$

Considering eq. (3.78) we get the following equations for the static displacement a_0, the vibration amplitude a and the phase angle φ between the vibration and the excitation

$$U_0 (a_0, a) = 0, \tag{3.83}$$

$$a = \frac{F_1}{\sqrt{(\lambda^2 - \omega^2)^2 + 4n^2\omega^2}}, \tag{3.84}$$

$$\tan\varphi = \frac{2n\omega}{\omega^2 - \lambda^2}, \tag{3.85}$$

where

$$\lambda^2 = \frac{c_d (a_0, a)}{m}, \quad n = \frac{b}{2m}, \quad F_1 = \frac{F}{m}. \tag{3.86}$$

The relationship between the static displacement a_0 and the vibration amplitude a due to eq. (3.83) is the same as in the case of free vibration. Comparing eqs. (3.81) and (3.86) it is easy to see that $\lambda (a_0, a)$ defines the frequency of free vibration of the amplitude a relative to static displacement a_0. By expressing a_0 in terms of a with the help of eq. (3.83) one can obtain the following explicit formula

$$\lambda^2 (a) = \frac{c_d}{m} = \frac{1}{\pi a m} \int_0^{2\pi} U_e [a_0 (a) + a \cos\psi] \cos\psi d\psi. \tag{3.87}$$

The dependence $\lambda (a)$ is referred to as the *backbone curve.* Inserting $\lambda (a)$ in eq. (3.84) one can determine the vibration amplitude a. The most convenient way of solving this equation is graphical. To this aim, we plot the graph $a (\lambda)$ considering λ in eq. (3.84) as an independent variable. If n

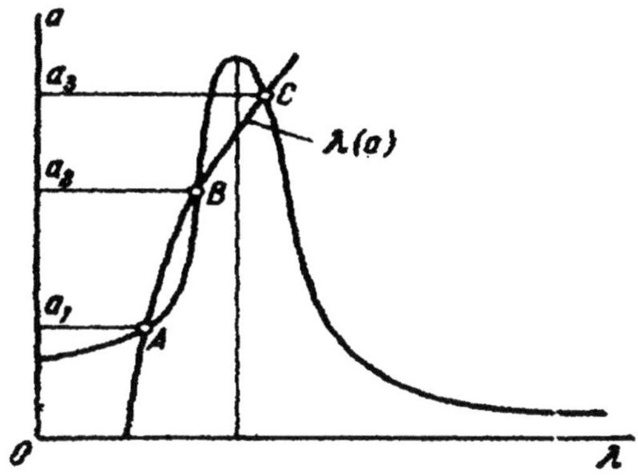

FIGURE 3.3.

is small, which is normally the case for the overwhelming majority of vibration protection systems, this graph takes the form shown in Fig. 3.3. The backbone curve $\lambda(a)$ is also plotted in this Figure. The points of the curves's intersection determine the solutions of eq. (3.84). Figure 3.3 shows that eq. (3.84) may have a number of solutions in the case of the nonlinear elastic restoring force. This implies that a number of different oscillatory regimes with different amplitudes and phases may occur, all the solutions being close to harmonic. The initial conditions are of crucial importance for establishing a particular regime. In practice however, the initial conditions are not strictly determined as they depend upon many random factors. For this reason one has to face the possibility of any of the above regimes.

Multiplicity of the steady-state solutions is one of the typical features of the non-autonomous nonlinear systems which we deal with.

The driving frequency in the vibration protection systems may vary. So, for example, the unbalanced mass can rotate with various angular velocities, the shaker frequency can be changed etc. The *amplitude-frequency characteristic* $a(\omega)$ can be obtained by computing the vibration amplitude from eq. (3.84) for any frequency ω. The graph of this curve is called the *response curve* or the *resonance curve*.

Before we proceed to analysing the form of the resonance curves let us note that in actual engineering systems the amplitude of the driving force F is usually a frequency-dependent function. For example, if $Q(t)$ is the component of a centrifugal force, then $F = \mu\omega^2$, where μ is the static moment. In other words, the amplitude of the vibrational action is proportional to the square of the frequency. The same dependence takes place in the case in which the object needs to be isolated from the base vibrating with amplitude ξ_0 which does not depend upon frequency ω. In this case $F_0 = m\xi_0\omega^2$. Other dependencies of F on the frequency ω are also possible. For this reason, the general form of the equation for the amplitude-frequency characteristic is given by

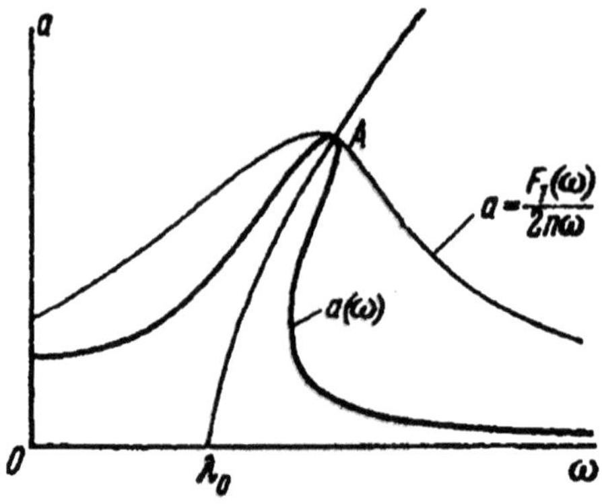

FIGURE 3.4.

$$a = \frac{F_1(\omega)}{\sqrt{\left[\lambda^2(a) - \omega^2\right]^2 + 4n^2\omega^2}}. \tag{3.88}$$

Let us determine the intersection points of the resonance (3.88) and backbone (3.87) curves. If we set $\lambda = \omega$ in eq. (3.88) (evidently, this holds true at the intersection points) we immediately obtain the following dependence of the amplitude on frequency

$$a = \frac{F_1(\omega)}{2n\omega}. \tag{3.89}$$

This equation determines the locus of the points of intersection of the resonance curves with the backbone curves in the plane (a, ω). This curve intersects the backbone curve at the same points at which the resonance curve intersects the backbone curve (e.g. point A in Fig. 3.4).

Since $F_1(\omega) > 0$ and the denominator of eq. (3.88) is larger than $2n\omega$ for any λ and ω the resonance curve lies under the curve (3.89) for the same values of ω.

Let us demonstrate that the curves (3.88) and (3.89) coincide at the point $\omega = \lambda$. With this in view, we find the derivative $da/d\omega$ considering eq. (3.88) as an implicit dependence a on ω. The result is

$$\begin{aligned} \frac{da}{d\omega} &= \frac{1}{\left(\lambda^2 - \omega^2\right)^2 + 4n^2\omega^2} \left\{ F_1'(\omega)\sqrt{\left(\lambda^2 - \omega^2\right)^2 + 4n^2\omega^2} \right. \\ &\left. -2F_1(\omega)\frac{\left(\lambda^2 - \omega^2\right)\left(\lambda\frac{d\lambda}{da}\cdot\frac{da}{d\omega} - 1\right) + 2n^2\omega}{\sqrt{\left(\lambda^2 - \omega^2\right)^2 + 4n^2\omega^2}} \right\}. \end{aligned} \tag{3.90}$$

Thus,

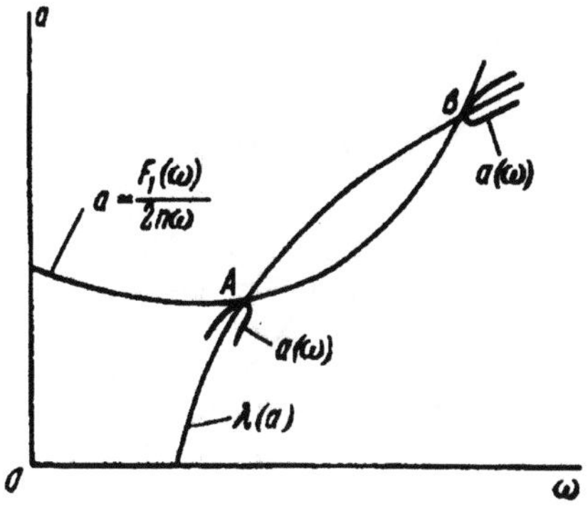

FIGURE 3.5.

$$\left(\frac{da}{d\omega}\right)_{\omega=\lambda} = \frac{F_1'(\lambda)\,2n\lambda - F_1(\lambda)\,2n}{4n^2\lambda^2} = \frac{F'(\lambda)\,\lambda - F_1(\lambda)}{2n\lambda^2},$$

which coincides with the derivative of function (3.89) with respect to ω provided that $\omega = \lambda$.

Hence, by plotting the backbone curve and curve (3.89) we can judge on the resonance curve in the vicinity of the intersection of the resonance and backbone curves (points A and B in Fig. 3.5).

The behaviour of the resonance curves far from these points essentially depends on the particular form of the backbone curve and function $F_1(\omega)$. Let us consider for example

$$F_1(\omega) = F_1 = const, \tag{3.91}$$

then curve (3.89) is a hyperbola. This hyperbola can intersect a hard backbone curve $(\lambda'(a) > 0)$ only once, see Fig. 3.6, while it can intersect a

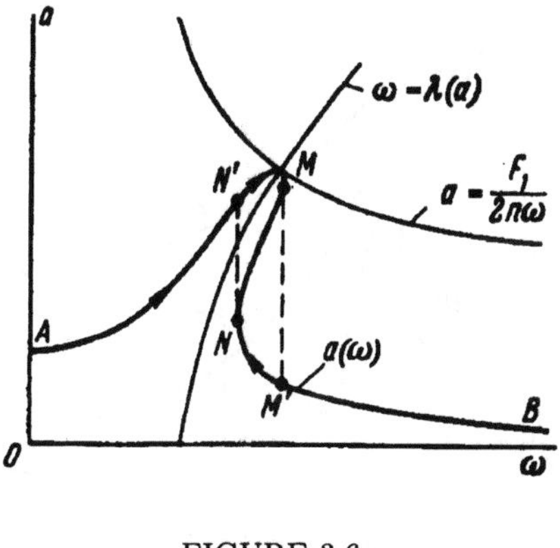

FIGURE 3.6.

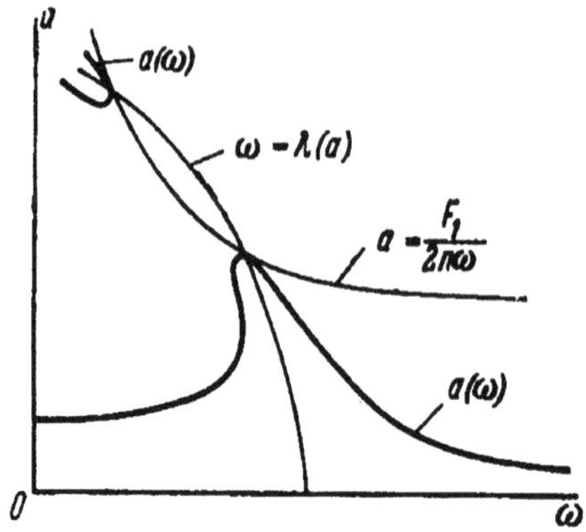

FIGURE 3.7.

soft backbone curve $(\lambda'(a) < 0)$ either twice, Fig. 3.7, or never, Fig. 3.8. Equating the derivative (3.90) to zero we find that the following frequency

$$\omega = \omega^* = \sqrt{\lambda^2 - 2n^2} \tag{3.92}$$

provides the extreme amplitude. If $2n^2 \gg \lambda^2$, then $\omega^* \approx \lambda$, that is the extremum values are achieved near the point of intersection of the resonance and the backbone curves.

Now it is easy to show that there are three types of resonance curve, see Figs. 3.6-3.8, depending on the number of intersection points of the backbone curve with the hyperbola (3.89). The case displayed in Fig. 3.6 corresponds to a hard backbone curve and is the most typical one. The cases shown in Figs. 3.7-3.8 occur much more seldom.

Assume that the driving frequency is so slowly varied that the free vibration response is neglected and only the steady-state vibration is considered. Determining the vibration amplitude a for any frequency we can plot the resonance curve. Increasing frequency ω, see Fig. 3.6, we can determine

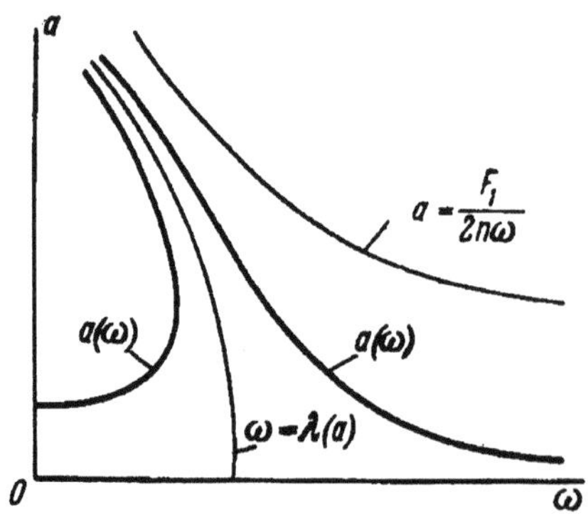

FIGURE 3.8.

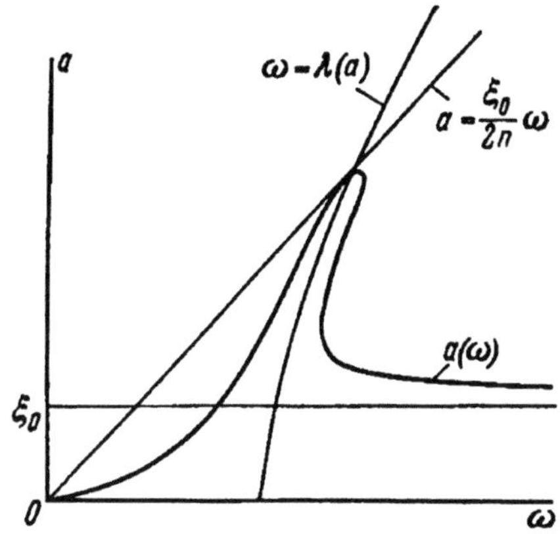

FIGURE 3.9.

the points of the resonance curve corresponding to the AM portion of the curve. A further increase in ω requires a jump from point M to point M' with an accompanying decrease in the amplitude. Starting at point B and decreasing ω the amplitude follows the BN portion of the curve and then jumps to the point N'. It is not possible to get the MN portion of the curve because the regimes corresponding to this portion are unstable. Jump phenomena are typical for nonlinear systems and often manifest themselves in practice.

Let us assume now that the amplitude of the driving force is proportional to the square of the frequency

$$F_1(\omega) = \xi_0 \omega^2. \tag{3.93}$$

In this case, the vibration amplitude takes the extremum values in the vicinity of the points of intersection of the resonance and the backbone curves. Indeed, equating the derivative $da/d\omega$ to zero yields

$$\omega^* = \frac{\lambda^2}{\sqrt{\lambda^2 - 2n^2}}, \tag{3.94}$$

so that $\omega^* \approx \lambda$ if $2n^2 \ll \lambda^2$.

Substituting eq. (3.93) into (3.89) we obtain the line equation

$$a = \frac{\xi_0}{2n}\omega. \tag{3.95}$$

This line can intersect a soft backbone curve at one point only (for $n \neq 0$) and a hard backbone curve at several points. The form of the resonance curve for the system with a hard backbone curve depends upon the number of the points of intersection, see [61]. The most typical forms are indicated in Figs. 3.9-3.11.

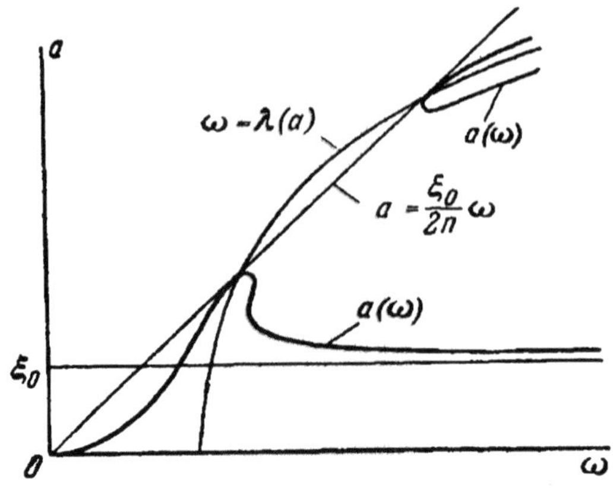

FIGURE 3.10.

Additional branches corresponding to large amplitude vibrations are typical feature of these resonance curves. These branches indicate that for a system with nonlinear elastic restoring force and linear dissipative force the harmonic driving force may cause large amplitude vibrations at the frequencies which considerably exceed the frequency of the small free vibration ("eigenfrequency" of the system). Thus, in these cases the vibration isolator amplifies the vibrational action rather than isolating the object from the vibrating support structure.

It is essential to note that similar phenomena occur for cases in which the estimations based on linear theory seem to ensure the converse conclusion. Indeed, let us consider a vibration isolator with a linear elastic element and symmetric elastic buffers. The elastic restoring force for this vibration isolator is given by eq. (3.1). The backbone curve for this case is as follows

$$\left.\begin{array}{ll} \lambda^2(a) = \lambda_0^2\left[1 + \left(\varkappa^2 - 1\right)\left(1 - \dfrac{2d}{\pi a}\sqrt{a^2 - d^2} - \dfrac{2}{\pi}\arcsin\dfrac{d}{a}\right)\right] & \text{if } a > d, \\ \lambda(a) = \lambda_0 & \text{if } a \le d, \end{array}\right\} \tag{3.96}$$

where

$$\lambda_0^2 = \frac{c}{m}.$$

This curve is plotted in Fig. 3.12.

Assume now that line (3.95) intersects the backbone curves at the three points indicated in Fig. 3.12. This takes place if the following condition holds

$$\frac{\xi_0}{2n} < \frac{d}{\lambda_0}. \tag{3.97}$$

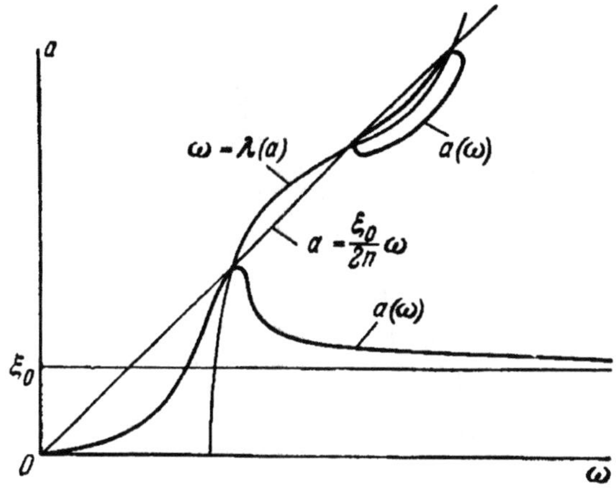

FIGURE 3.11.

The resonance curve is split into two branches. The entire lower branch lies in the zone of the linear elastic restoring force. Use of linear theory equations leads to the incorrect conclusion that the elastic buffers play no role since the vibration, even the resonance one, remains within the linear portion of the elastic characteristic. The existence of the additional closed branch BC is ignored in the framework of such an analysis. In reality, however, the system can exhibit dangerous vibrations of large amplitude with frequencies $\omega_1 < \omega < \omega_2$. This phenomenon, referred to as the "impact on buffers", is known to those dealing with vibration isolation systems.

It is worth noting that the system behaviour in the case in which both frequency and force amplitude are varied slowly can drastically differ from that in which the amplitude of the external force is held constant, $F = const$. For instance, upon performing an experiment with the system having the resonance curves of Figs. 3.11 and 3.12 the amplitude can follow the lower branch of the resonance curve if the frequency is varied slowly. The upper branch cannot be detected in the framework of this test. This branch however can manifest itself after a rather large occasional impact. In this case, large amplitude vibrations corresponding to the upper branch are excited. Thus, the large amplitude vibrations can be overlooked not only in calculations but also in tests.

In order to avoid the dangerous resonance phenomena the designer of the vibration protection system needs to ensure that no additional branches of the resonance curve appear. To this end, as Fig. 3.12 indicates, either the slope of the line (3.95) must be decreased or the resonance curve must be made steeper. However, reducing the slope of the line (3.95) (for given ξ_0) corresponds to an increase in factor n, i.e. increasing the dissipative forces. Increasing the steepness of the resonance curve inevitably leads to an

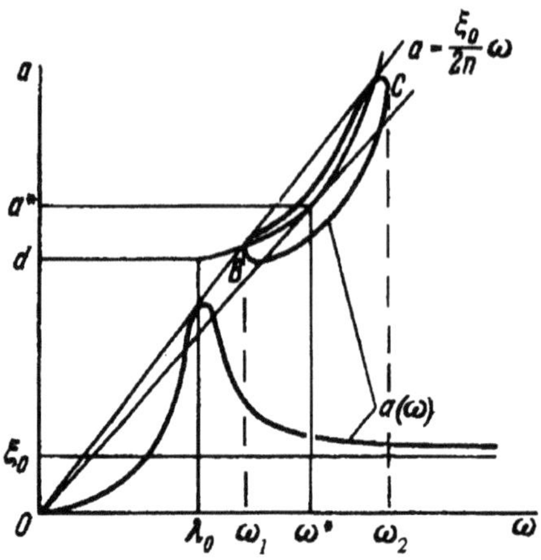

FIGURE 3.12.

increase in the gaps, which in turn results in an increase in the dimensions of the vibration isolation system.

Let us determine the value of the damping factor n^* which removes the threat of the occurrence of the additional branch in the system with elastic buffers. To this aim, we plot the tangent to the resonance curve which intersects the origin of the coordinate system, Fig. 3.12. If ω^* and a^* are the coordinates of the point at which this line touches the resonance curve, then

$$n^* = \frac{\xi_0 \omega^*}{2a^*}. \tag{3.98}$$

When $n > n^*$, the line (3.95) intersects the backbone curve only at one point, i.e. the additional branch no longer exists. This eliminates the threat of the occurrence of the "impact on buffers".

In many cases the elastic characteristic of the elastic buffers is not known *a priori*. Then, the designer is confronted with a question as to what will be the clearance d from the buffer and the damping factor n, so that no impact on the buffers takes place for any elastic characteristic of the buffers.

If the eigenfrequency λ_0 is prescribed, then the portion of the backbone curve corresponding to the linear part of the elastic characteristic is the vertical line AB in Fig. 3.13. The form of the backbone curve on the nonlinear part of the elastic characteristic depends on the elastic characteristic of the buffers. However, it is clear that the backbone curve lies above the line BC. This line can be said to correspond to rigid stops.

By means of the line (3.95) it is easy to verify that for the rigid stops there are two types of response curve. If this line does not intersect the broken line ABC (the case of small n), then the branches of the response curve are shown in Fig. 3.13, that is the impacts on buffers are possible at

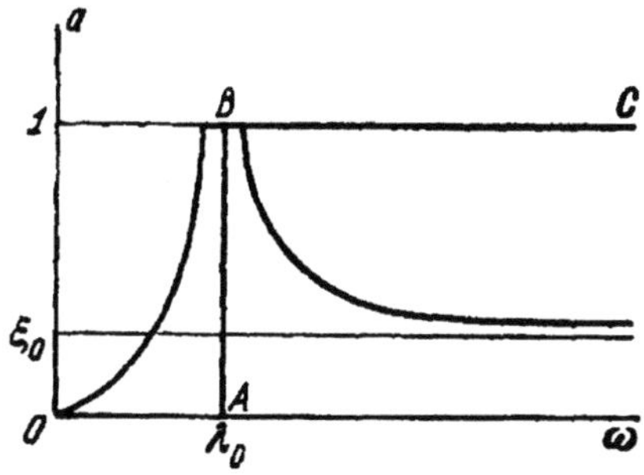

FIGURE 3.13.

any frequency higher than λ_0. But if the line (3.95) intersects the broken line ABC, an additional branch DC appears, see Fig. 3.14.

The driving amplitude in actual systems cannot be increased without bound. Thus, eq. (3.93) holds true only for those values of frequency ω which do not exceed a certain frequency $\omega_{\max}$. If the value of $\omega_{\max}$ is known, the posed problem is solved with ease. No impact on the buffers takes place when the abscissa of point D exceeds $\omega_{\max}$. This condition can be cast as follows

$$\frac{\xi_0}{2n} < \frac{d}{\omega_{\max}}. \tag{3.99}$$

This equation can be used both for the clearance d and the damping factor n.

It is evident that if frequency ω is not varied but is prescribed, $\omega = \omega_0$, then one must set $\omega_{\max} = \omega_0$ in eq. (3.99).

In order to eliminate impact on the buffers, one needs the clearance to be considerably greater than that due to the linear theory. We illustrate the above on the following numerical example.

Let a vibration protection system having the eigenfrequency 10 Hz ($\lambda_0 =$ 62,8 s^{-1}) undergo a harmonic excitation at frequency 100 Hz ($\omega = 628\ \mathrm{s}^{-1}$) and amplitude $\xi_0 = 0,1$ mm. The damping factor is $n = 6\ \mathrm{s}^{-1}$. Determine the clearance d in the case of symmetric buffers.

Linear theory renders the following formula for the amplitude of the vibration isolator

$$a = \frac{\xi_0\omega^2}{\sqrt{\left(\lambda_0^2 - \omega^2\right)^2 + 4n^2\omega^2}}, \tag{3.100}$$

which is obtained from eq. (3.84) by substituting $\lambda = \lambda_0$ and $F_1 = \xi_0\omega^2$.

Inserting the numerical values of λ_0, n and ω yields $a = 0,101$ mm. In order to avoid the buffer impacts one must take the clearance d greater than this amplitude. It seems that one can take, say, $d = 0,15 - 0,30\ mm$.

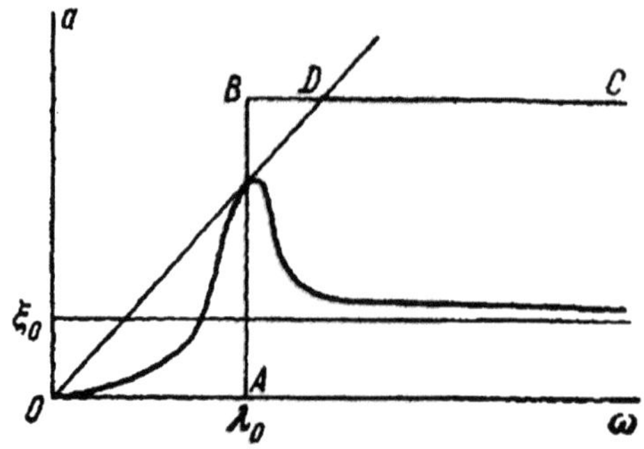

FIGURE 3.14.

This conclusion is however wrong. The required value of d must be determined by means of eq. (3.99) which gives for $\omega_{\max} = \omega_0$ the following result

$$d > \frac{\xi_0 \omega_0}{2n} = \frac{0,1 \cdot 628}{12} = 5,2 \; mm.$$

The required clearance is 50 times as large as that from eq. (3.100). If one has $d = 0,25$ mm, the dissipative force must be increased by 20 times in order to avoid any impact on the buffers.

In practice, the most challenging problem is to construct the backbone curve, especially, when the elastic restoring characteristic is not symmetric, i.e. $a_0 \neq 0$. Let us consider another example just to demonstrate the way to overcome these difficulties.

Let the object to be isolated be mounted on the linear vibration isolator, the mass of the object being 1 kg and the spring stiffness being 40 N/cm. The clearances from the equilibrium point to the upper and lower buffers are $d = 5$ mm and $d_1 = 3$ mm, respectively. Their stiffnesses are $c' = 1000$ N/cm and $c'' = 640$ N/cm, respectively. Determine the value of the damping factor n required to ensure no impact with the buffers.

Depending upon the values of a_0 and a the following vibrational regimes may take place in a system with nonsymmetric buffers:

I) vibrations without impacts on the buffers, i.e.

$$a_0 + a < d, \; a_0 - a > -d_1;$$

II) vibrations accompanied by impacts only on the upper buffer, i.e.

$$a_0 + a > d, \; a_0 - a > -d_1;$$

III) vibrations accompanied by impacts only on the lower buffer, i.e.

$$a_0 + a < d, \; a_0 - a < -d_1;$$

IV) vibrations with impacts on both buffers, i.e.

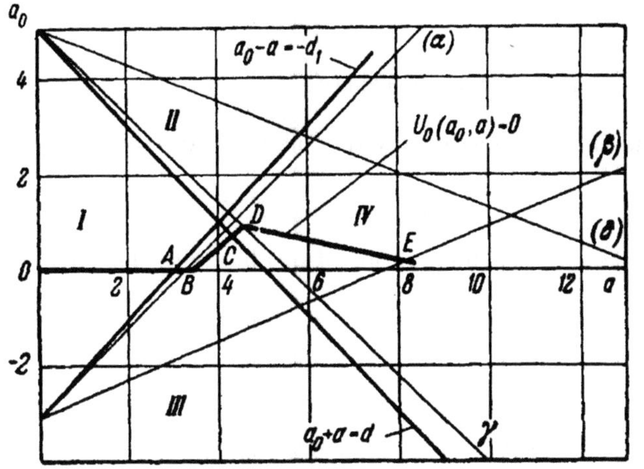

FIGURE 3.15.

$$a_0 + a > d,\ a_0 - a < -d_1.$$

For the sake of convenience, we plot the lines $a_0 + a = d$ and $a_0 - a = d_1$ in the plane (a_0, a). These lines split the half-plane $a > 0$ into four domains, in which the conditions (I)-(IV) hold, see Fig. 3.15.

Next, we plot the curve (3.83), to this end, we need a closed form expression for $U(a_0, a)$. An approximation derived in [65] by means of the method of linearisation by means of the distribution function is given by

$$\begin{aligned} &U_e(a_0 - 0,924a) + U_e(a_0 - 0,383a) \\ &+U_e(a_0 + 0,924a) + U_e(a_0 + 0,383a) = 0. \end{aligned} \tag{3.101}$$

When condition (I) is satisfied, it is apparent that $a_0 = 0$, i.e. the elastic characteristic is linear. For this reason, line (3.83) coincides with the abscissa axes, line OA, in the domain I. Increasing a we reach domain III, i.e. the impacts on the lower buffer occur. If $a < d_1/0,924$ all four points of the elastic characteristic in eq. (3.101) lie on the linear part. Therefore, when

$$d_1 < a < \frac{d_1}{0,924}\ (3 < a < 3,25)$$

line (3.83) coincides with the abscissa axes, line AB. Next, the first point $u = a_0 - 0,924a$ reaches the portion of the elastic characteristic corresponding to the lower buffer whereas the other points lie in the domain of linearity. Equation (3.101) takes the form

$$\begin{aligned} &c''(a_0 - 0,924a + d_1) - cd_1 + c(a_0 - 0,383a) \\ &+c(a_0 + 0,924a) + c(a_0 + 0,383a) = 0, \end{aligned}$$

or

$$a_0 (3c + c'') - 0,924a (c'' - c) + d_1 (c'' - c) = 0.$$

This is the line BC governed by the following equation

$$a_0 = 0,73a - 2,37. \tag{3.102}$$

The coordinates of the intersection of this line with the line $a_0 + a = d = 5$ mm

$$a = 4,25\,mm,\; a_0 = 0,75\,mm$$

determine the values of the amplitude and static displacement which cause impacts on the upper buffer. Increasing amplitude brings the line into domain (IV).

The following lines

$$a_0 - 0,924a = -d_1\,(\alpha), \quad a_0 + 0,924a = d\,(\gamma),$$
$$a_0 - 0,383a = -d_1\,(\beta), \quad a_0 + 0,383a = d\,(\delta)$$

are plotted in Fig. 3.15. Equation (3.101) changes when one of these lines intersects the line (3.83) because one of the four points determining the values of the elastic forces goes over to another part of the elastic characteristic.

The first break of line (3.83) takes place at point D

$$a = 4,45\,mm,\; a_0 = 0,88\,mm.$$

The next part is governed by the following equation

$$\begin{aligned} &c'' (a_0 - 0,924a + d_1) - cd_1 + c (a_0 - 0,383a) \\ &+c'(a_0 + 0,924a - d) + cd + c (a_0 + 0,383a) = 0, \end{aligned}$$

or

$$a_0(2c + c' + c'') + 0,924a(c' - c'') + c''d_1 - c'd + c (d - d_1) = 0.$$

Inserting numerical values yields the equation for the line DE

$$a_0 = -0,193a + 1,75, \tag{3.103}$$

wherein the coordinates of point E are given by

$$a = 8,25\,mm,\; a_0 = 0,11\,mm.$$

Further investigation of the form of curve (3.83) is not needed as the larger values of the amplitude are not required. Note that the line (3.83) has discontinuities at points B, D and E which can be explained by the approximate character of eq. (3.101).

The dependence $a_0\,(a)$ allows the backbone curve to be built. To this end, we use the result of [65] which is

$$\begin{aligned}\lambda^2 &= \frac{1}{2am}\{0,924\,[U_e\,(a_0+0,924a)-U_e\,(a_0-0,924a)]+\\ &+0,383\,[U_e\,(a_0+0,383a)-U_e\,(a_0-0,383a)]\}. \qquad (3.104)\end{aligned}$$

Thus, for a corresponding to the line OB

$$0<a<3,25\,mm,$$

we must take

$$\lambda^2=\frac{c}{m}=4000\ 1/\sec^2,\ \lambda=63,5\ 1/\sec=const.$$

Using eq. (3.102) we obtain for the line BD

$$\lambda^2=\frac{1}{2am}\{0,924[c\,(1,654a-2,37)-c''\,(-0,194a-2,37+3)+3c]$$

$$+0,383[c\,(1,113a-2,37)-c\,(0,347a-2,37)]\}=\frac{c}{m}\left(2,34-\frac{4,38}{a}\right).$$

Considering eq. (3.103) we obtain for the line DE

$$\lambda^2=\frac{1}{2am}\{0,924[c'\,(0,731a+1,75-5)+5c-c''\,(-1,117a+1,75+3)$$

$$+3c]+0,383[c\,(0,19a+1,75)-c\,(-0,576a+1,75)]\}$$

$$=\frac{c}{m}\left(16,68-\frac{68,5}{a}\right).$$

Finally, we obtain the backbone curve plotted in Fig. 3.16. The tangent to this curve which intersects the origin of the coordinate system is given in Fig. 3.16. This gives the required value of the damping factor n

$$n\geq 12,5\ 1/\sec.$$

Up until now we discussed the vibration isolator amplitude, in practice however the vibration isolator reaction is of crucial importance.

As pointed out in Chapter 1, the vibration isolation properties of the systems are described by means of the ratio of the reaction amplitude to the amplitude of the external excitation.

The dynamic component of the reaction consists of the elastic and dissipative forces, i.e.

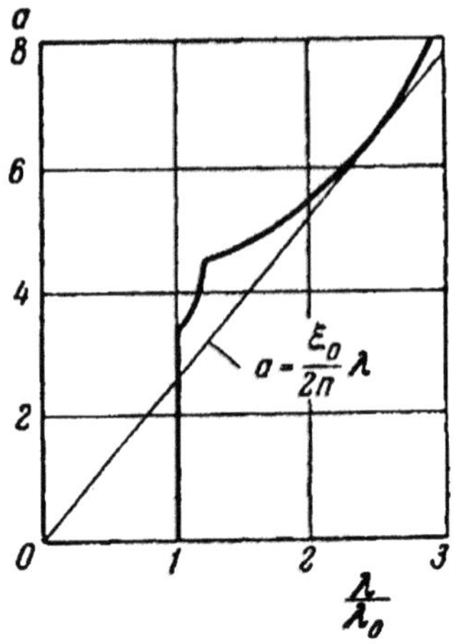

FIGURE 3.16.

$$R = -U(u, \dot{u}) = -b\dot{u} - U_e(u). \tag{3.105}$$

The linearised expression for the elastic force (3.79) can be used for the amplitude of the reaction. Taking into account that $U_0 = 0$ yields

$$R_{\max} \approx \left| b\dot{u} + c_d u^0 \right|_{\max} = a\sqrt{b^2\omega^2 + c_d^2} = ma\sqrt{\lambda^4 + 4n^2\omega^2}. \tag{3.106}$$

It should be kept in mind that the value of $R_{\max}$ due to eq. (3.106) is rather inaccurate. The explanation of this is that this equation determines only the amplitude of the first harmonic and ignores the higher harmonics. Neglecting amplitudes of the higher harmonics is allowed when one determines the kinematic parameters, but it is not the case when the nonlinear force is sought.

A better approximation to $R_{\max}$ can be obtained by explicit calculation of $R(t)$ due to the following formula

$$R(t) = ba\omega \sin\omega t - U_e(a_0 + a\cos\omega t). \tag{3.107}$$

When the dissipation is small, i.e. $ba\omega \gg U_{e\ \max}$ one can adopt that the amplitude value of the reaction coincides with that of the elastic force, i.e.

$$R_{\max} \approx U_e(a_0 \pm a). \tag{3.108}$$

On the right hand side of this equation the maximum of the two values is taken.

3.3 Forced vibrations under harmonic excitation and Coulomb friction

We consider a vibration isolator with a linear elastic restoring force. Let the only dissipative force in this vibration isolator be the Coulomb friction force. The equation of motion of the system under harmonic excitation is given by

$$m\ddot{u} + H\,\mathrm{sign}\dot{u} + U_e(u) = F\cos\omega t \quad (F = F(\omega)), \tag{3.109}$$

where the value of the frictional force H is assumed to be constant.

Seeking the approximate solution in the form (3.78) and applying the method of harmonic linearisation yields

$$U_e(u) \approx U_0 + c_d u^0, \quad H\mathrm{sign}\dot{u} \approx b_d \dot{u}. \tag{3.110}$$

The linearisation factors are due to eqs. (3.22)-(3.24), i.e. factor b_d is given by

$$b_d = \frac{-1}{\pi a\omega}\int_0^{2\pi} H\mathrm{sign}(-a\omega\sin\psi)\sin\psi d\psi = \frac{4H}{\pi a\omega}. \tag{3.111}$$

Solving the linearised equation

$$m\ddot{u}^0 + \frac{4H}{\pi a\omega}\dot{u}^0 + c_d u^0 + U_0 = F\cos\omega t, \tag{3.112}$$

we obtain the following equations for the solution's parameters

$$\left.\begin{aligned} &U_0(a_0, a) = 0, \\ &\left(c_d - m\omega^2\right) a = F\cos\varphi, \\ &-\frac{4H}{\pi} = F\sin\varphi. \end{aligned}\right\} \tag{3.113}$$

From these equations we find that

$$a = \frac{\sqrt{F_1^2(\omega) - h^2}}{\left|\lambda^2(a) - \omega^2\right|}, \tag{3.114}$$

$$\tan\varphi = \frac{h}{\sqrt{F_1^2 - h^2}}\,\mathrm{sign}\left(\omega^2 - \lambda^2\right), \tag{3.115}$$

where

$$F_1(\omega) = \frac{F(\omega)}{m}, \; h = \frac{4H}{\pi m}, \; \lambda^2 = \frac{c_d}{m}.$$

Expression (3.114) is in fact the amplitude-frequency characteristic given in an implicit form. Let us study the form of the resonance curve which is the graphic representation of eq. (3.114) in the plane (ω, a).

Analysis of eq. (3.114) indicates that if

$$F_1(\omega) < h, \tag{3.116}$$

then there exists no real value of a. This means that the method of harmonic linearisation does not yield an approximate harmonic solution. It is easy to understand the physical meaning of this result. When the frictional force H is larger than the amplitude of the harmonic force $F(\omega)$, the motion of the object relative to the base cannot begin. The vibration isolator is said to be "locked". It can be shown that this system returns to the equilibrium position provided that this system has nontrivial initial conditions. Indeed, let the motion of the system with some initial conditions be governed by the following function of time

$$u = \psi(t).$$

Then, the work due to the dry friction force between times t_1 and t_2 is given by

$$W_{df} = H \int_{t_1}^{t_2} \left|\dot{\psi}(t)\right| dt.$$

On the other hand, the work of the external force between the same times is

$$W_{ext} = F(\omega) \int_{t_1}^{t_2} \cos\omega t \dot{\psi}(t)\, dt.$$

But

$$F(\omega) < H, \; \cos\omega t \dot{\psi}(t) \leq \left|\dot{\psi}(t)\right|.$$

Thus, the dissipated energy exceeds the energy supplied into the system at any moment of time, for this reason the system must necessarily reach a state of relative equilibrium.

The condition of locking

$$F(\omega) < H \tag{3.117}$$

does not coincide with condition (3.117). No harmonic solution is obtained in the case in which

$$H < F(\omega) < \frac{4}{\pi} H. \tag{3.118}$$

The detailed analysis of this case indicates that there occurs very special periodic stick-slip motions. The system is in a state of relative equilibrium during a number of periodically repeated time intervals. As such vibrations occur very rarely in vibration isolation systems, they will not be analysed in what follows.

Assume now that

$$F_1(\omega) = \xi_0 \omega^2.$$

The system is locked if

$$\omega^2 < \frac{\pi h}{4\xi_0}. \tag{3.119}$$

The harmonic solutions are obtained provided that

$$\omega^2 > \frac{h}{\xi_0}.$$

When $\lambda = \omega$, the amplitude is unbounded, thus, the resonance curve does not intersect the backbone curve. The only possible exception is at frequency $\omega^* = \sqrt{h/\xi_0}$ because at this frequency the numerator and the denominator of (3.114) vanish.

Let us determine the frequencies ω' which yields extreme values of the amplitude. To this aim, we equate the following derivative to zero

$$\frac{d\left(a^2\right)}{d\left(\omega^2\right)} =$$

$$\frac{2\xi_0^2\omega^2\left(\lambda^2-\omega^2\right)^2 - \left(\xi_0^2\omega^4 - h^2\right)2\left(\lambda^2-\omega^2\right)\left[\frac{d\left(\lambda^2\right)}{d\left(a^2\right)}\cdot\frac{d\left(a^2\right)}{d\left(\omega^2\right)} - 1\right]}{\left(\lambda^2-\omega^2\right)^4} = 0.$$

This condition yields

$$\omega' = \frac{h}{\xi_0 \lambda(a)}. \tag{3.120}$$

Inserting eq. (3.120) into (3.114) we obtain the following equation for the extreme values of the amplitude

$$a' = \frac{h\xi_0}{\sqrt{h^2 - \xi_0^2\lambda^4(a')}}. \tag{3.121}$$

The solutions of this equation can be obtained graphically since they are the ordinates of the intersection points of the curve (3.121) with the backbone curve.

By introducing the following notation

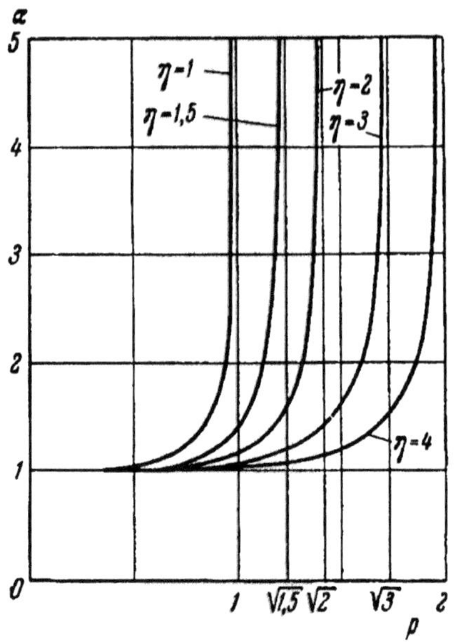

FIGURE 3.17.

$$\alpha = \frac{a'}{\xi_0}, \ \eta = \frac{h}{\xi_0 \lambda^2(0)}, \ p = \frac{\lambda(a)}{\lambda(0)}, \tag{3.122}$$

we can rewrite eq. (3.121) in the following nondimensional form

$$\alpha = \frac{\eta}{\sqrt{\eta^2 - p^4}}. \tag{3.123}$$

The plots of the dependence $\alpha(p)$ are shown in Fig. 3.17 for some values of η. All the curves start from the point $p = p, \alpha = 1$ and asymptotically tend to the lines $p = \sqrt{\eta}$.

Let us establish some properties of the resonance curve (3.114) for the case of $F_1(\omega) = \xi_0 \omega^2$.

i) Evidently, if $\xi_0 \omega^2 = h$, the amplitude a vanishes, and if $\xi_0 \omega^2 < h$, there are no real values of a. Therefore, the resonance curves start from the points $\omega = \sqrt{h/\xi_0}, a = 0$ and lie on the right hand side of the line $\omega = \sqrt{h/\xi_0}$.

ii) If $\xi_0 \omega^2 \gg h$, then approximately

$$a \approx \frac{\xi_0 \omega^2}{|\lambda^2 - \omega^2|}.$$

It is apparent that the resonance curves asymptotically approach the line $a = \xi_0$ from above as frequency increases.

iii) Solving eq. (3.114) for ω^2 yields

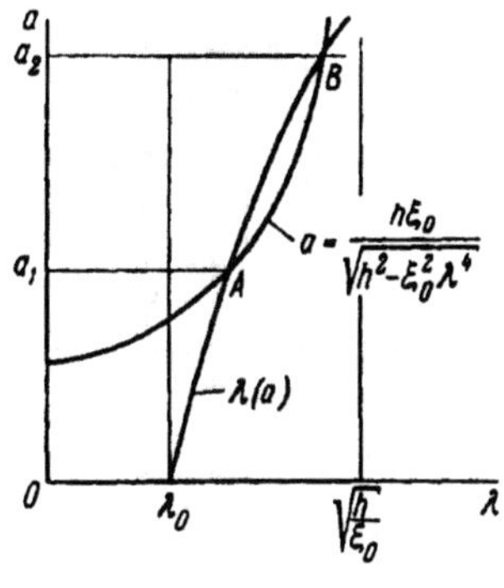

FIGURE 3.18.

$$\omega^2 = \frac{a^2\lambda^2 \pm \sqrt{a^2\left(\xi_0^2\lambda^4 - h^2\right) + h^2\xi_0^2}}{a^2 - \xi_0^2}.$$

Thus, there are no more than two real values of ω for any a. If

$$a^2\left(\xi_0^2\lambda^4 - h^2\right) + h^2\xi_0^2 < 0, \tag{3.124}$$

no real values of ω exist. The latter condition is equivalent to the following conditions

$$\xi_0\omega^2 < h, \tag{3.125}$$

$$a > \frac{h\xi_0}{\sqrt{h^2 - \xi_0^2\lambda^4}}. \tag{3.126}$$

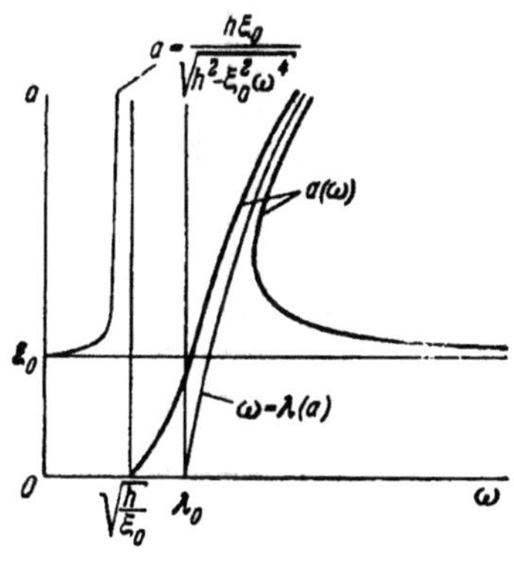

FIGURE 3.19.

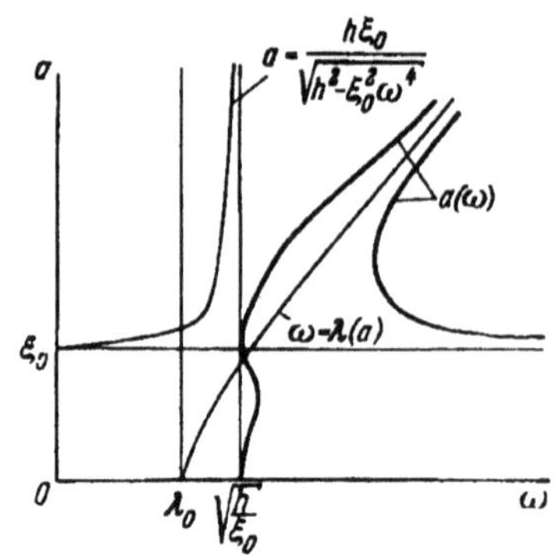

FIGURE 3.20.

The region which lies in Fig. 3.18 above the curve (3.121) satisfies these conditions. The resonance curve has no real points for $a_1 < a < a_2$.

It follows from this that the points at which the backbone curve intersects the resonance curve from below, e.g. point A in Fig. 3.18, correspond to the maxima of the resonance curves, whereas point B corresponds to the minima of the resonance curves.

These properties of the resonance curves allows us to define their form depending upon the number of intersection points of the curve (3.121) and the backbone curve.

If line (3.121) does not intersect the backbone curve, the resonance curve then takes on the form shown in Fig. 3.19 for $h < \xi_0\lambda^2(a)$, and in Fig. 3.20 for $h > \xi_0\lambda^2(a)$. If there is only one intersection point, the resonance curve is as shown in Fig. 3.21.

In the case of two intersection points which is possible only for the hard backbone curve and for $h > \xi_0\lambda^2(a)$ the resonance curve has two branches, see Fig. 3.22. The resonance curve corresponding to the case of three intersection points is displayed in Fig. 3.23. Introducing the Coulomb friction

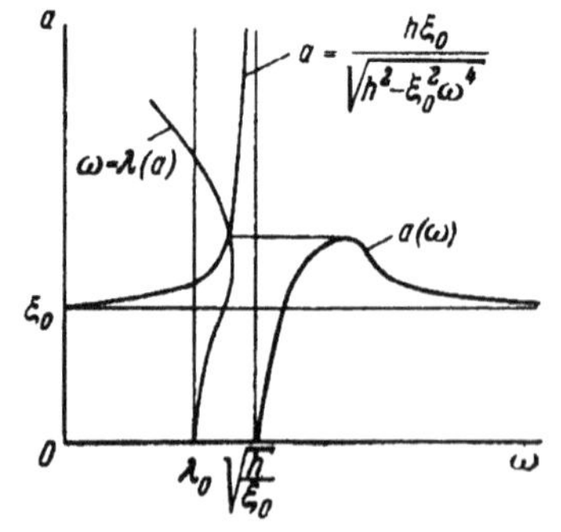

FIGURE 3.21.

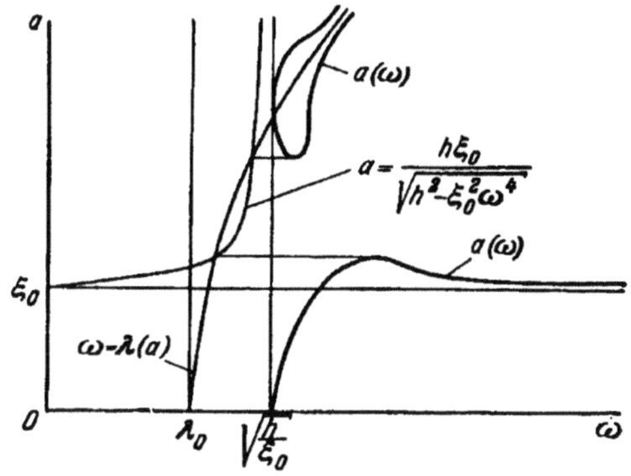

FIGURE 3.22.

does affect the form of the resonance curve provided that the amplitude of external excitation does not depend on frequency. Comparing eq. (3.114) for $F_1(\omega) = const$ with the equation for the resonance curve in the case of no friction, which is

$$a = \frac{F_1}{\left|\lambda^2(a) - \omega^2\right|}, \tag{3.127}$$

it is easy to note that introducing the Coulomb friction is equivalent to reduction of the external excitation amplitude by a factor of $F_1/\sqrt{F_1^2 - h^2}$. This reduction is considerable as h tends to F_1. On the other hand, an increase in the friction adversely affects the isolation properties of the vibration isolator. Indeed, the amplitude of the total reaction of the vibration isolator is given by the following expression

$$R_{\max} = \left|U_e(u) + H \operatorname{sign}\dot{u}\right|_{\max},$$

provided that $u \neq 0$, i.e. the vibration isolator is not "locked" by the Coulomb friction. Apparently, $R_{\max} \geq H$ and, therefore, for the vibration isolation to be efficient, the condition $H \ll F(\omega)$ must hold true.

Thus, when the friction is small the vibration isolation properties of the system are decreased because of the resonance phenomena, but when the friction is large the vibration isolation performance is poor as the vibration isolator reaction is large. It is reasonable to assume that there exists an optimum value of the friction which minimizes the dynamic magnification factor of the system.

Let us demonstrate the above for the example of the vibration isolator with a linear elastic characteristic. For this vibration isolator

$$\lambda(a) = \lambda(0) = \lambda_0.$$

Hence, if $F_1 = \xi_0\omega^2$

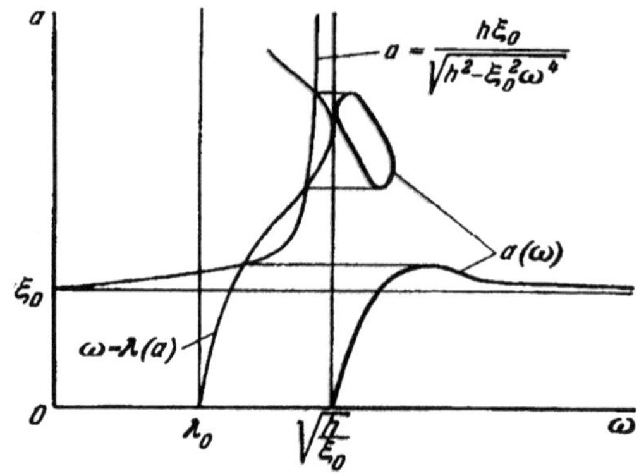

FIGURE 3.23.

$$a = \frac{\sqrt{\xi_0^2\omega^4 - h^2}}{\left|\lambda_0^2 - \omega^2\right|}. \tag{3.128}$$

The line $\lambda = \lambda_0$ intersects the curve (3.121) at a single point, if at all. Hence, the amplitude can have only one extreme value, which is maximum.

This value, due to eq. (3.121), is given by

$$a' = \frac{h\xi_0}{\sqrt{h^2 - \xi_0^2\lambda_0^4}}. \tag{3.129}$$

By virtue of eq. (3.120) the corresponding frequency is

$$\omega' = \frac{h}{\xi_0\lambda_0}.$$

The resonance curves plotted in the plane of the following coordinates

$$\alpha = \frac{a'}{\xi_0}, \quad z = \frac{\omega}{\lambda_0}$$

are shown in Fig. 3.24 for different values of $\eta = h/\xi_0\lambda_0^2$.

By using the above notation we can rewrite eq. (3.129) in the following form

$$\alpha = \frac{\eta}{\sqrt{\eta^2 - 1}}. \tag{3.130}$$

This dependence is plotted in Fig. 3.25. If $\eta \leq 1$ the maximum amplitude of the forced vibration can grow without bound as z tends to one. In practice this means that the impacts with the buffers is unavoidable for any clearance in the vibration isolator. Also, the lower branch of the resonance

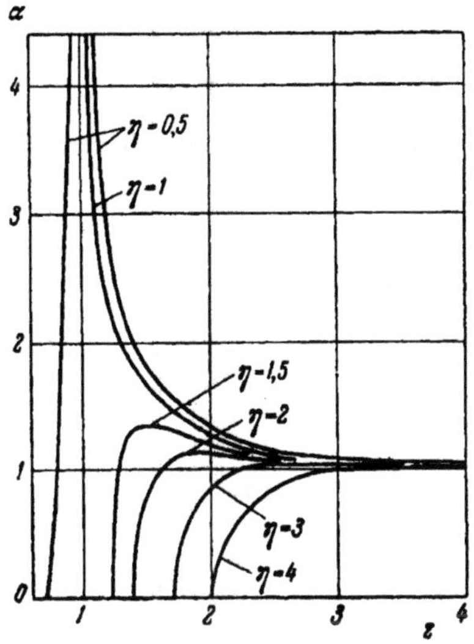

FIGURE 3.24.

curve corresponding to the linear part of the elastic restoring characteristic is no longer isolated but "joints" the additional branches of the resonance curve. A sharp decrease in α is observed in the region for which $\eta > 1$, e.g. $\alpha = 1,34$ for $\eta = 1,5$, additionally, α asymptotically tends to one with increasing η.

Therefore, increasing the friction force decreases the value of the maximum relative displacement of the vibration isolator at resonance which in turn decreases the amplitude of the elastic force and increases the dissipative force. Let us determine the value of the friction force which minimizes the maximum amplitude of the vibration isolator reaction

$$R_{\max} = \left| m\lambda_0^2 a \cos(\omega t + \varphi) - H \operatorname{sign} \sin(\omega t + \varphi) \right|_{\max} \approx m\lambda_0^2 a' + H.$$

Dividing by $m\lambda_0^2\xi_0$ yields the dynamic magnification factor

$$K = \frac{R_{\max}}{m\lambda_0^2\xi_0} = \frac{\eta}{\sqrt{\eta^2 - 1}} + \frac{\pi}{4}\eta. \tag{3.131}$$

The dependence K versus η is shown in Fig. 3.25. Equating the derivative of K with respect to η to zero, i.e. $dK/d\eta = 0$, yields the optimum value of η which ensures the minimum of K. The result is

$$\eta' = 1,47, \;\; K_{\min} = 2,52. \tag{3.132}$$

Hence, the optimum value of the Coulomb frictional force ensuring the vibration isolation from the harmonic driving of the amplitude $\xi_0\omega^2$ is

$$H = \frac{\pi}{4}\eta' m\lambda_0^2\xi_0 = 1,15 m\lambda_0^2\xi_0. \tag{3.133}$$

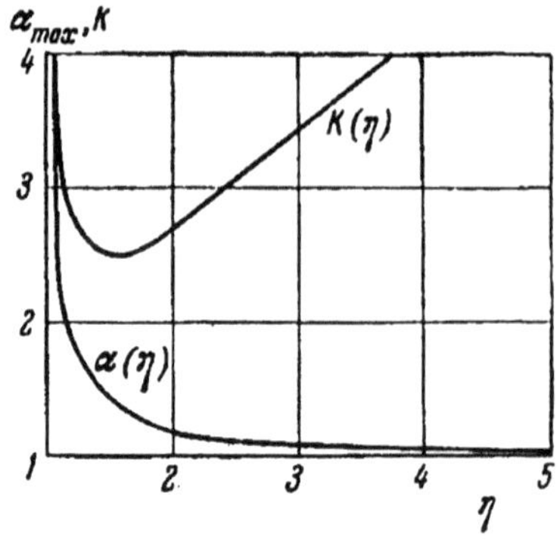

FIGURE 3.25.

Use is made of this expression while choosing the dissipative properties of the vibration isolator.

In the case of high frequency excitation the total reaction of the vibration isolator is actually equal to the Coulomb friction force. Indeed, the external excitation amplitude cannot grow without bounds, thus, ξ_0 decreases and tends to zero as frequency increases. The amplitude a decreases as well since $\lim_{\omega\to\infty} a = \xi_0$. The elastic force in the vibration isolator decreases which implies that the total reaction approaches the dry friction force.

3.4 Forced vibration in a system with internal friction

A vast amount of literature is concerned with investigating dissipation due to the internal friction in materials, see e.g. [96] and [95]. One can find a number of formulae expressing the dependence of the internal friction force on the amplitude of vibration. However, it is worth mentioning that dealing with the approximate analysis of the vibrational phenomena one is usually interested in the first harmonic of the force rather than in the entire periodic force.

With this in mind, the following formula is suggested in [96] for the internal friction force

$$U_{if}(u, \dot{u}) = \beta a^{\mu} \sqrt{1 - \frac{u^2}{a^2}} \,\mathrm{sign}\dot{u}, \tag{3.134}$$

with β and μ being the material parameters and a denoting the vibration amplitude.

It should be stressed that the above formula is applicable only for vibrations close to harmonic ones. Considering that

$$
\begin{aligned}
u &= a\sin\omega t, \ \sqrt{1-\frac{u^2}{a^2}} = \left|\cos\omega t\right|, \\
\dot{u} &= a\omega\cos\omega t = a\omega\sqrt{1-\frac{u^2}{a^2}}\ \mathrm{sign}\dot{u}
\end{aligned}
$$

we can rewrite eq. (3.134) in the form

$$
U_{if}\left(u,\dot{u}\right) = \frac{\beta a^{\mu-1}}{\omega}\dot{u}. \tag{3.135}
$$

This form implies that a harmonic linearisation of the dissipative force has already been performed.

Let us assume that the only dissipation is that due to the internal friction and harmonic vibration is sought. Substituting eqs. (3.135) and (3.79) into (3.74) yields the linearised equation

$$
m\ddot{u}^0 + \frac{\beta a^{\mu-1}}{\omega}\dot{u}^0 + c_d u^0 + U_0 = F\cos\omega t. \tag{3.136}
$$

Seeking the approximate solution of this equation in form (3.78) we obtain the following expression for the vibration amplitude

$$
a = \frac{F_1\left(\omega\right)}{\sqrt{\left[\lambda^2\left(a\right)-\omega^2\right]^2 + \beta_1^2 a^{2(\mu-1)}}}, \tag{3.137}
$$

where

$$
\beta_1 = \frac{\beta}{m}.
$$

Let us investigate the form of the resonance curves, to this aim we first find the extremum values a' of the amplitude. Equation (3.137) can be recast as follows

$$
a^2\left(\lambda^2-\omega^2\right)^2 + \beta_1^2 a^{2\mu} = F_1^2\left(\omega\right). \tag{3.138}
$$

We take the derivative of this expression with respect to ω^2 considering a^2 as a function of ω^2. The result is

$$
\frac{d\left(a^2\right)}{d\left(\omega^2\right)}\left(\lambda^2-\omega^2\right)^2 + 2a^2\left(\lambda^2-\omega^2\right)\left[\frac{d\left(\lambda^2\right)}{d\left(a^2\right)}\frac{d\left(a^2\right)}{d\left(\omega^2\right)} - 1\right] + \beta_1^2\mu a^{2(\mu-1)}\frac{d\left(a^2\right)}{d\left(\omega^2\right)}
$$
$$
= 2F_1\frac{dF_1}{d(\omega^2)}.
$$

Equating the derivative $d\left(a^2\right)/d\left(\omega^2\right)$ to zero yields

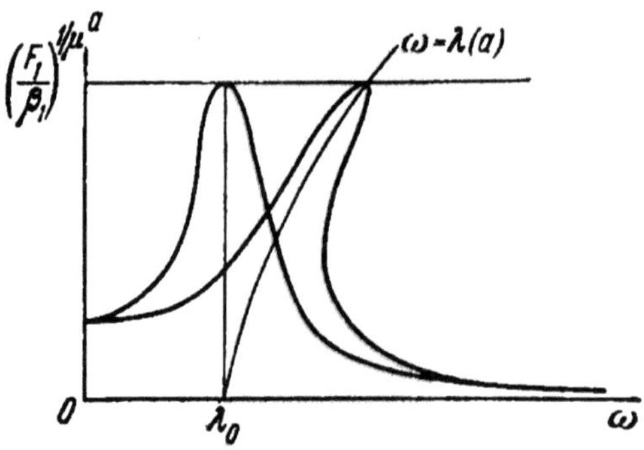

FIGURE 3.26.

$$(a')^2\left(\lambda^2-\omega^2\right)=-F_1\frac{dF_1}{d\left(\omega^2\right)}. \tag{3.139}$$

If $F_1 = const$, then $dF_1/d\left(\omega^2\right)=0$, thus

$$\lambda\left(a'\right)=\omega, \tag{3.140}$$

so that the amplitude achieves the extrema at the points of intersection of the resonance and backbone curves.

When $F_1=\xi_0\omega^2$, eq. (3.139) yields

$$\lambda^2\left(a'\right)=\omega^2\left[1-\frac{\xi_0^2}{(a')^2}\right]. \tag{3.141}$$

The extrema of a' are usually much greater than ξ_0. For this reason, we can accept that the extrema are achieved near the points of intersection of the resonance and backbone curves. The only exception is the point $a=0$, $\omega=0$ which is an extremum point, too.

In order to define the points of intersection of the resonance and backbone curves one needs to set $\lambda=\omega$ in eq. (3.137) that yields

$$a=\left[\frac{F_1(\omega)}{\beta_1}\right]^{1/\mu}. \tag{3.142}$$

Thus, it is sufficient to find the points of intersection of the backbone curve and the curve (3.142). The number of intersection points is determined by the number of branches of the resonance curve as it has been in all the above cases.

Equation (3.138) is solved for ω^2, to give

$$\omega^2=\lambda^2\pm\sqrt{\frac{F_1^2}{a^2}-\beta_1^2a^{2(\mu-1)}}. \tag{3.143}$$

If

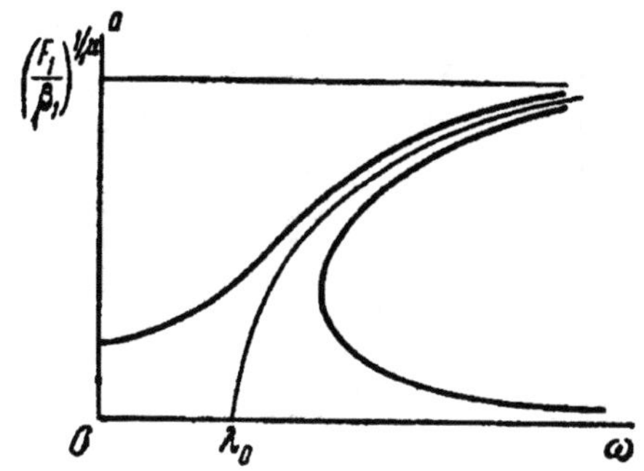

FIGURE 3.27.

$$a > \left[\frac{F_1(\omega)}{\beta_1} \right]^{1/\mu} \tag{3.144}$$

eq. (5.9) has no longer real-valued roots. This implies that the resonance curve has no points with such ordinates, i.e. the entire resonance curve lies under the curve (3.142).

The form of curve (3.142) depends on factor μ and the particular function $F_1(\omega)$. When $F_1(\omega) = const$, eq. (3.142) describes a horizontal line having not more than one intersection point with any backbone curve, see Fig. 3.26. This Figure also shows the form of the corresponding resonance curve. If $[F_1(\omega)/\beta]^{1/\mu}$ exceeds the value of the amplitude admissible by the vibration isolator dimensions, see Fig. 3.27, the line (3.142) does not intersect the backbone curve that implies the possibility of the resonance phenomena for any frequency $\omega > \lambda$.

If $F_1(\omega) = \xi_0 \omega^2$, the form of the curve (3.142) depends upon the value of the factor μ. For $0 < \mu < 2$ this curve is concave and tangential to the abscissa axis. This curve can have a different number of points of intersection with the backbone curve, namely, from zero up to three and more, see Fig. 3.28. In accordance with this number, there appear additional branches of

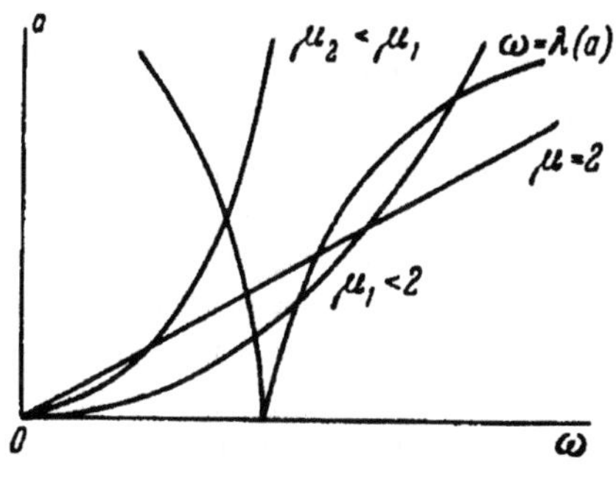

FIGURE 3.28.

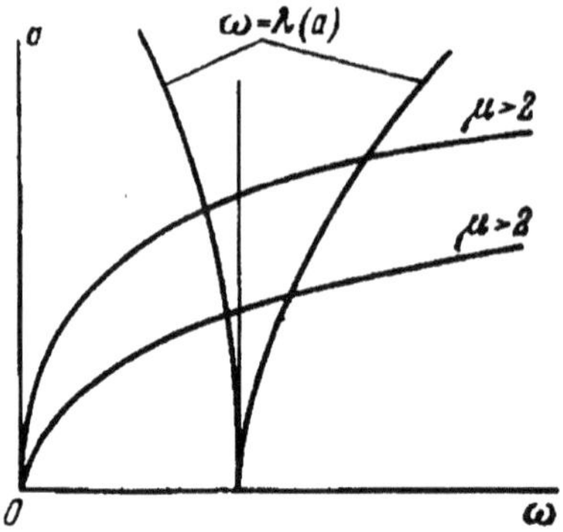

the resonance curve. For $\mu > 2$ the curve (3.142) is convex and is tangential to the ordinate axis. In this case, the number of intersection points can be different, however, only one intersection point is probable as μ increases, see Fig. 3.4.

When $\mu = 2$, the curve (3.142) is a line. The internal friction can then be said to be equivalent to linear damping from the perspective of the influence on the resonant vibration. This equivalence is, however, is not complete as the slope of the line corresponding to the linear damping is $\xi_0/(2n)$, whereas in the case of the internal friction the slope is $\sqrt{\xi_0/\beta_1}$.

When $\mu = 0$, the line (3.142) is parallel to the ordinate axis and governed by the following equation

$$\xi_0 \omega^2 = \beta_1.$$

This result shows that the internal friction turns out to be equivalent to the Coulomb friction in this particular case, and moreover $h = \beta_1$.

3.5 Comparative study of the various types of damping

We are now in a position to compare the various types of damping and draw some conclusion about the expediency of their use in vibration isolation systems. The influence of damping on the resonant vibration of the system has been proven to essentially depend on the form of the elastic characteristic (strictly speaking on the form of the backbone curve) and on the type of function $F_1(\omega)$. In what follows, a benchmark study is performed under the following assumptions which are typical for vibration isolation systems.

1) The backbone curve has one of the forms schematically depicted in Fig. 3.29a, b, c, d and e. In all these cases, the amplitude of the vibration isolator vibration is limited by the stops and cannot exceed Δ. It is easy to

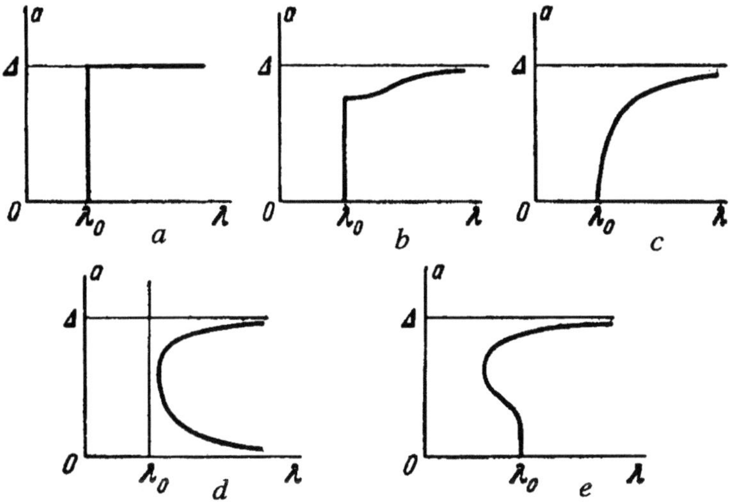

FIGURE 3.29.

understand from the above analysis that this fact is of crucial importance whereas the particular form of the elastic characteristic is not so important. The backbone curves shown in Fig. 3.29 correspond to the following types of elastic characteristic:

a) linear characteristic with symmetric rigid stops
b) linear characteristic with hardening elastic stops
c) nonlinear symmetric characteristic of the sort of eq. (3.5)
d) nonlinear asymmetric characteristic of the sort of eq. (3.6)
e) characteristic with an initial compressive force and stops.

2) We consider the following cases for $F_1(\omega)$

$$F_1(\omega) = F_1 = \text{const} \quad \text{if} \quad \omega_{\min} \le \omega \le \omega_{\max}, \tag{3.145}$$

$$F_1(\omega) = \xi_0 \omega^2 \quad \text{if} \quad \omega_{\min} \le \omega \le \omega_{\max}. \tag{3.146}$$

In each of the above variants we need to consider two cases: in the first case $\omega_{\min}$ is smaller than or of the order of the eigenfrequency λ_0, while in the second case $\omega_{\min} \gg \lambda_0$.

The damping has been mentioned to be necessary only for suppressing the large amplitude resonant vibration. It plays a negative part at small amplitudes increasing the total reaction in the vibration isolator and deteriorating the vibration isolation. With this in mind, it becomes evident that the best performance will be achieved by means of such a damping which ensures a rapid increase in the dissipative force with increasing amplitude. The internal friction in the case of large μ appears to be the most suitable one. Dry friction delivers the worst result as the Coulomb friction force does not depend upon the vibration amplitude at all.

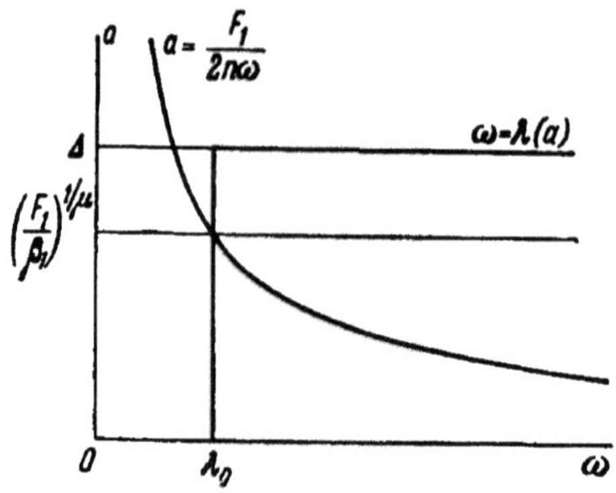

FIGURE 3.30.

If the dependence of the excitation amplitude upon frequency is determined by eq. (3.145) then the line (3.89) (or (3.142)) intersects the backbone curve only once, see Fig. 3.30. In order to avoid the impacts with the stops it is necessary to require that the intersection point lies on the elastic part of the characteristic, i.e. the following condition must be satisfied

$$\frac{F_1}{2n\lambda_0} < \Delta \quad \text{or} \quad \left(\frac{F_1}{\beta_1}\right)^{1/\mu} < \Delta, \tag{3.147}$$

where Δ denotes the gap between the equilibrium position and the stop.

These conditions are absolutely necessary when $\omega_{\min} \leq \lambda_0$. In the case in which $\omega_{\min} \gg \lambda_0$ it is sufficient that the abscissa of the intersection point is smaller than $\omega_{\min}$.

In this case, dry friction should not be used at all. Using dry friction leads either to the "locked" vibration isolator (if $F_1 < h$) which implies no vibration isolation or to impacts with the stops at any frequency (if $F_1 > h$). This conclusion immediately follows from the analysis of Section 3.3.

Let us compare now the influence of the internal friction and linear damping on the vibration isolation properties of the vibration isolator at $\omega \gg \lambda_0$.

At high frequencies, i.e. $\omega \gg \lambda_0$, one can approximately take

$$a \approx \frac{F_1}{\omega^2}.$$

The amplitude of the velocity-proportional resistance force (linear friction) is

$$|U_{lf}|_{\max} = 2mn\omega a \approx \frac{2mnF_1}{\omega}, \tag{3.148}$$

while the amplitude of the internal friction is given by

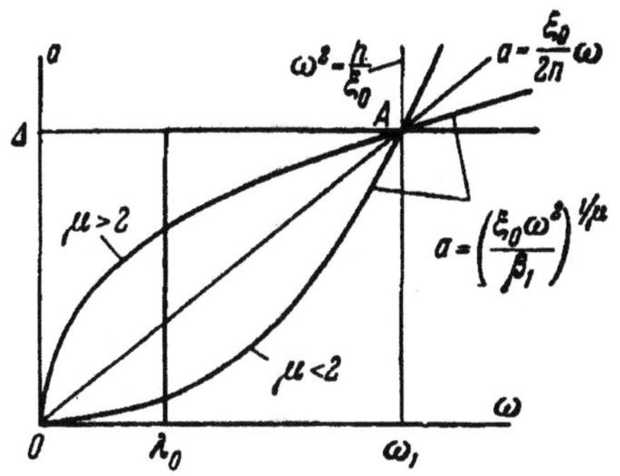

FIGURE 3.31.

$$|U_{if}|_{\max} = m\beta_1 a^{\mu} = \frac{m\beta_1 F_1^{\mu}}{\omega^{2\mu}}. \tag{3.149}$$

From these equations it is clear that at high frequencies and $\mu > 1/2$, the amplitude of the internal friction is smaller than that of the linear damping, i.e. vibration isolation in the first case is better for the same amplitude.

We proceed now to the case in which $F_1(\omega)$ is given by eq. (3.146). The character of the resonance curves is determined by the number of points of intersection of the backbone curve with curve (3.95) for linear damping, curve (3.142) for internal friction and line $\omega^2 = h/\xi_0$ for Coulomb friction. Figure 3.31 displays these curves and the backbone curve of the linear elastic characteristic with rigid stops. One can avoid impacts with the stops only by ensuring that the point of intersection of the above curves with the line $a = \Delta$ has the abscissa larger than $\omega_{\max}$, i.e.

$$\omega_1 > \omega_{\max}.$$

In the case of Coulomb friction this can be provided only if the vibration isolator is locked at any frequency up to $\omega_{\max}$, i.e. under the following condition

$$h > \xi_0\omega_{\max}^2. \tag{3.150}$$

In the case of linear damping we obtain the necessary condition (3.99). Finally, in the case of internal friction the following condition must hold

$$\Delta > \left(\frac{\xi_0\omega_{\max}^2}{\beta_1}\right)^{1/\mu}. \tag{3.151}$$

This condition is more easily satisfied for large μ and β_1.

Provided that conditions (3.99) and (3.151) are met and $\lambda_0 \ll \omega < \omega^2$, one obtains the amplitudes of the dissipative forces by substituting $F_1 = \xi_0\omega^2$ into eqs. (3.148) and (3.149). The result is

$$\left|U_{lf}\right|_{\max} = 2mn\xi_0\omega, \tag{3.152}$$

$$\left|U_{if}\right|_{\max} = m\beta_1\xi_0^{\mu}. \tag{3.153}$$

The vibration isolation system utilising internal friction is seen to have better performance for any value of μ.

Summarising we can state that the best way to struggle against resonance vibrations in vibration protection systems is to use of internal friction dampers. Design of internal friction dampers is a very challenging task, for this reason, in practice, one has to use other types of damping, like dry friction. For the overwhelming majority of the structural materials the values of β is very small.

The most effective material in this regard turns out to be rubber. Nevertheless, the internal friction proves to be insufficient to suppress resonant vibrations in the vibration isolation systems of aircraft.

The efficiency of the internal friction is often described by the so called *absorption factor* ψ which is equal to the ratio of the hysteretic loop area to the maximum potential strain energy, [96]. For elements with linear elastic characteristic and nonlinear dissipative force describing internal friction one obtains

$$\psi = \frac{2\pi\beta a^{\mu}}{ca^2} = 2\pi\frac{\beta_1}{\lambda_0^2}a^{\mu-1}, \tag{3.154}$$

with c being the stiffness of the elastic element. By virtue of (3.154) eq. (3.137) can be recast as follows

$$a = \frac{F_1(\omega)}{\sqrt{\left(\lambda_0^2 - \omega^2\right)^2 + \left(\psi/2\pi\right)^2\lambda_0^4}}, \tag{3.155}$$

since $\lambda = \lambda_0$ in the linear system. At $\omega = \lambda_0$ we obtain

$$a' = \frac{F_1(\lambda_0)}{\lambda_0^2\dfrac{\psi}{2\pi}}. \tag{3.156}$$

The value of ψ does not exceed 0,3-0,5 for conventional structural materials. When $F_1(\omega) = \xi_0\omega^2$ eq. (3.156) yields

$$a' \geq \frac{\xi_0}{0,08} = 12,5\xi_0,$$

i.e. the dynamic magnification factor is considerable at resonance. The amplitude ξ_0 can achieve 1-1,5 mm in systems of passive vibration isolation. Provided that internal friction is utilised this means that the "free running" of the vibration isolator must be increased up to 25-40 mm which is frequently inadmissible from the perspective of the absorber's dimensions.

Combining various types of damping allows the problem of struggling with the resonant vibrations to be solved successfully. We limit our analysis to the consideration of the combined action of internal and dry friction. In this case

$$U_{if+df}\left(u,\dot{u}\right)=H\mathrm{sign}\dot{u}+\beta a^{\mu-1}\frac{\dot{u}}{\omega} \tag{3.157}$$

and, thus,

$$b_d=m\left(\frac{h}{a\omega}+\frac{\beta_1 a^{\mu-1}}{\omega}\right). \tag{3.158}$$

The governing equation for the amplitude-frequency characteristic is given by

$$a=\frac{\sqrt{F_1^2\left(\omega\right)-h^2}}{\sqrt{\left[\lambda^2\left(a\right)-\omega^2\right]^2+\beta_1 a^{2(\mu-1)}}}. \tag{3.159}$$

It is easy to show that the amplitude reaches the extremum near the point of intersection of the resonance curve (3.159) with the backbone curve. To define this point we put $\lambda=\omega$ in eq. (3.159), to get

$$a'=\left(\frac{\sqrt{F_1^2\left(\omega\right)-h^2}}{\beta_1}\right)^{1/\mu}. \tag{3.160}$$

This is sufficient to determine the points of intersection of the curve (3.160) and the backbone curve $\omega=\lambda\left(a\right)$. Consider the case in which the backbone curve has the form shown in Fig. 3.29a and $F_1\left(\omega\right)=\xi_0\omega^2$. In this case the curves (3.160) start at the point

$$\omega=\sqrt{\frac{h}{\xi_0}},\quad a=0$$

and have the form shown in Fig. 3.32 for different values of μ. If $h>\xi_0\lambda_0^2$, the curves (3.160) intersect the backbone curve at only one point, namely at point B, see Fig. 3.32 , and then it is only necessary to require the abscissa of the intersection point to be greater than $\omega_{\max}$. When this condition holds true, the resonant vibration of the system is suppressed.

Considering the condition under which the resonant vibration in the system with rigid stops occurs, we assumed that this vibration is nearly harmonic. In general, this assumption is incorrect as the higher harmonics play an important role in modelling impacts on stops. For this reason, the results of the previous Sections should be deemed as a guide. A more rigorous investigation of the phenomenon of impact with the stops, which also accounts for the energy dissipation at impact, is given below.

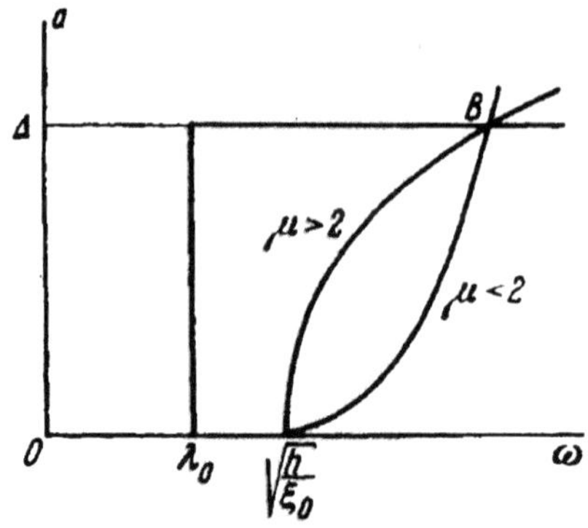

FIGURE 3.32.

3.6 Forced vibration under polyharmonic excitation

In many problems of vibration protection one has to deal with polyharmonic vibration excitations. The problem of vibration isolation from machines with reciprocating parts, internal combustion engines, piston compressors etc. requires the higher harmonics of the periodic driving force to be taken into account and the conditions for resonance due to the higher harmonics to be established. Vibrations of vehicles, vessels and aircraft are always polyharmonic, therefore, the systems of vibration isolation for the equipment of these structures must be properly designed in order to reduce the transmission of polyharmonic external excitation.

Let us assume that the right hand side of eq. (3.74) is a polyharmonic function of time, i.e.

$$Q(t) = \sum_{i=1}^{N} F_i \cos(\omega_i t + \varphi_i). \tag{3.161}$$

This function may be periodic under the condition that all the harmonics are excited by a single vibration source or several synchronous vibration sources. This function can be nearly periodic if a there are a number of independent sources exciting vibrations with the incommensurable frequencies. In both cases, the steady-state (regardless of being periodic or nearly periodic) solution of eq. (3.74) must be polyharmonic.

As eq. (3.74) is nonlinear, its solution must contain in general not only the harmonics with frequencies ω_i, but also the ultraharmonics of frequencies $p\omega_i$, p being an integer, harmonics with combination frequencies $\sum p_i\omega_i$ and so on.

In the present Section the approximate solution will be sought in the form

$$u = a_0 + \sum_{i=1}^{N} a_i \cos(\omega_i t + \varphi_i), \tag{3.162}$$

retaining the harmonics only of those frequencies which are present in the driving. This choice is natural provided that eq. (3.74) is close to being linear because eq. (3.162) implies a solution of a linear equation. Use will be made of the method of harmonic linearisation using the distribution function covered in the previous Chapter.

To begin with, we consider a few particular examples.

3.6.1 System with nonlinear elastic force and linear damping

We analyse the behaviour of a system with a nonlinear elastic force and a velocity proportional resistance force. The equation of motion governing the system can be written in the following form

$$m\ddot{u} + b\dot{u} + U_e(u) = \sum_{i=1}^{N} F_i \cos(\omega_i t + \varphi_i). \tag{3.163}$$

Let us linearise the nonlinear elastic force

$$U_e(u) = U_0 + c_d u^0, \tag{3.164}$$

with $u^0 = u - a_0$ and the factors U_0 and c_d being defined according to eqs. (3.42) and (3.43).

Choosing the probability density in the form of a linear combination of three Dirac's delta-functions and restricting our analysis to approximating the first four moments, yields the following formulae (see [65] for detail)

$$U_0 = \frac{1}{\varepsilon}\left[\frac{1}{2}U_e\left(a_0 - \sqrt{\varepsilon}\sigma_u\right) + \frac{1}{2}U_e\left(a_0 + \sqrt{\varepsilon}\sigma_u\right) + (\varepsilon - 1)\, U_e(a_0)\right], \tag{3.165}$$

$$c_d = \frac{1}{2\sqrt{\varepsilon}\sigma_u}\left[U_e\left(a_0 + \sqrt{\varepsilon}\sigma_u\right) - U_e\left(a_0 + \sqrt{\varepsilon}\sigma_u\right)\right], \tag{3.166}$$

with ε being given by eq. (3.50). Inserting eq. (3.164) into eq. (3.163) we obtain the linearised equation. Equating the constant terms and the factors in $\cos(\omega_i t + \psi_i)$ and $\sin(\omega_i t + \psi_i)$ leads to the following system of equations

$$U_0(a_0, \sigma_u, \varepsilon) = 0, \tag{3.167}$$

$$\left[\lambda^2(a_0, \sigma_u, \varepsilon) - \omega_i^2\right] a_i = A_i \cos(\varphi_i - \psi_i) \quad (i = 1, ..., N), \tag{3.168}$$

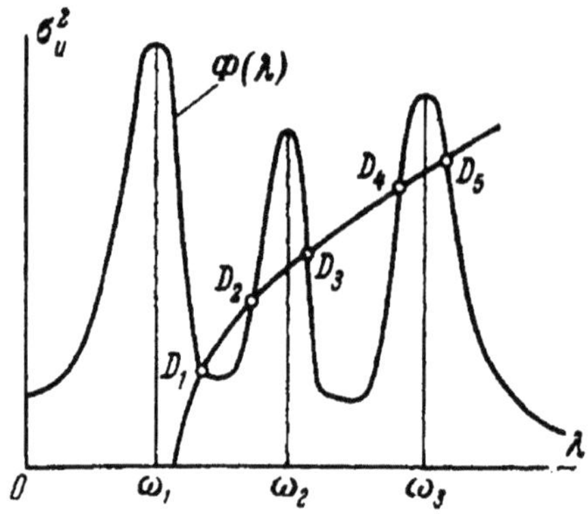

FIGURE 3.33.

$$2na_i\omega_i = A_i \sin(\varphi_i - \psi_i) \;\; (i = 1, ..., N), \tag{3.169}$$

where

$$\lambda^2 = \frac{c_d}{m}, \;\; 2n = \frac{b}{m}, \;\; A_i = \frac{F_i}{m}.$$

It follows from eqs. (3.168) and (3.169) that

$$a_i = \frac{A_i}{\sqrt{\left(\lambda^2 - \omega_i^2\right)^2 + 4n^2\omega_i^2}}, \quad \tan(\varphi_i - \psi_i) = \frac{2n\omega_i}{\lambda^2 - \omega_i^2}. \tag{3.170}$$

Thus,

$$\sigma_u^2 = \frac{1}{2}\sum_{i=1}^{N} a_i^2 = \frac{1}{2}\sum_{i=1}^{N} \frac{A_i^2}{\left(\lambda^2 - \omega_i^2\right)^2 + 4n^2\omega_i^2} = \Phi(\lambda), \tag{3.171}$$

$$\begin{aligned} \varepsilon &= \frac{M_4^0}{\sigma_u^4} = \frac{3}{2} + \frac{3}{8\sigma_u^4} \sum_{i,j=1,i\neq j}^{N} a_i^2 a_j^2 \\ &= \frac{3}{8\sigma_u^4} \sum_{i,j=1,i\neq j}^{N} \frac{A_i^2 A_j^2}{\left[\left(\lambda^2 - \omega_i^2\right)^2 + 4n^2\omega_i^2\right]\left[\left(\lambda^2 - \omega_j^2\right)^2 + 4n^2\omega_j^2\right]}. \end{aligned} \tag{3.172}$$

Equations (3.167), (3.171) and (3.172) allows us to determine the unknown a_0, σ_u^2 and ε. This in turn permits us to determine λ^2 and to find the amplitudes and phases of the solution's harmonics by virtue of eq. (3.170). The parameters a_0, σ_u^2 and ε may be determined by the method of successive approximation, see [65] for detail.

We intend now to show a way of defining the damping factor n which suppresses the resonance vibrations of the system. From the prescribed value of ε we can obtain a_0 as a function of σ_u from eq. (3.167). Substituting this function into the expression for λ yields the dependence $\lambda\left(\sigma_u^2\right)$ whose locus in the plane (λ, σ) is a sort of backbone curve. Using eqs. (3.165) and (3.166) it is not difficult to plot a family of such curves for different values of ε. Dependence $\Phi(\lambda)$, due to eq. (3.171), is shown in Fig. 3.33 together with the curve $\lambda\left(\sigma_u^2\right)$ plotted for a prescribed value of ε. The intersection points give the first approximation to the values of σ_u and λ. Substituting them into eq. (3.172) yields a more accurate value of ε which allows us to obtain the second approximation for σ_u and λ etc.

The diagrams for $\Phi(\lambda)$ for not very large values of n have the form depicted schematically in Fig. 3.33. The curve $\lambda\left(\sigma_u^2\right)$ may intersect the curve $\Phi(\lambda)$ at a number of points, $D_1, ..., D_s$, yielding a number of steady-state solutions of the nonlinear equation of concern. One of these solution can be nonresonant, e.g. point D_1 in Fig. 3.33, while the others describe the resonant frequencies ω_i.

In order to suppress the resonant vibrations, it is sufficient to take that value of n which ensures that the peaks of the curve $\Phi(\lambda)$ lie below the curve $\lambda\left(\sigma_u^2\right)$. For small values of n the maxima of the curve $\Phi(\lambda)$ are located near the lines $\lambda = \omega_i$. Based on this observation the following strategy for determining factor n may be suggested. Taking a value of ε we plot curve $\lambda\left(\sigma_u^2\right)$ and determine the ordinates of this curve corresponding to the values $\lambda = \omega_i$, Fig. 3.34. For the hard curves it is sufficient to study $\omega_i > \lambda_0$. Next, we set successively $\sigma_u^2 = \left(\sigma_u^2\right)_i$ and $\lambda = \omega_i$ in eq. (3.171) and calculate every time the value of n. Then we choose the minimum value of n from the values obtained. For this particular value of n the curve $\Phi(\lambda)$ intersects the curve $\lambda\left(\sigma_u^2\right)$ in the vicinity of the point $\lambda = \omega_i$. Letting now $\sigma_u^2 = \Phi(\omega_i)$, $\lambda = \omega_i$ we get a more accurate value of ε due to eq. (3.172) and plot the new curve $\lambda\left(\sigma_u^2\right)$ corresponding to the new value of ε. If this curve lies below the previous one, it is necessary to increase the value of n so that the value of $\Phi(\omega_i)$ is smaller than the ordinate of the curve $\lambda\left(\sigma_u^2\right)$ at $\lambda = \omega_i$.

We consider now the following numerical example. Let the forced vibration in the vibration isolation system be governed by the following equation

$$\ddot{u} + 2n\dot{u} + \frac{2\lambda_0^2\Delta}{\pi}\tan\frac{\pi u}{2\Delta} = A\left(\cos z_1\lambda_0 t + \cos z_2\lambda_0 t\right), \tag{3.173}$$

where

$$\Delta = \frac{A}{\lambda_0^2}, \; z_1 = \sqrt{5}, \; z_2 = 2\sqrt{2}.$$

Required is the value of the ratio n/λ_0 which excludes the occurrence of the resonance vibrations.

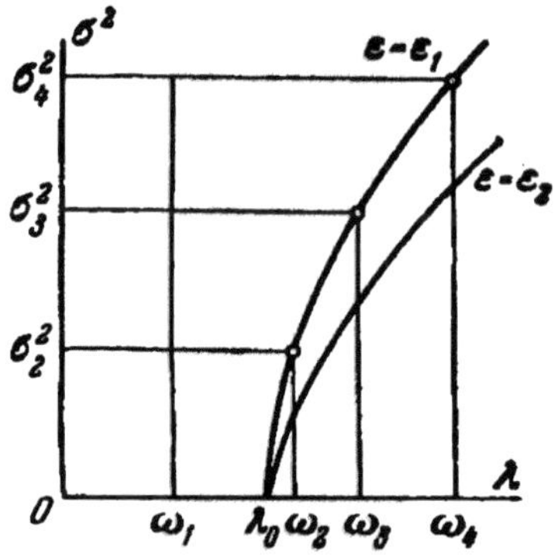

FIGURE 3.34.

We seek the solution in the form

$$u = a_1 \cos\left(z_1 \lambda_0 t + \psi_1\right) + a_2 \cos\left(z_2 \lambda_0 t + \psi_2\right). \tag{3.174}$$

As the nonlinear function on the left hand side of eq. (3.173) is odd, no constant term is observed in this equation. Equation (3.166) yields

$$\lambda^2\left(\varepsilon, \sigma_u\right) = \frac{2\lambda_0^2 \Delta}{\pi\sqrt{\varepsilon}\sigma_u} \tan\left(\frac{\pi}{2\Delta}\sqrt{\varepsilon}\sigma_u\right). \tag{3.175}$$

Let us take

$$\varepsilon_1 = 1{,}875,$$

since $3/2 < \varepsilon < 3$ as mentioned in Section 3.1. Inserting ε_1 into eq. (3.175) yields the dependence $\lambda\left(\sigma_u\right)$. Next, we find the values of σ_u^2 for $\lambda = z_1\lambda_0$ and $\lambda = z_2\lambda_0$. To this end, we solve the following equations

$$\left.\begin{aligned} z_1^2\lambda_0^2 &= \frac{2\lambda_0^2 \Delta}{\pi\sqrt{\varepsilon_1}\sigma_u} \tan\left(\frac{\pi}{2\Delta}\sqrt{\varepsilon_1}\sigma_u\right), \\ z_2^2\lambda_0^2 &= \frac{2\lambda_0^2 \Delta}{\pi\sqrt{\varepsilon_1}\sigma_u} \tan\left(\frac{\pi}{2\Delta}\sqrt{\varepsilon_1}\sigma_u\right). \end{aligned}\right\} \tag{3.176}$$

Denoting

$$\frac{\pi\sqrt{\varepsilon_1}\sigma_u}{2\Delta} = s$$

allows us to recast eqs. (3.176) in the form

$$\tan s = z_1^2 s, \ \tan s = z_2^2 s.$$

Solving these equations yields

$$s_1 = 1,432; \; s_2 = 1,487.$$

Equation (3.171) can be rewritten as follows

$$\sigma_u^2 = \frac{4s^2\Delta^2}{\pi^2\varepsilon_1} = \frac{A^2}{2}\left[\frac{1}{\left(\lambda^2 - z_1^2\lambda_0^2\right)^2 + 4n^2\lambda_0^2 z_1^2} + \frac{1}{\left(\lambda^2 - z_2^2\lambda_0^2\right)^2 + 4n^2\lambda_0^2 z_2^2}\right].$$

Substitution of the numerical values leads to the equation

$$s^2 = 2,31\left[\frac{1}{\left(p^2 - z_1^2\right)^2 + 4\nu^2 z_1^2} + \frac{1}{\left(p^2 - z_2^2\right)^2 + 4\nu^2 z_2^2}\right], \tag{3.177}$$

where

$$p = \frac{\lambda}{\lambda_0}, \; \nu = \frac{n}{\lambda_0}.$$

Inserting $p = z_1$, $s = s_1$ and $p = z_2$, $s = s_2$ into eq. (3.177) yields respectively the following values of ν

$$\nu_1 = 0,248, \; \nu_2 = 0,211.$$

The larger value of ν, i.e. ν_1, corresponds to the resonance at frequency $z_1\lambda_0$, thus, it is necessary to make a more accurate estimation for this regime. Equation (3.172) renders

$$\varepsilon_2 = \frac{3}{2} + \frac{3}{8}\cdot\frac{\pi^4\varepsilon_1^2}{16\Delta^4 s_1^4}\cdot\frac{\lambda_0^8\Delta^4}{4\nu_1^2 z_1^2\left[\left(z_1^2 - z_2^2\right)^2 + 4\nu_1^2 z_2^2\right]\lambda_0^8} = 1,637. \tag{3.178}$$

Taking $\varepsilon_1 = \varepsilon_2$ we obtain the new value of ν_1 which changes only the factor on the right hand side of eq. (3.177), but not the value of s_1. The solution of this equation is

$$\nu_1 = 0,246.$$

This value of ν_1 is so close to the previous one, i.e. $\nu_1 = 0,248$, that there is no necessity for further approximation.

The example considered was relatively simple which allows us to avoid the graphical construction associated with more complex cases.

3.6.2 System with linear elastic force and dry friction

We study now a system with linear elastic restoring force and dry friction subject to polyharmonic external excitation.

The equation of motion is given by

$$m\ddot{u} + H\operatorname{sign}\dot{u} + cu = \sum_{i=1}^{N} F_i \cos\left(\omega_i t + \varphi_i\right). \tag{3.179}$$

The method of linearisation by means of the distribution function is applied to linearize the dry friction force. The result is

$$b_d = \frac{1}{\sigma_v^2} \int_{v_1}^{v_2} yH\operatorname{sign}y \cdot w\left(y\right) dy = \frac{H}{\sigma_v^2} \int_{v_1}^{v_2} \left|y\right| w\left(y\right) dy, \tag{3.180}$$

where $w\left(y\right)$ is the distribution function of the polyharmonic process $v = \dot{u}\left(t\right)$, and σ_v is the standard deviation of $v\left(t\right)$.

We prove now that the linearisation factor b_d can be cast in the form

$$b_d = \frac{\eta}{\sigma_v}, \tag{3.181}$$

where η is a factor whose value depends only upon the ratio of the odd central moments to the corresponding power of the standard deviation σ_v. Indeed, any symmetric distribution law $w\left(y\right)$ can be represented to the required accuracy as a linear combination of Dirac's delta-functions

$$w\left(y\right) = 2\sum_{i=0}^{s} \alpha_i \delta\left(y - y_i\right). \tag{3.182}$$

Inserting this equation into (3.180) yields

$$b_d = \frac{2H}{\sigma_v^2} \sum_{i=0}^{s} \alpha_i \left|v_i\right| = \frac{2H}{\sigma_v} \sum_{i=0}^{s} \alpha_i k_i = \frac{\eta}{\sigma_v}, \tag{3.183}$$

where

$$k_i = \frac{\left|v_i\right|}{\sigma_v}, \quad \eta = 2H \sum_{i=0}^{s} \alpha_i k_i.$$

The values of α_i and k_i are determined from the following equations

$$2\sum_{i=0}^{s} \alpha_i = 1, \quad 2\sum_{i=0}^{s} \alpha_i v_i^{2p} = M_{2p}^0 \quad \left(p = 1, 2, \ldots\right).$$

Dividing all these equation except the first one by σ_v^{2p} and considering eq. (3.183) leads to the conclusion that factor η depends only on the ratio M_{2p}^0 / σ_v^{2p}.

By linearising eq. (3.179) we obtain

$$\ddot{u} + \frac{\gamma}{\sigma_v}\dot{u} + \lambda_0^2 u = \sum_{i=1}^{N} A_i \cos(\omega_i t + \varphi_i), \tag{3.184}$$

where

$$\gamma = \frac{\eta}{m}, \quad \lambda_0^2 = \frac{c}{m}, \quad A_i = \frac{F_i}{m}.$$

Solving the linearised equation renders

$$u = \sum_{i=1}^{N} a_i \cos(\omega_i t + \psi_i) = \sum_{i=1}^{N} \frac{A_i \cos(\omega_i t + \psi_i)}{\sqrt{\left(\lambda_0^2 - \omega_i^2\right)^2 + \frac{\gamma^2}{\sigma_v^2}\omega_i^2}}, \tag{3.185}$$

wherein

$$\psi_i = \varphi_i + \arctan\frac{\gamma\omega_i}{\sigma_v\left(\lambda_0^2 - \omega_i^2\right)}. \tag{3.186}$$

We construct now the equation for σ_v^2 which is

$$\sigma_v^2 = \frac{1}{2}\sum_{i=1}^{N} a_i^2\omega_i^2 = \frac{1}{2}\sum_{i=1}^{N} \frac{A_i^2\omega_i^2}{\left(\lambda_0^2 - \omega_i^2\right)^2 + \frac{\gamma^2}{\sigma_v^2}\omega_i^2}. \tag{3.187}$$

The values of σ_v^2 which satisfy this equation may be obtained by means of the graphical construction. To this end, we plot the graph of the following function

$$\sigma_v^2(z) = \frac{1}{2}\sum_{i=1}^{N} \frac{A_i^2\omega_i^2}{\left(\lambda_0^2 - \omega_i^2\right)^2 + z\omega_i^2} \tag{3.188}$$

and the hyperbola

$$\sigma_v^2 = \frac{\gamma^2}{z}. \tag{3.189}$$

The point of intersection of the curves (3.188) and (3.189) determines the value of σ_v^2 satisfying eq. (3.187), see Fig. 3.35. Clearly, we need the value of γ to plot the curve (3.189). Limiting our consideration to the first six moments of the distribution function for the process $v(t)$ yields

$$\gamma = \frac{2H}{n}\left(\alpha_1 k_1 + \alpha_2 k_2\right). \tag{3.190}$$

The factors α_1, α_2, k_1 and k_2 can be expressed in terms of ε and μ. Prescribing some values of ε and μ we can apply the method of successive approximations to obtain more accurate values of these parameters.

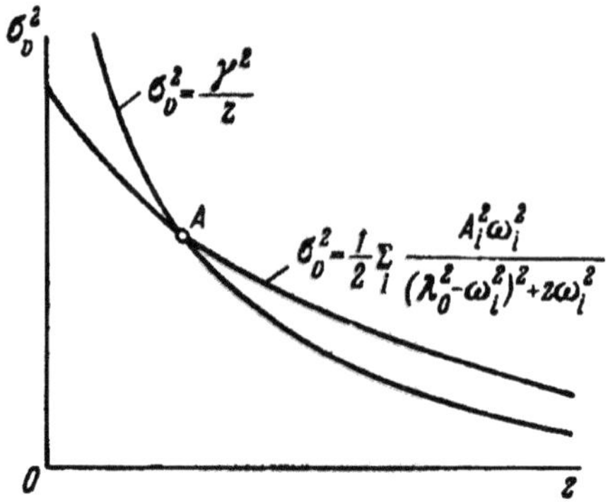

FIGURE 3.35.

The curve (3.188) may have no intersection points with the hyperbola (3.189) which implies that there exists no solution in the form (3.185). Two cases must be studied under these circumstances.

a) The curve (3.188) lies under the hyperbola (3.189). (Formally, these curves intersect at the point $\sigma_v^2 = 0, z = \infty$.) This means that the external driving is not sufficiently strong to overcome the dry friction force. Either the vibration isolator is locked or stick-slip motion occurs.

b) The curve (3.188) lies above the hyperbola (3.189). This case can be realised only in the case in which λ coincides with one of the frequencies ω_i, otherwise σ_v^2 is bounded at $z = 0$ and the curve (3.188) cannot lie above the hyperbola (3.189). Hence, σ_v^2 is unbounded and the resonant vibrations, typical for a linear system with dry friction, are observed.

Let us investigate eq. (3.187) in the case of $N = 2$. One obtains

$$2\sigma_v^2 = \frac{A_1^2\omega_1^2}{\left(\lambda_0^2 - \omega_1^2\right)^2 + \dfrac{\gamma^2}{\sigma_v^2}\omega_1^2} + \frac{A_2^2\omega_2^2}{\left(\lambda_0^2 - \omega_2^2\right)^2 + \dfrac{\gamma^2}{\sigma_v^2}\omega_2^2}. \tag{3.191}$$

Manipulating this equation leads to the quadratic equation in σ_v^2

$$\begin{aligned} \sigma_v^4 &+ \left[\left(\gamma^2 - \frac{A_1^2}{2}\right)\frac{\omega_1^2}{\left(\lambda_0^2 - \omega_1^2\right)^2} + \left(\gamma^2 - \frac{A_2^2}{2}\right)\frac{\omega_2^2}{\left(\lambda_0^2 - \omega_2^2\right)^2}\right]\sigma_v^2 \\ &+ \left(\gamma^2 - \frac{A_1^2 + A_2^2}{2}\right)\frac{\gamma^2\omega_1^2\omega_2^2}{\left(\lambda_0^2 - \omega_1^2\right)\left(\lambda_0^2 - \omega_2^2\right)} = 0. \end{aligned} \tag{3.192}$$

It is easy to see that this equation has a solution $\sigma_v^2 > 0$ provided that

$$\gamma^2 < \frac{A_1^2 + A_2^2}{2}. \tag{3.193}$$

When

$$\gamma^2 > \frac{A_1^2 + A_2^2}{2},$$

then

$$\gamma^2 > \frac{A_1^2}{2}, \; \gamma^2 > \frac{A_2^2}{2},$$

i.e. both factors in eq. (3.192) are positive and this equation possesses no positive roots. Equation (3.193) is therefore the necessary condition for existing solutions of the type (3.185).

Special attention needs to be paid to the case in which one frequency of the polyharmonic driving coincides with the eigenfrequency of the system. Let, for instance, $\omega_1 = \lambda_0$, then eq. (3.188) takes the following form

$$\sigma_v^2 = \frac{1}{2}\left(\frac{A_1^2}{z} + \frac{A_2^2\omega_2^2}{\left(\lambda_0^2 - \omega_2^2\right)^2 + z\omega_2^2}\right). \tag{3.194}$$

It is evident that if

$$\frac{A_1^2}{2} > \gamma^2,$$

the curve (3.194) lies above the hyperbola (3.189), i.e. σ_v^2 grows without bounds. If conversely

$$\frac{A_1^2}{2} < \gamma^2, \tag{3.195}$$

then either the entire curve (3.194) lies under the hyperbola (3.189) or these curves intersect at a point corresponding to a finite value of σ_v^2. Thus, condition (3.195) ensures the absence of resonant vibrations regardless of the value of amplitude A_2. This conclusion has been confirmed by numerical simulation.

3.6.3 General solution for the problem of polyharmonic driving

We consider now the problem of seeking the polyharmonic solutions in general. Let the governing equation for the system be given as follows

$$m\ddot{u} + U_{if}\left(u, \dot{u}\right) + U_e\left(u\right) = \sum_{i=1}^{N} F_i \cos\left(\omega_i t + \varphi_i\right), \tag{3.196}$$

then the linearised equation takes the form

$$m\ddot{u} + b_d \dot{u} + U_0 + c_d u^0 = \sum_{i=1}^{N} F_i \cos(\omega_i t + \varphi_i). \tag{3.197}$$

The factors U_0, c_d and b_d are functions of a_0, σ_u^2 and σ_v^2 provided that our analysis is limited to the approximation of the first and second order moments while determining the joint probability function for the processes u and $\dot{u}$. One can use eqs. (11.9)-(11.11) from [75] to obtain the factors U_0, c_d and b_d for any particular system under consideration.

Looking for the solution of the linearised equation in the form of eq. (3.172) yields

$$a_i = \frac{A_i}{\sqrt{\left[\lambda^2\left(\sigma_u^2\right) - \omega_i^2\right]^2 + 4n_d\left(\sigma_u^2, \sigma_v^2\right)\omega_i^2}}, \tag{3.198}$$

where

$$\lambda^2 = \frac{c_d}{m}, \quad n_d = \frac{b_d}{2m}.$$

Hence

$$\sigma_u^2 = \frac{1}{2}\sum_{i=1}^{N} \frac{A_i^2}{\left[\lambda^2\left(\sigma_u^2\right) - \omega_i^2\right]^2 + 4n_d^2\left(\sigma_u^2, \sigma_v^2\right)\omega_i^2} = \Phi(\lambda), \tag{3.199}$$

$$\sigma_v^2 = \frac{1}{2}\sum_{i=1}^{N} \frac{A_i^2\omega_i^2}{\left[\lambda^2\left(\sigma_u^2\right) - \omega_i^2\right]^2 + 4n_d^2\left(\sigma_u^2, \sigma_v^2\right)\omega_i^2} = \Psi(\lambda). \tag{3.200}$$

Solving these equations for σ_u^2 and σ_v^2 is rather laborious. However, while dealing with the vibration protection systems one needs to know if the resonance vibrations can occur in the system or the damping present in the system suppresses them. If the analysis is limited only to these aspects, one can apply the approach analogous to that of the first example.

Equations (3.199) and (3.200) determine σ_u^2 and σ_v^2 as functions of λ and n_d. For weak damping, i.e. for small values of n_d, the functions $\Phi(\lambda)$ and $\Psi(\lambda)$ can be adopted to achieve the maxima at $\lambda = \omega_i$. Using diagram $\lambda\left(\sigma_u^2\right)$ and taking $\lambda\left(\sigma_u^2\right) = \omega_i$ we find the maximum values of σ_u^2 denoted by $\left(\sigma_u^2\right)_i$. For each values of $\left(\sigma_u^2\right)_i$ we can find n_{di} from the following equation

$$\left(\sigma_u^2\right)_i = \sum_{j=1}^{N} \frac{A_j^2}{\left(\omega_i^2 - \omega_j^2\right)^2 + 4n_{di}^2\omega_j^2}. \tag{3.201}$$

Figure 3.36 shows that for the value of n_{di} from the latter equation the curve $\Phi(\lambda)$ is tangential with the backbone curve at the point having

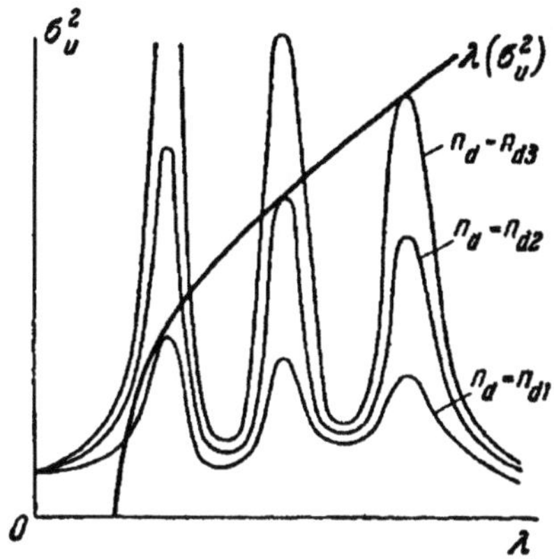

FIGURE 3.36.

abscissa $\lambda = \omega_i$. Inserting the obtained value of n_{di} into eq. (3.200) leads to the following equation for $\left(\sigma_v^2\right)_i$

$$\left(\sigma_v^2\right)_i = \sum_{j=1}^{N} \frac{A_j^2 \omega_j^2}{\left(\omega_i^2 - \omega_j^2\right)^2 + 4n_{di}^2 \omega_j^2}. \tag{3.202}$$

Let us determine the value of n_{di}^* which corresponds to $\left(\sigma_u^2\right)_i$ and $\left(\sigma_v^2\right)_i$ with the help of the following equation

$$n_{di}^* = \frac{b_d\left[\left(\sigma_u^2\right)_i, \left(\sigma_v^2\right)_i\right]}{2m}. \tag{3.203}$$

If $n_{di}^* > n_{di}$, then $\Phi\left(\omega_i\right) < \left(\sigma_u^2\right)_i$, i.e. the curve $\Phi\left(\lambda\right)$ lies below the backbone curve in the vicinity of the point $\lambda = \omega_i$. In this case, the harmonic of frequency ω_i cannot cause resonant vibration. Conversely, i.e. when $n_{di}^* < n_{di}$, the danger of resonance remains.

Concluding this Section let us remember that the equation for the linearisation factors used in this Section are valid only under the assumption that the frequencies do not meet the following conditions

$$2\omega_i = \omega_j, \omega_i + \omega_j = \omega_k, \omega_i - \omega_j = \omega_k, \tag{3.204}$$

$$3\omega_i = \omega_j, 2\omega_i - \omega_j \pm \omega_k = 0, \omega_i \pm \omega_j - \omega_k \pm \omega_l = 0. \tag{3.205}$$

If this is the case, it is necessary to take into account the dependence of the third and forth moments on the phases ψ_i.

3.7 Subharmonic resonances in vibration protection systems

The general theory of resonances which may occur in a single-degree-of-freedom system undergoing some periodic external excitation is considered in the present Section.

In physics, oscillations of large amplitude caused by relatively weak external actions is referred to as resonance. In order to understand the nature of the resonance phenomena we recall first some typical features of resonance in a linear single-degree-of-freedom system. Its governing equation is

$$m\ddot{u} + b\dot{u} + cu = Q(t). \tag{3.206}$$

Let

$$\frac{b}{m} = 2n, \ \frac{c}{m} = \lambda_0^2, \ \frac{1}{m}Q(t) = F_0 \cos \omega t,$$

which implies that the external excitation is harmonic with the frequency coinciding with the eigenfrequency of the undamped system. Seeking the solution in the form

$$u = a \cos(\lambda_0 t + \theta), \tag{3.207}$$

and inserting the latter equation into eq. (3.206) we obtain

$$-a\lambda_0^2 \cos(\lambda_0 t + \theta) - 2na\lambda_0 \sin(\lambda_0 t + \theta) + a\lambda_0^2 \cos(\lambda_0 t + \theta) = F_0 \cos \lambda_0 t. \tag{3.208}$$

Evidently, the elastic force is identically equal to the product of the mass and acceleration with the opposite sign. This is also the case in which free vibration in the undamped system is considered. Thus, the external action must be equal to the resistance force at any moment of time, i.e.

$$-2na\lambda_0 \sin(\lambda_0 t + \theta) \equiv F_0 \cos \lambda_0 t,$$

which immediately yields

$$a = \frac{F_0}{2n\lambda_0}, \ \theta = -\frac{\pi}{2}. \tag{3.209}$$

Hence, when the dissipation is weak, i.e. $n \ll \lambda_0$, large amplitude vibration may occur in the system. The amplitude of the elastic force and the inertia force are much greater than the external excitation amplitude. Indeed,

$$|\ddot{u}|_{\max} = \left|\lambda_0^2 u\right|_{\max} = a\lambda_0^2 = \frac{F_0\lambda_0}{2n} \gg F_0. \tag{3.210}$$

The above allows the resonance in a linear single-degree-of-freedom system to be viewed as free vibration supported by external excitation. The very existence of resonance is caused by the fact that harmonic vibration of an arbitrarily large amplitude occurs in conservative system without dissipation subject to the appropriate initial conditions. Free vibrations in the real system decay since the nonconservative damping force continually dissipates energy. The external harmonic action can sustain that level of free vibration at which the influence of the dissipation is completely compensated.

An analogous physical phenomenon can take place in nonlinear systems. Again, let us consider the single-degree-of-freedom system governed by eq. (3.74) assuming the external action $Q(t)$ be a periodic function of time with period T. Considering the vibration isolator reaction as a sum of elastic and dissipative forces we can recast eq. (3.74) in the form

$$m\ddot{u} + U_e(u) = -U_{if}(u, \dot{u}) + Q(t). \tag{3.211}$$

We generalise the concept of "resonance" to nonlinear systems as follows. Vibration is said to be resonant if the amplitude of the elastic force and inertial force considerably exceed the amplitude of the external action. In this case, the forces on the right hand side of eq. (3.211) are small. This allows us to assume that the periodic solution of this equation will be close to the periodic solution of the following equation

$$m\ddot{u}_0 + U_e(u_0) = 0, \tag{3.212}$$

i.e. the free vibration of the system without dissipation. Therefore, the forced vibration of a nonlinear system may also be deemed to be a "free vibration" supported by external excitation.

To find the periodic solutions of eq. (3.211) we utilise the Galerkin approach, see [65]. For small forces U_d and $Q(t)$ the system governed by eq. (3.211) is "close" to the conservative system (3.212) and the condition for existence of the periodic solutions can be written as follows

$$\int_0^T \{-U_{if}[u_0(t+\alpha), \dot{u}_0(t+\alpha)] + Q(t)\}\, \dot{u}_0(t+\alpha)\, dt = 0. \tag{3.213}$$

Equation (3.211) may have periodic solutions with period T or pT where p is an integer, then, by virtue of the perturbation method, we seek the periodic solutions of eq. (3.212) with the same period.

Assume that eq. (3.212) has a periodic solution with period pT, then, as the system governed by eq. (3.212) is autonomous it has a family of solution of the following sort

$$u_0 = \varphi(t+\alpha, pT), \tag{3.214}$$

with α being an arbitrary real parameter.

The solution (3.214) can be expanded as a Fourier series which contains only cosines because substituting $-t$ instead of t does not affect eq. (3.212). Hence,

$$u_0 = \sum_{i=0}^{\infty} a_i \cos i \frac{2\pi}{pT} (t + \alpha) . \tag{3.215}$$

The value of parameter α defining the phase angle between the external action and the forced vibration must be determined from eq. (3.213) which can be rewritten in the form

$$P(\alpha) = -\int_0^{pT} U_{if} (u_0, \dot{u}_0) \dot{u}_0 dt + \int_0^{pT} Q(t) \dot{u}_0 dt = 0. \tag{3.216}$$

The first integral represents the work of the dissipative force through the free vibration of the conservative system. Its value does not depend on parameter α and is equal to the area of the hysteresis loop

$$W_c = \int_0^{pT} U_{if} (u_0, \dot{u}_0) \dot{u}_0 dt = \oint U_{if} (u, \dot{u}) du. \tag{3.217}$$

In the case of linear damping

$$W_c = \int_0^{pT} b\dot{u}_0\dot{u}_0 dt = \frac{4\pi^2 b}{p^2T^2} \int_0^{pT} \left[\sum_{i=1}^{\infty} a_i i \sin \frac{2\pi}{pT} (t + \alpha) \right]^2 dt = \frac{2b\pi^2}{pT} \sum_{i=1}^{\infty} a_i i^2. \tag{3.218}$$

For dry friction

$$W_c = \int_0^{pT} H \operatorname{sign}\dot{u}_0 \dot{u}_0 dt = \int_0^{pT} H |\dot{u}_0| dt = 4Ha, \tag{3.219}$$

where a denotes the amplitude of free vibration. While obtaining eq. (3.219) it was assumed that the velocity reverses its sign only twice during the vibration period.

In practice, an approximate value of W_c can be obtained under the assumption that the free vibration is close to being harmonic. In the case of a linear damping this assumption yields

$$W_c \approx \frac{2b\pi^2}{pT} a_1^2, \tag{3.220}$$

where a_1 is the amplitude of the first harmonic.

Expressions for W_c for other types of damping can be derived by analogy.

Let us analyse the second integral in eq. (3.216) which expresses the work of the external vibration excitation through the free vibration of the conservative system. To this end, we expand the periodic external force as a Fourier series, to get

$$Q(t) = \sum_{j=1}^{\infty} \left(Q_{cj} \cos j \frac{2\pi}{T} t + Q_{sj} \sin j \frac{2\pi}{T} t \right), \tag{3.221}$$

where Q_{cj} and Q_{sj} are the Fourier coefficients. The constant component in force $Q(t)$ is assumed to be absent.

Inserting eqs. (3.215) and (3.221) into eq. (3.216) and noting that

$$\int_0^{pT} \cos j \frac{2\pi}{T} t \sin i \frac{2\pi}{pT} (t+\alpha) = \begin{cases} 0 & \text{if } i \neq jp, \\ \dfrac{pT}{2} \sin j \dfrac{2\pi\alpha}{T} & \text{if } i = jp, \end{cases}$$

$$\int_0^{pT} \sin j \frac{2\pi}{T} t \sin i \frac{2\pi}{pT} (t+\alpha) = \begin{cases} 0 & \text{if } i \neq jp, \\ \dfrac{pT}{2} \cos j \dfrac{2\pi\alpha}{T} & \text{if } i = jp, \end{cases}$$

yields

$$\begin{aligned} W_b &= \int_0^{pT} Q(t)\, \dot{u}_0(t)\, dt = \\ &= -\pi p \sum_{j=1}^{\infty} j a_{pj} \left(Q_{cj} \sin j \frac{2\pi}{T} \alpha + Q_{sj} \cos j \frac{2\pi}{T} \alpha \right), \end{aligned} \tag{3.222}$$

where index pj means the product of p and j. Hence, parameter α must be determined from the following equation

$$W_c + \pi p \sum_{j=1}^{\infty} j a_{pj} \left(Q_{cj} \sin j \frac{2\pi}{T} \alpha + Q_{sj} \cos j \frac{2\pi}{T} \alpha \right) = 0, \tag{3.223}$$

which has a simple physical meaning, namely, the work of the external vibrational action through one cycle must be equal to that of the dissipative work.

By analysing expression (3.223) it is easy to understand the principal difference between linear and nonlinear systems with respect to resonant vibration. Free vibration in a linear single-degree-of-freedom system is always a harmonic vibration with an eigenfrequency of the system, i.e.

$$u_0 = a \cos \lambda (t + \alpha),$$

where a and α can be chosen arbitrarily. Expression (3.222) is equal to zero unless $j2\pi/T = \lambda$, that is if the frequency of one of harmonics in the external excitation coincides with the eigenfrequency of the system. Only this harmonic can perform the nontrivial work through free vibration and support free vibration by compensating the energy dissipation due to the dissipative forces.

Free vibration in the nonlinear system is not harmonic. Thus, the external harmonic action is also able to perform a nontrivial work in the case in which its frequency is equal to p times the eigenfrequency (p is any integer), i.e. the frequency of the external harmonic excitation coincides with the frequency of one of the higher harmonics.

Resonant vibration whose period is p times the least period of the driving is referred to as the *subharmonic resonance* of order p or the fractional resonance. With this in view, resonant vibrations with the same period as the external excitation can be termed the first order resonances. They are also called the *principal* resonances.

In general, when the frequency of $r - th$ harmonic of the external excitation coincides with the frequency of the $p - th$ harmonic of free vibration, eq. (3.222) gains a nontrivial term. This also implies that in this case a resonant vibration is observed. Its frequency is equal to the frequency of external excitation multiplied by r/p. Such resonances are termed as ultrasubharmonic ones of order r/p, r and p being relatively prime integers, [127]. For the resonant vibration to have a period equal to or a multiple period of the external excitation, it is necessary that ratio p/r is an integer.

Consider now the condition for the occurrence of subharmonic resonances for the case of harmonic external excitation

$$Q(t) = F \cos \omega t. \tag{3.224}$$

As the excitation is harmonic, only subharmonic resonances of order p can occur. Since $j = 1$ eq. (3.223) takes the form

$$P(\alpha) = W_c + \pi p a_p F \sin \omega \alpha = 0. \tag{3.225}$$

In order to obtain a real value of α, the following condition must hold true

$$|\sin \omega a| = \left| \frac{W_c}{\pi p a_p F} \right| < 1. \tag{3.226}$$

Another form of this condition is

$$|W_c| < |\pi p a_p F| . \tag{3.227}$$

This inequality is the condition of occurrence of the subharmonic resonance of order p. For this reason, let us analyse it in detail.

The left hand side of this inequality depends on the intensity and type of damping. Increasing the intensity of the dissipative forces violates inequality (3.227), i.e. there is a critical damping which make subharmonic

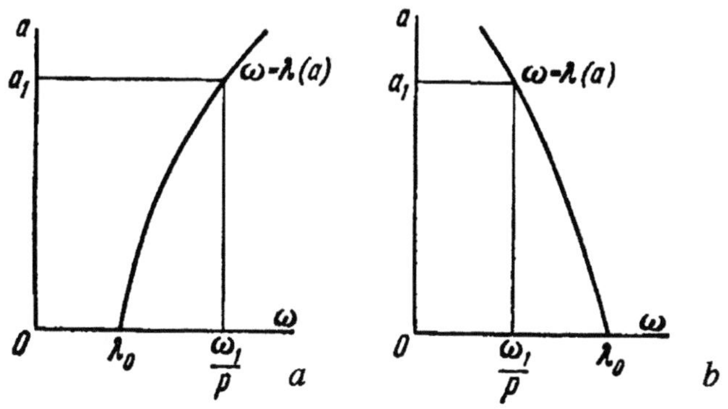

FIGURE 3.37.

resonance impossible. The right hand side of the inequality is proportional to pa_p, a_p being the Fourier factor for the $p - th$ harmonic of the free vibration. The value of pa_p decreases as p increases provided that function $U_e(u)$ has a continuous first derivative. In general, the smoother is the elastic characteristic of the vibration isolator (in other words, the higher the order of the continuous derivatives of $U_e(u)$), the faster the decrease in value of a_p which in turn means that the occurrence of subharmonic resonant vibrations becomes less and less probable. Subharmonic resonant vibrations are most probable in vibration protection systems involving rigid stops since their elastic characteristic has a discontinuous derivative $U_e'(u)$. The likelihood of the occurrence of subharmonic resonances increases with the growth of the driving force amplitude F.

As the subharmonic resonant vibrations are close to the free vibrations, solution (3.214) can be used not only for establishing the condition for subharmonic resonances, but also for estimation of their amplitudes. As a first approximation, one can adopt the amplitude of the subharmonic vibration to coincide with the amplitude of free vibration of the same frequency, which is the case for principal resonant vibrations. In practice, in order to estimate the amplitude of the subharmonic resonance of order p it is convenient to use the backbone curve $\lambda(a)$, Fig 3.37. Dividing the driving frequency ω_1 by p we find the value of the amplitude a. It is easy to see from Figs. 3.37a and b that the subharmonic resonances of order p are possible in systems with hardening and softening backbone curves if $\omega_1 > p\lambda_0$ and $\omega_1 < p\lambda_0$, respectively.

We proceed now to the investigation of the conditions for the existence of subharmonic resonances in some particular cases. To begin with, we write down condition (3.227) for different types of damping.

In the case of linear damping, we insert eq. (3.227) into (3.227), to have

$$2n < \frac{p^2 |a_p| F_1}{\omega \sum\limits_{i=1}^{\infty} i^2 a_i^2}, \tag{3.228}$$

where

$$n = \frac{b}{2m}, \ F_1 = \frac{F}{m}.$$

Condition (3.228) can be simplified by using eq. (3.220) for the work of the dissipative force

$$2n < \frac{p^2 |a_p| F_1}{\omega a_1^2}. \tag{3.229}$$

In particular, from this equation we can establish the condition for the existence of the following principal resonant vibrations

$$p = 1, \ a_p = a_1. \tag{3.230}$$

This gives

$$2n < \frac{F_1}{\omega a_1}. \tag{3.231}$$

This result fully agrees with that obtained in Section 3.2. Formula (3.231) points out that the resonant vibration of amplitude a^* and frequency ω can occur if the curve (3.89) intersects the backbone curve above point a^*.

The case of dry friction yields by virtue of eq. (3.219)

$$H < \frac{\pi p |a_p| F}{4a}$$

or

$$h < \frac{p |a_p| F_1}{a}, \tag{3.232}$$

where

$$h = \frac{4H}{\pi m}.$$

We obtain for the principal resonance that

$$h < F_1,$$

coinciding with the condition for existence of solutions close to harmonic ones, see Section 3.3.

Finally, eq. (3.135) can be utilised for the analysis of internal friction in the system under harmonic excitation. The work of the dissipative force through the period of free vibration is the area of the hysteresis loop for the amplitude a, i.e.

$$W_c = \pi\beta a^{\mu+1}. \tag{3.233}$$

Hence, for subharmonic resonances to exist, the following condition must hold true

$$\beta_1 = \frac{\beta}{m} < \frac{pa_p F_1}{a^{\mu+1}}. \tag{3.234}$$

In fact, using the results obtained leads to determination of those damping parameters which ensure absence of subharmonic resonant vibrations. In other words, the following inequality must hold

$$|W_c| > |\pi p a_p F|,$$

see eq. (3.227). The most challenging part of the analysis is to obtain the Fourier coefficient for the $p-th$ harmonic of the free vibration. We proceed now to some particular examples which demonstrate the way in which this coefficient is obtained.

For the sake of convenience we introduce nondimensional time

$$\tau = \lambda_0 t, \tag{3.235}$$

where λ_0 denotes the eigenfrequency of the vibration isolator. The nondimensional eigenfrequency ν and the nondimensional frequency of the external excitation z are introduced as follows

$$\nu = \frac{\lambda}{\lambda_0}, \quad z = \frac{\omega}{\lambda_0}. \tag{3.236}$$

The first example is that of the vibration isolator with symmetric cubic elastic characteristic. Its free vibration is governed by the equation

$$m\ddot{u} + cu + eu^3 = 0. \tag{3.237}$$

This equation rewritten in terms of nondimensional time τ takes the form

$$\frac{d^2u}{d\tau^2} + u + \gamma u^3 = 0, \tag{3.238}$$

where

$$\gamma = \frac{e}{c}.$$

This equation allows a closed form solution. To obtain this solution, we use the first integral of this equation which expresses the law of conservation of energy

$$\frac{1}{2}\left(\frac{du}{d\tau}\right)^2+\frac{u^2}{2}+\gamma\frac{u^4}{4}=\frac{a^2}{2}+\gamma\frac{a^4}{4}, \tag{3.239}$$

where a is the amplitude of free vibration.

There follows from eq. (3.239) that

$$\frac{du}{d\tau}=\sqrt{\left(a^2-u^2\right)+\frac{\gamma}{2}\left(a^4-u^4\right)}=a\sqrt{1+\frac{\gamma}{2}a^2}\sqrt{\left(1-\zeta^2\right)\left(1+\varepsilon^2\zeta^2\right)}, \tag{3.240}$$

where

$$\varepsilon^2=\frac{\gamma a^2}{2+\gamma a^2},\ \zeta=\frac{u}{a}.$$

Integrating eq. (3.240) yields

$$\begin{aligned}\tau+\tau_0 &= \int_0^u \frac{du}{\sqrt{\left(a^2-u^2\right)+\frac{\gamma}{2}\left(a^4-u^4\right)}}= \\ &= \frac{1}{\sqrt{1+\frac{\gamma}{2}a^2}}\int_0^\zeta \frac{d\zeta}{\left(1-\zeta^2\right)\left(1+\varepsilon\zeta^2\right)}.\end{aligned} \tag{3.241}$$

Solution of eq. (3.237) is therefore the elliptic function

$$u=\operatorname{cn}\left[\sqrt{1+\gamma a^2}\left(\tau+\tau_0\right)\right]. \tag{3.242}$$

By means of the known expansion of the elliptic function in Fourier series we obtain

$$u=a\frac{2\pi}{\varkappa K\left(\varkappa\right)}\sum_{p=0}^{\infty}\frac{q^{p+1/2}}{1+q^{2p+1}}\cos\left(2p+1\right)\psi, \tag{3.243}$$

where $\varkappa=\sqrt{\gamma a^2/\left(2+2\gamma a^2\right)}$ is the modulus of the elliptic integral (3.241), $K\left(\right)$ denotes the complete elliptic integral of the first kind, $\psi=\nu\left(\tau+\tau_0\right)$ and q is given by

$$q=\exp\left\{-\pi\frac{K\left(\sqrt{1-\varkappa^2}\right)}{K\left(\varkappa\right)}\right\}.$$

Expansion (3.243) allows the Fourier coefficients to be expressed in terms of the amplitude of free vibration

$$\left.\begin{aligned} a_p &= a\frac{2\pi}{\varkappa K(\varkappa)}\frac{q^{p/2}}{1+q^p} && \text{for odd } p \\ a_p &= 0 && \text{for even } p. \end{aligned}\right\} \tag{3.244}$$

It is also necessary to find the relationship between the frequency and the amplitude of free vibration. To this aim, we integrate eq. (3.241) with respect to u from 0 to a, and take $\tau_0 = 0$, to obtain

$$\frac{T}{4} = \frac{K(\varkappa)}{\lambda_0\sqrt{1+\gamma a^2}} \text{ or } \nu = \frac{\pi\sqrt{1+\gamma a^2}}{2K(\varkappa)}. \tag{3.245}$$

Now we can plot the backbone curve which relates the nondimensional frequency of free vibration to the amplitude, and then by virtue of eq. (3.244) determine the amplitude of the harmonic of order p. It goes without saying that the exact expression for the backbone curve (3.245) can be replaced by an approximation obtained by the method of harmonic linearisation if an approximate solution is sought.

Let us define now the value of dry friction ensuring suppression of the subharmonic resonances of third order. Substituting eq. (3.244) into eq. (3.245) yields

$$h > \frac{6\pi}{\varkappa K(\varkappa)}\frac{q^{3/2}}{1+q^3}F_1. \tag{3.246}$$

If the frequency of the external force is equal to ω, we obtain for the third harmonics that $\lambda = \omega/3$ and $\nu = z/3$. The latter expression renders ν which allows us to find a from eq. (3.245) and h from eq. (3.246).

We note at this stage that the approximate analysis based on obtaining the higher harmonics by means of the method of harmonic balance (the so-called improved first approximation, [65]) would give a rather inaccurate result in this case. For example, for $\nu = 1$ and $z = 3$ the value of the dry friction force would be infinite because ratio a_3/a_1 tends to infinity (in this case a_1 implies the amplitude of the first harmonic).

Let us consider now a vibration isolator having symmetric elastic buffers. Its characteristic is given by eq. (3.1) under the assumption that $\Delta = \Delta_1$. Using the nondimensional time τ allows the equation for free vibration to be written in the following form

$$\left.\begin{aligned} &\frac{d^2u}{d\tau^2} + u = 0 && \text{if } |u| \le d, \\ &\frac{d^2u}{d\tau^2} + \varkappa^2 u = d\left(\varkappa^2 - 1\right)\,\mathrm{sign}\,u && \text{if } |u| > d. \end{aligned}\right\} \tag{3.247}$$

Let us determine the free vibration whose frequency is equal to ν. To this end, we utilise the method of fitting which is frequently used in the theory of nonlinear vibrations while studying piecewise linear systems.

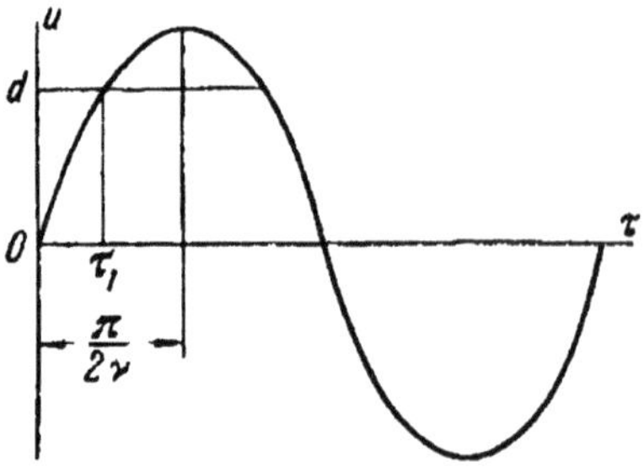

FIGURE 3.38.

As the phase of free vibration can be chosen arbitrarily, we take, for the sake of definiteness, that $u = 0$, $du/d\tau > 0$ at $\tau > 0$. Due to the symmetry of the elastic characteristic, the free vibration must have the form shown in Fig. 3.38, i.e. each half-wave must be symmetric about its maximum ordinate. Thus, it is sufficient to find the law of motion within the interval $0 \leq \tau \leq \pi/2\nu$.

Integrating the linear equations in (3.247) yields

$$\left.\begin{aligned} u_1 &= A_1 \sin\tau + B_1 \cos\tau && \text{if } 0 \leq \tau \leq \tau_1, \\ u_2 &= A_2 \sin\varkappa\tau + B_2 \cos\varkappa\tau + d\left(\varkappa^2 - 1\right) && \text{if } \tau_1 \leq \tau \leq \pi/2\nu, \end{aligned}\right\} \tag{3.248}$$

wherein τ_1 denotes the time moment corresponding to $u = d$. In order to determine the unknown parameters A_1, B_1, A_2, B_2 and τ_1 we use the initial condition and the condition for fitting the curves which implies the continuity conditions for u and $du/d\tau$ at $\tau = \tau_1$.

The initial condition is: $\tau = 0$, $u_1 = 0$.

The fitting conditions are: 1) $\tau = \tau_1$, $u_1 = d$; 2) $\tau = \tau_1$, $u_1 = u_2$; and 3) $\tau = \tau_1$, $du_1/d\tau = du_2/d\tau$. Also, we use the condition for a maximum at $\tau = \pi/2\nu$ which is

$$\tau = \frac{\pi}{2\nu}, \ \frac{du_2}{d\tau} = 0.$$

Determining the unknown parameters from these equations gives

$$A_1 = \frac{d}{\sin\tau_1}, \ B_1 = 0,$$

$$A_2 = \frac{d}{\varkappa^2} \frac{\sin\dfrac{\varkappa\pi}{2\nu}}{\cos\varkappa\left(\dfrac{\pi}{2\nu} - \tau_1\right)}, \ B_2 = \frac{d}{\varkappa^2} \frac{\cos\dfrac{\varkappa\pi}{2\nu}}{\cos\varkappa\left(\dfrac{\pi}{2\nu} - \tau_1\right)}. \tag{3.249}$$

The value of τ_1 is obtained from the following equation

$$\tan\tau_1 \cdot \tan\left(\frac{\pi}{2\nu} - \tau_1\right) = \varkappa. \tag{3.250}$$

Figure 3.39 illustrates dependence of τ_1 on ν for different values of $\varkappa$. Inserting eq. (3.249) into (3.248) yields

$$u_1 = d\frac{\sin\tau}{\sin\tau_1}, \tag{3.251}$$

$$u_2 = \frac{d}{\varkappa^2}\left[\frac{\cos\varkappa\left(\dfrac{\pi}{2\nu}-\tau\right)}{\cos\varkappa\left(\dfrac{\pi}{2\nu}-\tau_1\right)} + \varkappa^2 - 1\right]. \tag{3.252}$$

The vibration amplitude is apparently equal to the value of u_2 at $\tau = \pi/2\nu$, hence

$$a = \frac{d}{\varkappa^2}\left[\frac{1}{\cos\varkappa\left(\dfrac{\pi}{2\nu}-\tau_1\right)} + \varkappa^2 - 1\right]. \tag{3.253}$$

We can now define the Fourier coefficients for function $u(\tau)$. The coefficients in front of the even harmonics vanish due to symmetry whereas the coefficients in front of the odd harmonics are as follows

$$\begin{aligned} a_p &= \frac{\nu}{\pi}\int_0^{\pi/2\nu} u\cdot\cos p\nu\tau\, d\tau = \\ &= \frac{4\nu}{\pi}\left(\int_0^{\tau_1} u_1\cos p\nu\tau\, d\tau + \int_{\tau_1}^{\pi/2\nu} u_2\cos p\nu\tau\, d\tau\right) = \\ &= \frac{4d}{\pi}\frac{\left(\varkappa^2-1\right)\left(\cos p\nu\tau_1\cdot\tan\tau_1 - p\nu\sin p\nu\tau_1\right)}{p\left(p^2\nu^2-1\right)\left(p^2\nu^2-\varkappa^2\right)\tan\tau_1}. \end{aligned} \tag{3.254}$$

Let us take, for example, $\varkappa = 2$, the amplitude of the external force Fm, the nondimensional frequency $z = 4,5$ and determine the intensity of the damping force which is necessary for suppressing resonant vibration.

Considering Fig. 3.39 we find that $1 \le \nu \le 2$ for $\varkappa = 2$. The only integer p which satisfies the condition $z = p\nu$ is $p = 3$, thus, only the third order subharmonic resonance is possible in the system, and, moreover, $\nu = 1,5$.

For $\nu = 1,5$ we obtain from Fig. 3.39 that $\tau_1 = 0,35$. Equations (3.253) and (3.254) yield

$$a = 2,24d, \ a_3 = 0,41d.$$

Now, by means of eq. (3.229) we find the linear damping factor which guarantees that the condition for resonant vibrations does not hold

$$n > \frac{9\cdot 0,41d\cdot F}{4,5\lambda_0\cdot 2,24^2 d^2} = 0,16\frac{F}{\lambda_0 d}.$$

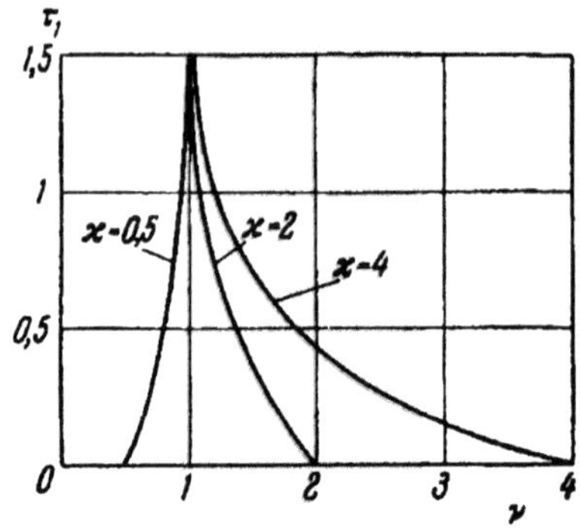

FIGURE 3.39.

By virtue of eq. (3.232) we obtain the following condition for the friction

$$h > \frac{3 \cdot 0,41d \cdot F}{2,24d} = 0,55F.$$

3.8 Subharmonic resonance in systems with rigid stops

As shown above, the higher the buffer stiffness and the sharper the change in the elastic characteristics of the elastic element and the buffer, the higher the probability of occurrence of subharmonic resonances. This is the main reason for investigating systems with rigid stops. In most cases, one can assert that if the dissipative forces suffice for suppression of the subharmonic resonances in the system with rigid stops, they then will suffice in the case of elastic buffers having any elastic characteristic, but the same gap.

Consider the free vibration of a system with a linear elastic characteristic and rigid stops.

The distinction will be made in what follows between the "upper" stop ($u = d$) and the "lower" stop ($u = -d_1$), therefore

$$U_e(u) = cu \text{ if } -d_1 < u < d. \tag{3.255}$$

An assumption about the nature of the impact on stops must be made. As we intend to consider free vibration we adopt the assumption of an ideal elastic impact, i.e. no energy disappears. At the time instant of the impact the velocity reverses its sign retaining its absolute value, i.e. the law of motion has the form schematically depicted in Fig. 3.40. By introducing the nondimensional time we can write down the governing differential equation for the time interval between impacts

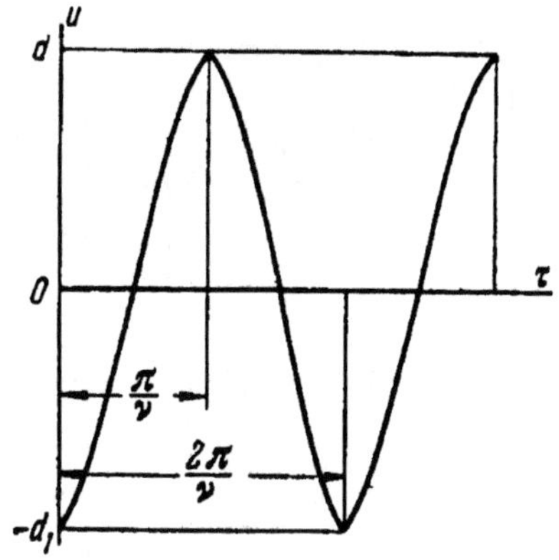

FIGURE 3.40.

$$\frac{d^2u}{d\tau^2} + u = 0. \tag{3.256}$$

Let us denote the period of vibration equal to the time interval between the impacts on the same stop as $2\pi/\nu$, where ν is the nondimensional frequency. The law of motion within the period is symmetric about the ordinate $\tau = \pi/\nu$ which implies that it is sufficient to find the law of motion within the interval $0 \le \tau \le \pi/\nu$.

Integration of eq. (3.256) yields

$$u = A\cos\tau + B\sin\tau. \tag{3.257}$$

The boundary conditions

$$\tau = 0,\; u = -d;\; \tau = \frac{\pi}{\nu},\; u = d$$

allow the integration constants A and B to be determined

$$A = -d_1,\; B = \frac{d + d_1\cos\frac{\pi}{\nu}}{\sin\frac{\pi}{\nu}}. \tag{3.258}$$

Hence,

$$u = \frac{1}{\sin\frac{\pi}{\nu}}\left[d\sin\tau - d_1\sin\left(\frac{\pi}{\nu} - \tau\right)\right] \; \left(0 \le \tau \le \frac{\pi}{\nu}\right). \tag{3.259}$$

We can now find the $p-th$ order Fourier coefficient

$$a_p = \frac{2\nu}{\pi}\int_0^{\pi/\nu} u\cos p\nu\tau d\tau =$$

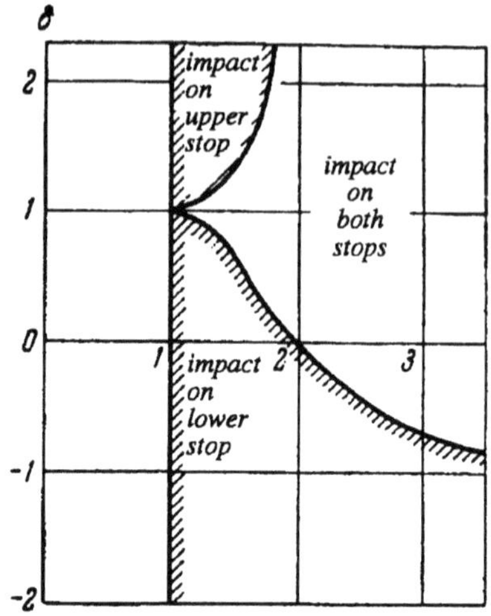

FIGURE 3.41.

$$= \frac{2\nu}{\pi}\int_0^{\pi/\nu}\left[d\frac{\sin\tau}{\sin\frac{\pi}{\nu}} + d_1\frac{\sin\left(\frac{\pi}{\nu}-\tau\right)}{\sin\frac{\pi}{\nu}}\right]\cos p\nu\tau\, d\tau =$$

$$= \left\{\begin{array}{ll} \dfrac{2\nu\,(d-d_1)}{\pi\,(1-p^2\nu^2)}\tan\dfrac{\pi}{2\nu} & \text{for even } p, \\ \dfrac{2\nu\,(d+d_1)}{\pi\,(1-p^2\nu^2)}\cot\dfrac{\pi}{2\nu} & \text{for odd } p. \end{array}\right\} \quad (3.260)$$

If $d = d_1$, i.e. in the case of symmetric stops, the coefficients of the even harmonics vanish.

Impacts on both stops have been assumed up until now. However this regime occurs not for all values of d, d_1 and ν. For $0 \le \tau \le \pi/\nu$ one has

$$\frac{du}{d\tau} = \frac{1}{\sin\frac{\pi}{\nu}}\left[d\cos\tau + d_1\cos\left(\frac{\pi}{\nu}-\tau\right)\right]. \quad (3.261)$$

For an impact on the lower stop to occur at $\tau = 0$, $du/d\tau$ must be nonnegative ($du/d\tau \ge 0$), otherwise the system cannot achieve this stop. Analogously, the condition $du/d\tau \ge 0$ at $\tau = \pi/\nu$ must hold to ensure that an impact occurs on the upper stop. Considering (3.261) yields the condition for impact on the upper stop

$$\cos\frac{\pi}{\nu} \ge -\frac{d_1}{d}, \quad (3.262)$$

as well as that for the lower stop

$$\cos\frac{\pi}{\nu} \ge -\frac{d}{d_1}. \quad (3.263)$$

Figure 3.41 displays the regions of impacts only on the upper stop, only on the lower stop and on both stops in the plane of ν and $\delta = d_1/d$. Negative values of δ imply that both stops lie on one side of the equilibrium position, i.e. from the equilibrium position the system is driven into one of the stops. In the case of impacts on the upper stop only, the integration constants in eq. (3.257) are determined from the following conditions (see Fig. 3.42a)

$$\tau = \frac{\pi}{\nu}, \; u = d; \; \tau = 0, \; \frac{du}{d\tau} = 0.$$

We obtain then

$$u = d\frac{\cos\tau}{\cos\dfrac{\pi}{\nu}}. \tag{3.264}$$

From this equation it is easy to obtain the following

$$a_p = \frac{2\nu}{\pi}\int\limits_0^{\pi/\nu} d\frac{\cos\tau}{\cos\dfrac{\pi}{\nu}}\cos p\nu\tau d\tau = (-1)^p \cdot \frac{2d\nu}{\pi\left(p^2\nu^2 - 1\right)}\tan\frac{\pi}{\nu}. \tag{3.265}$$

The vibration amplitude in this case is given by

$$2a = u\left(\frac{\pi}{\nu}\right) - u\left(0\right) = d\left(1 - \frac{1}{\cos\dfrac{\pi}{\nu}}\right). \tag{3.266}$$

In the case of impacts on the lower stop only, Fig 3.42b, we have the following conditions

$$\tau = 0, \; u = -d_1; \; \tau = \frac{\pi}{\nu}, \; \frac{du}{d\tau} = 0.$$

Thus,

$$u = -d_1\frac{\cos\left(\dfrac{\pi}{\nu} - \tau\right)}{\cos\dfrac{\pi}{\nu}}. \tag{3.267}$$

$$a_p = \frac{2\nu}{\pi}\int\limits_0^{\pi/\nu} -d_1\frac{\cos\left(\dfrac{\pi}{\nu} - \tau\right)}{\cos\dfrac{\pi}{\nu}}\cos p\nu\tau d\tau = \frac{2d_1\nu}{\pi\left(p^2\nu^2 - 1\right)}\tan\frac{\pi}{\nu}, \tag{3.268}$$

and consequently

$$2a = u\left(\frac{\pi}{\nu}\right) - u\left(0\right) = -d_1\left(\frac{1}{\cos\dfrac{\pi}{\nu}} - 1\right). \tag{3.269}$$

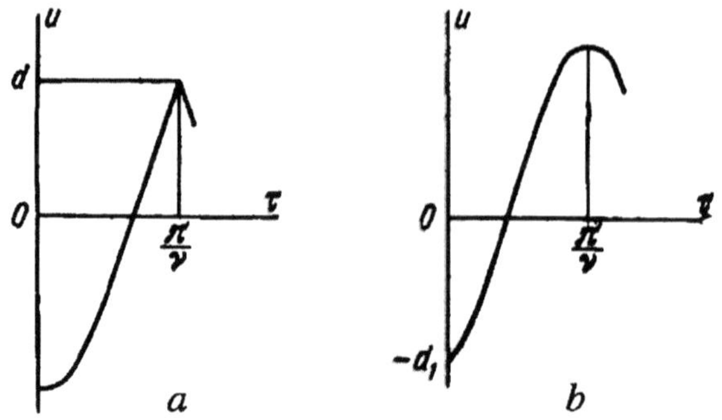

FIGURE 3.42.

Now it is not difficult to find the value of the dissipative force required for suppressing the subharmonic resonances.

Let us consider the following problem which can serve as a particular example. Let the system be subject to the excitation whose amplitude and nondimensional frequency are mF and ω/λ_0, respectively. The distance from the lower stop is $d_1 = 0,5d$. Determine the dry friction force sufficient for suppressing the subharmonic resonances.

As $z = p\nu$ and $\nu > 1$, the following subharmonic resonances may occur in the system

$$p = 2 \quad \nu = 2,5;$$
$$p = 3 \quad \nu = 1,67;$$
$$p = 4 \quad \nu = 1,25.$$

Considering Fig 3.41 and taking into account that $\delta = d_1/d = 0,5$ we find that in the first and the second case impacts on both stops take place while in the third case only impacts with the lower stop take place. Using eq. (3.260) we obtain

$$p = 2 \quad \nu = 2,5, \qquad a_2 = \frac{5 \cdot 0,5d}{\pi\,(1-25)} \tan\frac{\pi}{5} = 0,024d;$$

$$p = 3 \quad \nu = 1,67, \quad a_3 = \frac{3,53 \cdot 1,5d}{\pi\,(1-25)} \cot\frac{\pi}{3,33} = 0,047d.$$

In both cases, $a = (d + d_1)/2 = 0,75d$.

By virtue of eq. (3.232) we obtain

$$p = 2, \quad h > \frac{2 \cdot 0,024}{0,75} F = 0,064F,$$

$$p = 3, \quad h > \frac{3 \cdot 0,047}{0,75} F = 0,188F.$$

For $p = 4$ eqs. (3.268) and (3.269) yield

$$a_4 = -\frac{2 \cdot 0,5d \cdot 1,25}{\pi\,(25-1)} \tan\frac{\pi}{1,25} = 0,0303d,$$

$$a = -0,25d\left(\frac{1}{\cos\dfrac{\pi}{1,25}} - 1\right) = 0,56d.$$

Thus, the following inequality

$$h > \frac{4 \cdot 0,0303}{0,56} F = 0,214F$$

must hold. Comparing the values of h obtained we choose the maximum value, i.e. $h = 0,214F$. This value of h ensures that all subharmonic resonances are suppressed. The present example indicates that the subharmonic resonances of higher order ($p = 4$ in this particular case) can be more dangerous provided that the distances from the equilibrium position to the stops are quite distinct.

Up to now, we have not considered that impacts on the stops in actual systems are always accompanied by energy dissipation. The loss can be estimated by means of the Carnot theorem which in the case of the impact of the material point on the stop is reduced to the following formula

$$\Delta W = \frac{m\left(1-k^2\right)}{2}\dot{u}_i^2. \tag{3.270}$$

Here ΔW is the energy loss during the impact, $\dot{u}_i$ is the material point velocity immediately before the collision and k is the restitution coefficient whose particular value depends on the mechanical properties of the colliding bodies, $0 < k < 1$.

We are confronted with a question as to whether we can calculate the energy loss ΔW by means of the solution of eq. (3.212), i.e. by substituting the value of velocity $\dot{u}_i$ from the equation for the free vibration into (3.270). As shown in [62], this is allowed in the cases for which the restitution coefficient is close to one, i.e. in the case of a nearly elastic impact. The energy loss ΔW must be added to the work of the dissipative forces during the vibration period. Condition (3.227) must be replaced by the following condition for occurrence of the subharmonic resonances in a system with stops

$$\left|W_s + \Delta W\right| < \left|\pi p a_p F\right|. \tag{3.271}$$

In the case of impacts on both stops, eq. (3.261) yields

$$\dot{u}_l = \lambda_0\left(\frac{du}{d\tau}\right)_{\tau=0} = \frac{\lambda_0}{\sin\pi/\nu}\left(d + d_1\cos\frac{\pi}{\nu}\right),$$

$$\dot{u}_u = \lambda_0 \left(\frac{du}{d\tau}\right)_{\tau=\pi/\nu} = \frac{\lambda_0}{\sin \pi/\nu} \left(d \cos\frac{\pi}{\nu} + d_1\right).$$

where $\dot{u}_l$ and $\dot{u}_u$ are velocities of the impact on the lower and upper stop, respectively. It follows that

$$\begin{aligned} \Delta W &= \frac{m}{2}\left(1-k^2\right)\left(\dot{u}_l^2 + \dot{u}_u^2\right) = \\ &= \frac{m}{2}\left(1-k^2\right)\left[\left(d^2+d_1^2\right)\left(1+\cos^2\frac{\pi}{\nu}\right) + 4dd_1\cos\frac{\pi}{\nu}\right]\lambda_0^2. \end{aligned} \tag{3.272}$$

If

$$|\Delta W| > |\pi p a_p F|, \tag{3.273}$$

i.e. the energy loss due to the impact exceeds the work of driving through the free vibration, then the corresponding subharmonic resonance cannot occur. The resonance is suppressed by the impacts on the stops in this case. Note that the energy loss due to impact is proportional to the square of the distances between the stops whereas the right hand side of eq. (3.273) is linearly dependent on the distances. Thus, by moving the stops apart one can always succeed in eliminating the subharmonic resonant vibrations.

In the case of impacts only on the upper stop

$$\Delta W = \frac{1}{2} m \left(1-k^2\right) \lambda_0^2 d^2 \tan^2\frac{\pi}{\nu}, \tag{3.274}$$

while in the case of impacts only on the lower stop

$$\Delta W = \frac{1}{2} m \left(1-k^2\right) \lambda_0^2 d_1^2 \tan^2\frac{\pi}{\nu}. \tag{3.275}$$

Let us now determine, for example, the value of the restitution coefficient which guarantees suppression of the subharmonic resonance of the fourth order in the system considered above. Using eq. (3.275) we obtain condition (3.273) in the form

$$\frac{1}{2} m \left(1-k^2\right) \lambda_0^2 d_1^2 \cdot 0,725^2 > \pi \cdot 4 \cdot 0,0606 d_1 F,$$

since $\nu = 1,25$. Recasting this equation yields

$$k^2 < 1 - 0,345 \frac{F}{d_1 \lambda_0^2}.$$

Taking for example $d_1 = 5$ mm, $\lambda_0 = 62,8$ 1/s and $f = 4g \approx 40$ m/s^2 we obtain

$$k^2 < 0,31;\ k < 0,55.$$

3.9 Forced random vibrations

There exists a broad class of problems of vibration protection in which it is convenient to consider the external excitation as a stationary random process. We list here a few of them: (i) vibration isolation of vehicles travelling along a road with "random"profi le, (ii) vibration isolation of items of equipment mounted on board of aircraft and spacecraft, the latter having a large number of independent sources of vibration, e.g. jet engine, (iii) vibration protection of devices and mechanisms in the roll of ships and many others.

In all these problems, the factors which determine the characters of the vibrational excitation, e.g. the road profile, physical processes in the engines, height of sea waves etc., are not deterministic. For this reason, these processes are also viewed as random processes rather than as deterministic functions of time.

In the majority of cases, the probability characteristics of the vibration excitation cannot be obtained *a priori*, e.g. by means of analysis of physical properties of the vibration source. As a rule, such an analysis turns out to be unrealisable in practice. The probabilistic properties of the random process are determined by means of statistical processing of some finite number of realisations over some finite time interval. It is worth noting that in this case the probabilistic approach to the problem could be omitted and one could limit the investigation to the behaviour of the vibration protection system subject to deterministic driving corresponding to each realisation. In this case the information about the external driving would be used in full. However this approach is associated with considerable computational difficulties. The probabilistic approach to the problem proves to be a convenient way to simplify the problem solving. For the cases in which only the probabilistic properties of the external excitation are given, this approach is the only way to handle the problem.

On the other hand, knowledge of a finite number of realisations of the random process is insufficient for the unique determination of all probabilistic properties. The information incompleteness forces one to utilise such postulates of the theory of random processes as ergodicity and stationarity.

Forced vibration in vibration protection systems subjected to stationary random external excitation is also a stationary random process. Analysis of the forced vibration reduces to obtaining its probabilistic characteristics in terms of the probabilistic properties of the excitation.

We proceed to investigate the forced vibration of a vibration protection system with one degree-of-freedom. Let $Q(t)$ in eq. (3.74) be a centered Gaussian random process. In order to determine the probabilistic properties of the displacement in the vibration isolator we apply the method of statistical linearisation covered in Section 3.1. For the statistical linearisation of function $U(u, \dot{u})$ one needs the distribution laws for the random processes $u(t)$ and $\dot{u}(t)$. Assuming Gaussian distributions leads to the following expressions for the linearisation factors, see e.g. [65] and [108]

$$U_0 = \frac{1}{2\pi\sigma_u\sigma_v} \int_{-\infty}^{\infty}\int_{-\infty}^{\infty} U(u,v) \exp\left[-\frac{(u-m_u)^2}{2\sigma_u^2} - \frac{v^2}{2\sigma_v^2}\right] dudv, \quad (3.276)$$

$$c_d = \frac{1}{2\pi\sigma_u^3\sigma_v} \int_{-\infty}^{\infty}\int_{-\infty}^{\infty} (u-m_u) U(u,v) \exp\left[-\frac{(u-m_u)^2}{2\sigma_u^2} - \frac{v^2}{2\sigma_v^2}\right] dudv, \quad (3.277)$$

$$b_d = \frac{1}{2\pi\sigma_u\sigma_v^3} \int_{-\infty}^{\infty}\int_{-\infty}^{\infty} vU(u,v) \exp\left[-\frac{(u-m_u)^2}{2\sigma_u^2} - \frac{v^2}{2\sigma_v^2}\right] dudv. \quad (3.278)$$

Here σ_u^2 and σ_v^2 denote the variances of $u(t)$ and $\dot{u}(t)$, respectively, and m_u is the mean value of $u(t)$.

The linearised equation (3.74) can be written as follows

$$m\ddot{u}^0 + c_d u^0 + b\dot{u}_d^0 + U_0 = Q(t), \quad (3.279)$$

with $u^0 = u - m_u$ being centered random process.

Expressions for the linearisation factors are simplified in the case in which the nonlinear function U can be represented in the form

$$U(u,\dot{u}) = U_e(u) + U_d(\dot{u}). \quad (3.280)$$

In this case

$$U_0 = \frac{1}{\sqrt{2\pi}\sigma_u} \int_{-\infty}^{\infty} U_e(u) \exp\left[-\frac{(u-m_u)^2}{2\sigma_u^2}\right] du, \quad (3.281)$$

$$c_d = \frac{1}{\sqrt{2\pi}\sigma_u^3} \int_{-\infty}^{\infty} U_e(u)(u-m_u) \exp\left[-\frac{(u-m_u)^2}{2\sigma_u^2}\right] du, \quad (3.282)$$

$$b_d = \frac{1}{\sqrt{2\pi}\sigma_v^3} \int_{-\infty}^{\infty} U_d(v) \exp\left[-\frac{v^2}{2\sigma_v^2}\right] dv. \quad (3.283)$$

The linearised equation allows the moments of u and $\dot{u}$ to be defined in terms of the spectral density or the correlation function of the external excitation $Q(t)$. We find first the spectral density of u and $\dot{u}$

$$S_u(\omega) = \frac{S_Q(\omega)}{|c_d - m\omega^2 + b_d i\omega|^2}, \tag{3.284}$$

$$S_v(\omega) = \frac{\omega^2 S_Q(\omega)}{|c_d - m\omega^2 + b_d i\omega|^2}. \tag{3.285}$$

Next, we obtain the variances σ_u^2 and σ_v^2

$$\sigma_u^2 = \frac{1}{2\pi}\int_{-\infty}^{\infty} S_u(\omega)\,d\omega = \frac{1}{2\pi}\int_{-\infty}^{\infty} \frac{S_Q(\omega)\,d\omega}{(c_d - m\omega^2)^2 + b_d^2\omega^2}, \tag{3.286}$$

$$\sigma_v^2 = \frac{1}{2\pi}\int_{-\infty}^{\infty} S_v(\omega)\,d\omega = \frac{1}{2\pi}\int_{-\infty}^{\infty} \frac{\omega^2 S_Q(\omega)\,d\omega}{(c_d - m\omega^2)^2 + b_d^2\omega^2}. \tag{3.287}$$

Finally, equating the constant component on the left hand side of eq. (3.279) to zero yields

$$U_0 = 0. \tag{3.288}$$

Expressing in eqs. (3.286)-(3.288) the linearisation factors U_0, c_d and b_d in terms of σ_u^2, σ_v^2 and m_u leads to a system of three equations from which the unknown moments can be obtained.

The probabilistic properties of the vibration isolator reaction is of primary importance for the estimation of the quality of the vibration protection system. From the linearised equation we get

$$U(u, \dot{u}) = U_0 + c_d u^0 + b_d \dot{u}^0. \tag{3.289}$$

This renders the mean value and the spectral density

$$\left.\begin{aligned} m_U &= U_0 = 0, \\ S_U(\omega) &= \frac{|b_d \cdot i\omega + c_d|^2 \cdot S_Q(\omega)}{|c_d - m\omega^2 + b_d \cdot i\omega|^2} = \frac{(c_d^2 + b_d^2\omega^2)\, S_Q(\omega)}{(c_d - m\omega^2)^2 + b_d^2\omega^2}. \end{aligned}\right\} \tag{3.290}$$

Hence,

$$\sigma_U^2 = \frac{1}{2\pi}\int_{-\infty}^{\infty} S_U(\omega)\,d\omega = \frac{1}{2\pi}\int_{-\infty}^{\infty} \frac{(c_d^2 + b_d^2\omega^2)\, S_Q(\omega)}{(c_d - m\omega^2)^2 + b_d^2\omega^2} d\omega. \tag{3.291}$$

We proceed now to solving some particular examples. Let the spectral density of the normal external excitation be prescribed by the following expression

$$S_Q(\omega) = \frac{2\eta\beta\sigma_Q^2}{|\eta - \omega^2 + \beta i\omega|^2}. \tag{3.292}$$

Such a spectrum is observed, for example, in the case in which a broad-band process which is similar to white noise is filtered by a second order linear system. For instance, an elastic support structure can play the part of such a filter. The same spectrum is observed in the case of several sources of vibration, one source dominating and producing vibrations close to being harmonic. Parameter η in eq. (3.292) is approximately equal to the square of the dominating frequency whereas parameter β determines the filter bandwidth.

Inserting eq. (3.292) into (3.286) yields

$$\sigma_u^2 = \frac{1}{2\pi}\int_{-\infty}^{\infty} 2\beta\eta\sigma_Q^2 \frac{d\omega}{|\eta - \omega^2 + \beta i\omega|^2 \, |c_d - m\omega^2 + b_d i\omega|^2}. \tag{3.293}$$

Such integrals are encountered in many books, e.g. [105]. Estimating the integrals we obtain

$$\sigma_u^2 = \sigma_Q^2 \frac{m^2\beta\eta + b_d c_d + b_d\beta(b_d + m\beta)}{b_d c_d\left[(c_d - m\eta)^2 + (b_d + m\beta)(\beta c_d + b_d\eta)\right]}, \tag{3.294}$$

$$\begin{aligned}
\sigma_U^2 &= \frac{1}{2\pi}\int_{-\infty}^{\infty} \frac{\left(c_d^2 + b_d^2\omega^2\right)\cdot 2\beta\eta\sigma_Q^2 \cdot d\omega}{|\eta - \omega^2 + \beta i\omega|^2 \, |c_d - m\omega^2 + b_d i\omega|^2} = \\
&= \sigma_Q^2 \frac{\left[c_d\left(b_d c_d + m^2\beta\eta\right) + b_d(b_d + m\beta)(b_d\eta + \beta c_d)\right]}{b_d\left[(c_d - m\eta)^2 + (b_d + m\beta)(\beta c_d + b_d\eta)\right]},
\end{aligned} \tag{3.295}$$

$$\begin{aligned}
\sigma_v^2 &= \frac{1}{2\pi}\int_{-\infty}^{\infty} 2\beta\eta\sigma_Q^2 \frac{\omega^2 d\omega}{|\eta - \omega^2 + \beta i\omega|^2 \, |c_d - m\omega^2 + b_d i\omega|^2} = \\
&= \sigma_Q^2 \frac{\eta(b_d + m\beta)}{b_d\left[(c_d - m\eta)^2 + (b_d + m\beta)(\beta c_d + b_d\eta)\right]}.
\end{aligned} \tag{3.296}$$

Thus, the dynamic magnification factor K which is ratio of the standard deviation of the vibration isolator reaction to the standard deviation of the external force is given by the following expression

$$K^2 = \frac{c_d\left(b_d c_d + m^2\beta\eta\right) + b_d(b_d + m\beta)(b_d\eta + \beta c_d)}{b_d\left[(c_d - m\eta)^2 + (b_d + m\beta)(c_d\beta + b_d\eta)\right]}. \tag{3.297}$$

For small β and b_d the maximum value of K occurs when $\eta \approx c_d/m$. Then,

$$K^2 = K_*^2 = 1 + \frac{c_d m}{b_d (b_d + \beta m)}, \tag{3.298}$$

$$\sigma_u^2 = (\sigma_{u*})^2 = \sigma_Q^2 \frac{(c_d + b_d\beta)\, m}{b_d c_d^2 (b_d + m\beta)} \approx \sigma_Q^2 \frac{m}{b_d c_d (b_d + m\beta)}. \tag{3.299}$$

We study now the action of the vibration excitation (3.292) on a system with symmetric elastic buffers and linear damping.

Assuming $m_Q = 0$ and taking into account the symmetry of the elastic characteristic we obtain

$$m_u = 0, \; U_0 = 0.$$

Moreover, it is evident that $b_d = b = const.$

Inserting $U_e(u)$ for the system considered into (3.282) and estimating the integrals we obtain

$$c_d = c\left\{1 + \varkappa^2 \left[1 - 2\Phi\left(\frac{d}{\sigma_u}\right)\right]\right\}, \; \left(\varkappa^2 = \frac{c' - c}{c}\right), \tag{3.300}$$

where $\Phi(u)$ is the error function

$$\Phi(u) = \frac{1}{\sqrt{2\pi}} \int\limits_0^u \exp\left(-\frac{u^2}{2}\right) du. \tag{3.301}$$

By virtue of eq. (3.299) we obtain

$$(\sigma_{u*})^2 = \frac{\mu}{1 + \varkappa^2 \left[1 - 2\Phi\left(\dfrac{d}{\sigma_{u*}}\right)\right]}, \; \left(\mu = \frac{\sigma_Q^2 m}{b\,(b + m\beta)\, c}\right). \tag{3.302}$$

Solving this equation for σ_{u*} and substituting the solution into (3.300) we find first c_d and then K_*^2 from eq. (3.298)

$$K_*^2 = 1 + \frac{cm}{b\,(b + m\beta)^2} \left\{1 + \varkappa^2 \left[1 - 2\Phi\left(\frac{d}{\sigma_{u*}}\right)\right]\right\}. \tag{3.303}$$

Figures 3.43 and 3.44 display the dependencies $\sigma_{u*}/\sqrt{\mu}$ and K_* versus nondimensional parameter $d/\sqrt{\mu}$ for the following value of β

$$\beta = \frac{b}{m} = 0,2\sqrt{\frac{c}{m}}.$$

These Figures show that if the gap is large then even rather stiff buffers do not influence the probabilistic characteristics of the forced vibration of the object to be isolated. Provided that the following condition is satisfied

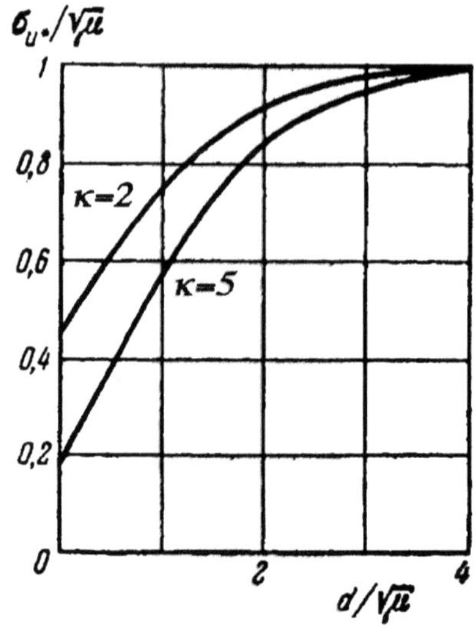

FIGURE 3.43.

$$d > 3\sqrt{\mu},$$

the probabilistic properties of the system with and without buffers coincide very accurately. The graph of Fig. 3.44 indicates that an increase in gap leads to a decrease in K_*, which improves the vibration isolating properties of the system at "resonance" despite the fact that the standard deviation of deflection increases.

Consider now the forced vibration caused by the same excitation in a system having a linear elastic characteristic and dry friction. In this case $c_d = c = const$ and eq. (3.283) yields

$$b_d = \frac{q}{\sigma_v} = \sqrt{\frac{2}{\pi}\frac{H}{\sigma_v}}, \tag{3.304}$$

where H denoted the dry friction value.

Inserting eq. (3.304) into (3.296) the following equation for σ_v is found

$$\sigma_v^2 = \frac{\sigma_Q^2 \cdot \eta\,(q + \beta m \sigma_v) \cdot \sigma_v^2}{q\left[\sigma_v^2\,(c - m\eta)^2 + (q + \beta m \sigma_v)\,(\beta c \sigma_v + q\eta)\right]}. \tag{3.305}$$

This equation is reduced to the following quadratic equation

$$\sigma_v^2\left[(c - m\eta)^2 + cm\beta^2\right] + \sigma_v\left(\frac{q^2 - \sigma_Q^2}{q}\beta m\eta + q\beta c\right) + \left(q^2 - \sigma_Q^2\right)\eta = 0.$$

Only one of the roots is positive and therefore is meaningful. Hence

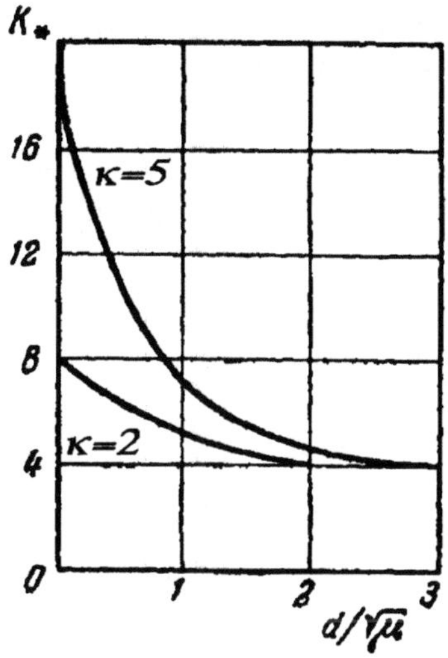

FIGURE 3.44.

$$\sigma_v = \frac{\left(\sigma_Q^2 - q^2\right)\beta m\eta - q^2\beta c}{2q\left[(c-m\eta)^2 + cm\beta^2\right]} + \frac{\sqrt{\left[\left(\sigma_Q^2 - q^2\right)\beta m\eta + q^2\beta c\right]^2 + 4q^2\eta\left(\sigma_Q^2 - q^2\right)(c-m\eta)^2}}{2q\left[(c-m\eta)^2 + cm\beta^2\right]}. \tag{3.306}$$

A positive value of σ_v exists if

$$\sigma_Q > q. \tag{3.307}$$

At resonance, i.e. if $c = m\eta$, the standard deviation is bounded in contrast to the case of deterministic harmonic excitations. Inserting $c = m\eta$ into eq. (3.306) gives

$$(\sigma_{v*})^2 = \frac{\sigma_Q^2 - q^2}{qm\beta}, \quad b_{d*} = \frac{q}{\sigma_{v*}} = \frac{q^2}{\sigma_Q^2 - q^2} m\beta. \tag{3.308}$$

Using these relations we find

$$(\sigma_{u*})^2 = \sigma_Q^2 \frac{(c+\beta b_d)\, m}{b_d c^2 (b_d + m\beta)} = \frac{\sigma_Q^2 - q^2}{q^2 c^2 \beta^2 m^2}\left[c\left(\sigma_Q^2 - q^2\right) + m\beta^2 q^2\right], \tag{3.309}$$

$$K_*^2 = 1 + \frac{cm}{b_d (b_d + m\beta)} = 1 + \frac{c}{m\beta^2}\left(\frac{\sigma_Q^2}{q^2} - \frac{q^2}{\sigma_Q^2}\right). \tag{3.310}$$

If parameter η (which is the square of the dominating frequency of the external excitation) is sufficiently large, so that $m\eta \gg c$, one can obtain from eq. (3.306)

$$\sigma_v \approx \frac{\sqrt{\left(\sigma_Q^2 - q^2\right) m^2\eta^2\beta^2 + 4m^2q^2\eta^3\left(\sigma_Q^2 - q^2\right)}}{2qm^2\eta^2} \approx \frac{\sqrt{\sigma_Q^2 - q^2}}{m\sqrt{\eta}},$$

with small terms being neglected. Hence

$$b_d \approx \frac{q\sqrt{\eta}m}{\sqrt{\sigma_Q^2 - q^2}}.$$

Inserting this value into eq. (3.295) and letting $\eta \to \infty$ we have

$$\lim_{\eta\to\infty} \sigma_U^2 = q^2.$$

This means that the standard deviation of the vibration isolator reaction is smaller by a factor of $\sqrt{\pi/2}$ than the dry friction force provided that the spectrum of vibration lies at high frequencies.

4

Nonlinear passive multi-degree-of-freedom systems

4.1 Static analysis of nonlinear elastic suspensions

We consider here the general case of an object which is suspended on elastic elements.

Here and throughout the rest of the present Chapter, except Section 4.7, the suspended object is assumed to be rigid body. This assumption is acceptable if the fundamental eigenfrequency of the object is much higher than the maximum frequency of the external excitation.

Let the object be attached to the base by means of N elastic isolators. To simplify the notation, it is advantageous to denote the components of the isolator deflection along the principal directions u_1, v_1, w_1, ..., u_N, v_N, w_N as u_1, u_2, ...u_{3N}. This is equivalent to replacing each isolator by three uniaxial elastic elements attached to the object at the same point, see Fig. 4.1. The axis of each uniaxial elastic element coincides with one of the principal directions of the isolator. The reaction of any uniaxial elastic element is assumed to depend only on the deflection and its time derivative. Therefore, the static and dynamic characteristics of elastic isolators may be cast as follows

$$U_{si} = U_{si}(u_{si}), \tag{4.1}$$

$$U_i = U_i(u_i, \dot{u}_i) \; (i = 1, ..., 3N). \tag{4.2}$$

The uniaxial elements are conditionally depicted as springs, see Fig. 4.1, however they cannot be identified with conventional elastic springs. In par-

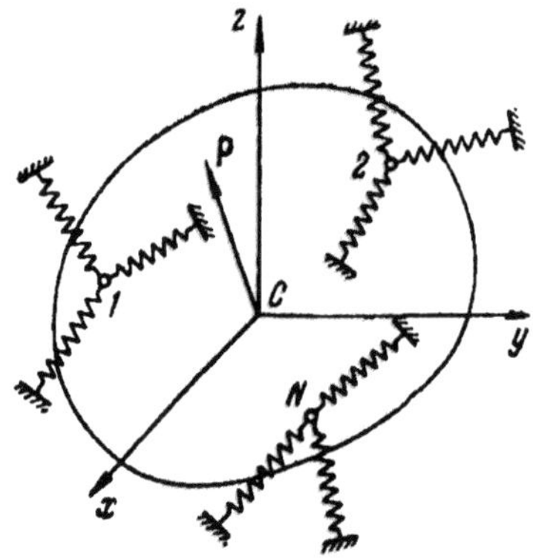

FIGURE 4.1.

ticular, one must bear in mind that the displacement of the attachment point in any direction orthogonal to the element axis does not affect the reaction which is often the case for elastic springs.

We proceed now to some aspects of the static analysis of the elastic suspension which involves determination of the static forces U_{si} and static deflections u_{si}. Let us place the origin of the coordinate system xyz at the center of inertia of the suspended object. Let α_i, β_i and γ_i denote cosines of the angles between the axis of $i-th$ element and the x, y and z directions, respectively. The static loadings acting on the object to be protected are ordinarily composed of the weight and the inertia force caused by the constant component of the base acceleration. The resultant of these forces P acts upon the center of inertia. Let x_i, y_i and z_i be the coordinates of the point of attachment of the $i-th$ element to the object. Resolving forces along and about the x, y and z axes gives the following static equations

$$\sum_{i=1}^{3N} U_{si}\alpha_i = P_x, \quad \sum_{i=1}^{3N} U_{si}\beta_i = P_y, \quad \sum_{i=1}^{3N} U_{si}\gamma_i = P_z, \tag{4.3}$$

$$\left.\begin{aligned} &\sum_{i=1}^{3N} U_{si}\left(y_i\gamma_i - z_i\beta_i\right) = 0, \quad \sum_{i=1}^{3N} U_{si}\left(z_i\alpha_i - x_i\gamma_i\right) = 0, \\ &\sum_{i=1}^{3N} U_{si}\left(x_i\beta_i - y_i\alpha_i\right) = 0. \end{aligned}\right\} \tag{4.4}$$

These 6 equations contain $3N$ unknown parameters which implies that the system is statically indeterminate if $N > 2$.

Let u_{si}^0 denote the deflections of the uniaxial elements in the unloaded state, i.e. when the static loads are absent. These deflections are conventionally termed as mounting deflections. After the static loads have been applied to the object, its displacement can be described in terms of the translation of the center of inertia $\boldsymbol{\delta}$ and the angle of rotation $\boldsymbol{\psi}$. In systems of vibration protection the isolators' displacements are much smaller than the distances between the points of attachment of the isolators which allows us to presume a small angle of rotation. In this case, the displacement of the $i-th$ attachment point is given by

$$\boldsymbol{\delta}_i = \boldsymbol{\delta} + \boldsymbol{\psi} \times \mathbf{r}_i, \tag{4.5}$$

where $\mathbf{r}_i$ denotes the radius vector of the attachment point. The deflection of the $i-th$ element is

$$u_{si} = u_{si}^0 + \boldsymbol{\delta}_i \cdot \mathbf{n}_i \ \ (i = 1, ..., 3N). \tag{4.6}$$

Here $\mathbf{n}_i$ is the unit vector along the axis of the $i-th$ element. Expanding expression (4.6) yields

$$\begin{aligned} u_{si} &= u_{si}^0 + \delta_{ix}\alpha_i + \delta_{iy}\beta_i + \delta_{iz}\gamma_i = \\ &= u_{si}^0 + (\delta_x + \psi_y z_i - \psi_z y_i)\,\alpha_i + (\delta_y + \psi_z x_i - \psi_x z_i)\,\beta_i + \\ &\quad (\delta_z + \psi_x y_i - \psi_y x_i)\,\gamma_i \ \ (i = 1, ..., 3N). \end{aligned} \tag{4.7}$$

Solving eqs. (4.1), (4.3), (4.4) and (4.7) gives $6N+6$ unknown parameters, namely, static forces U_{si}, static deflections u_{si} and 6 kinematic parameters δ_x, δ_y, δ_z, ψ_x, ψ_y and ψ_z defining the kinematics of the rigid body. Solving this system of equations is a challenging problem because the static characteristics (4.1) may be nonlinear.

Fortunately, there is no need to solve this problem. The explanation for this is that the elastic elements are usually chosen by means of the static analysis. This means that the characteristics of the elastic elements are unknown during the static design. One can choose an appropriate isolator only after the values of the static load in all principal directions have been calculated.

This circumstance simplifies the static design which can be carried out as follows:

1) One arbitrarily imposes $3N-6$ linear conditions on the values of the static forces U_{si}. Then all $3N$ forces can be determined from these conditions and static equations (4.3) and (4.4).

2) Using the forces obtained one chooses the elastic isolators. This means that their static characteristics, eq. (4.1), are defined.

3) By means of the static characteristics one finds the values of u_{si} corresponding to the previously determined forces.

4) There remains to determine the mounting deflections from eq. (4.7). In general, this part of the design is the most laborious because it is necessary to dispose the values of δ_x, δ_y, δ_z, ψ_x, ψ_y and ψ_z so that the static forces corresponding to the mounting deflections satisfy the static equation for

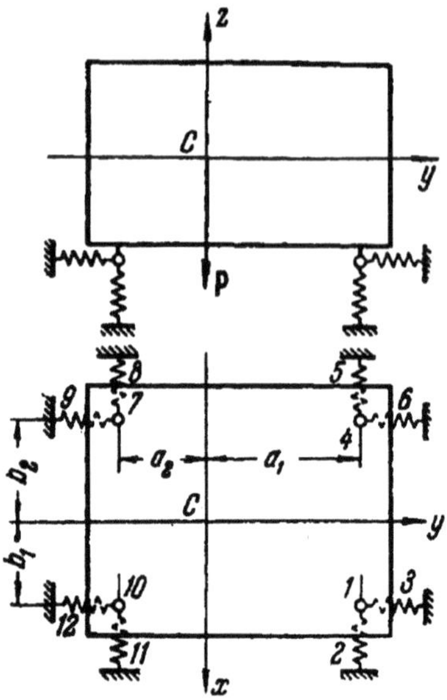

FIGURE 4.2.

$\mathbf{P} = 0$. It will be shown in what follows that some simple methods can be suggested for the majority of the important cases.

The calculated mounting deflections of the isolators must be ensured when assembling by shifting the points at which the isolator is attached to the base. The actual values of the static forces and deflections can only coincide with the computed ones when the mounting deflections are kept. Only this can guarantee the coincidence of the actual and design dynamic characteristics, the latter essentially depending upon the static deflections.

Consider some simple examples.

1. The object to be protected is mounted on four isolators which are equivalent to 12 uniaxial elastic elements, see Fig. 4.2. Force P is directed along axis z, i.e. $P_x = P_y = 0$ and $P_z = P$. The principal directions of the isolators are parallel to the coordinate axes, i.e.

$$\begin{aligned}
\alpha_1 &= \beta_1 = \alpha_4 = \beta_4 = \alpha_7 = \beta_7 = \alpha_{10} = \beta_{10} = \beta_2 = \gamma_2 = \beta_5 = \gamma_5 = \beta_8 \\
&= \gamma_8 = \beta_{11} = \gamma_{11} = \alpha_3 = \gamma_3 = \alpha_6 = \gamma_6 = \alpha_9 = \gamma_9 = \alpha_{12} = \gamma_{12} = 0 \\
\gamma_1 &= \alpha_2 = \beta_3 = \gamma_4 = \alpha_5 = \beta_6 = \gamma_7 = \alpha_8 = \beta_9 = \gamma_{10} = \alpha_{11} = \beta_{12} = 1
\end{aligned}$$

The following conditions

$$U_{s2} = U_{s3} = U_{s5} = U_{s6} = U_{s8} = U_{s9} = U_{s11} = U_{s12} = 0, \tag{4.8}$$

do not contradict the static equations, therefore they can be imposed on the static forces. Conditions (4.8) imply that the reactions of each isolator are parallel to the z axis.

Since the static loads and the static reactions of the isolators are parallel, this static loading is referred to as *unidirectional loading* and the number of the static equations reduces to three. These equations can be written as follows

$$\left.\begin{aligned} U_{s1} + U_{s4} + U_{s7} + U_{s10} &= P, \\ (U_{s1} + U_{s4})\, a_1 - (U_{s7} + U_{s10})\, a_2 &= 0, \\ (U_{s1} + U_{s10})\, b_1 - (U_{s4} + U_{s7})\, b_2 &= 0. \end{aligned}\right\} \tag{4.9}$$

As the number of unknowns exceeds the number of equations, the system is statically indeterminate.

Let us impose, for example, the following condition on forces U_{s1}, U_{s4}, U_{s7} and U_{s10}

$$U_{s1}a_1b_1 - U_{s4}a_1b_2 + U_{s7}a_2b_2 - U_{s10}a_2b_1 = 0. \tag{4.10}$$

Solving the system of linear equations (4.9) and (4.10) yields values of U_{s1}, U_{s4}, U_{s7} and U_{s10}. It can be proven that all these forces are positive under the above assumption. Next, we choose the elastic isolators so that the obtained forces lie in the bounds admissible for these isolators and determine u_{s1}, u_{s4}, u_{s7} and u_{s10}.

We proceed now to eq. (4.7). In the case of unidirectional loading the following condition must hold

$$\delta_x = \delta_y = \psi_x = \psi_y = \psi_z = 0, \tag{4.11}$$

otherwise eq. (4.8) is not satisfied. In other words, in equilibrium the object must be mounted without skewing and have the translation only in the direction of the z axis. For instance, the mounting deflections can be provided by means of shims or spacers.

Inserting eq. (4.11) into (4.7) yields

$$\left.\begin{aligned} &u^0_{s1} = u_{s1} - \delta_z, \quad u^0_{s4} = u_{s4} - \delta_z, \\ &u^0_{s7} = u_{s7} - \delta_z, \quad u^0_{s10} = u_{s10} - \delta_z, \\ &\delta_z = (u_{s1} + u_{s4} + u_{s7} + u_{s10})\,/4. \end{aligned}\right\} \tag{4.12}$$

Condition (4.10) can be replaced, for instance, by the following one

$$U_{s4} = \frac{P}{4},$$

which results in another values of the static forces. This means that another isolator might be required. The values of the static deflections satisfying condition (4.11) can be found in this case by analogy. The arbitrariness of the additional conditions allows the design requirements to be met.

2. Consider unidirectional loading of a system having N isolators, see Fig. 4.3. The static equations are given by

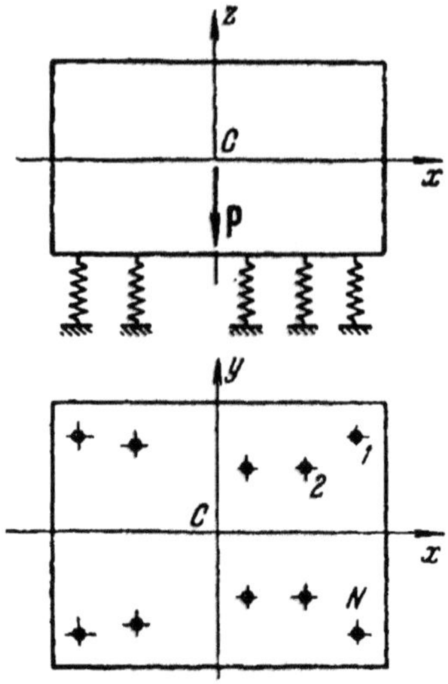

FIGURE 4.3.

$$\sum_{i=1}^{N} U_{si} = P, \sum_{i=1}^{N} U_{si} x_i = 0, \sum_{i=1}^{N} U_{si} y_i = 0. \tag{4.13}$$

Here U_{si} denotes the force in $i-th$ isolator in the principal direction which is parallel to the z axis. Let us take the additional condition in the following form

$$U_{s1} = \dots = U_{sN}. \tag{4.14}$$

In other words, we require the reactions of the isolators to be coincident. The number of conditions in eq. (4.14), $N-1$, exceeds the number of excessive unknown variables in eq. (4.13), i.e. these conditions can be satisfied only under certain relationships between the coordinates of the attachment points. Substituting eq. (4.14) into (4.13) we obtain

$$U_{s1} = \dots = U_{sN} = \frac{P}{N}, \sum_{i=1}^{N} x_i = 0, \sum_{i=1}^{N} y_i = 0. \tag{4.15}$$

Provided that the isolators are all identical, their static deflections coincide and no levelling is needed.

3. Consider the loading depicted in Fig. 4.4. The loading is not unidirectional as force P and the isolators' reactions parallel to this force cause a moment which in turn produces forces parallel to the y axis.

Considering the symmetry of the system about plane ycz enables the additional conditions to be chosen in the form

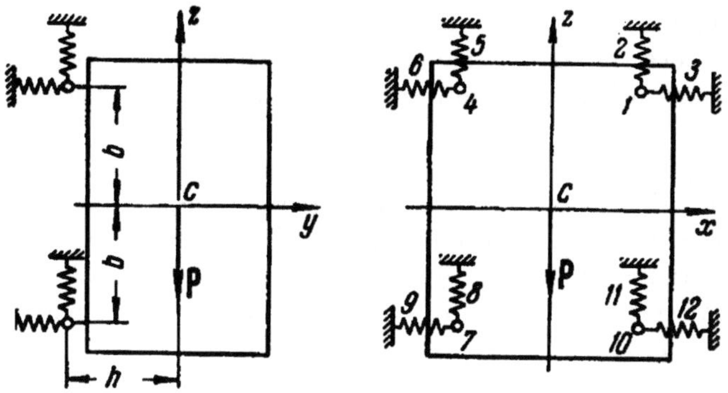

FIGURE 4.4.

$$\left.\begin{array}{c}U_{s3}=U_{s6}=U_{s9}=U_{s12}=0,\quad U_{s2}=U_{s5}=U_{s8}=U_{s11},\\ U_{s1}=U_{s4},\quad U_{s7}=U_{s10}.\end{array}\right\}\tag{4.16}$$

The static equations are partly satisfied identically, the rest yield

$$\left.\begin{array}{c}U_{s2}=U_{s5}=U_{s8}=U_{s11}=P/4,\\ U_{s1}=U_{s4}=-U_{s7}=-U_{s10}=Pb/(2h).\end{array}\right\}\tag{4.17}$$

Apparently, the isolators 1 and 4 work in tension whereas the isolators 7 and 10 are compressed. In order to get rid of scewness one needs additional shims for the isolators 7 and 10, their thickness being equal to the difference of the static deflections of the isolators 1 and 7, i.e.

$$u_{s7}^{0}=u_{s10}^{0}=u_{s1}-u_{s7}.\tag{4.18}$$

4.2 Small vibration of elastically suspended rigid body

Let us consider small vibrations of the object to be protected. We assume that the dynamic characteristics of the elastic elements can be linearised, to give

$$U_i=c_iu_i+b_i\dot{u}_i\ \ (i=1,...,3N),\tag{4.19}$$

where u_i are the deflections of the elastic elements relative to the equilibrium while c_i and b_i denote the stiffness and the damping factor of the $i-th$ element, respectively. The above assumption leads to a system of linear differential equations of motion with constant factors. These equations can be analysed using classical vibration theory. Here we restrict ourselves to a short description of some of the basic concepts of linear theory, the entire theory being adequately covered in many books, e.g. [3], [32], [81] etc.

The origin of the reference system at rest is assumed to coincide with the center of inertia of the object in its equilibrium position. The principal central axes of inertia of the object x, y and z are chosen to be the moving reference frame, the axes ξ, η and ζ of the reference system at rest are assumed to coincide with the x, y and z axes of the moving reference frame at equilibrium.

The system under consideration possesses six degrees of freedom. Let us consider the coordinates describing the displacement of the object relative to the base as the generalised coordinates. The relative and absolute displacements of the object are coincident for the cases in which the base must be protected from vibrating object. In other cases in which the object must be protected from the vibrating base one can use the kinematic equations of motion relative to moving reference frame to describe the effect of the vibrating base on the object. This can be done by introducing the forces of inertia in the relative motion and the inertia forces due to the Coriolis acceleration. The Coriolis acceleration is proportional to the product of small relative velocity and small relative angular velocity, i.e. it has the second order of smallness, and therefore can be neglected in the case of small vibrations.

The displacement of the protected object relative to the equilibrium position is thus described in terms of the deflection vector $\mathbf{s}$ and the vector of small angle of rotation $\boldsymbol{\theta}$.

The kinetic energy of the relative motion of the object is as follows

$$T = \frac{1}{2} m \left(\dot{s}_x^2 + \dot{s}_y^2 + \dot{s}_z^2 + \rho_x^2 \dot{\theta}_x^2 + \rho_y^2 \dot{\theta}_y^2 + \rho_z^2 \dot{\theta}_z^2 \right), \tag{4.20}$$

where m is the mass of the object, and ρ_x, ρ_y and ρ_z denote the radii of inertia about axes x, y and z, respectively.

Let us proceed to the potential energy. To this end, we find the displacement of the point of attachment of the $i-th$ uniaxial element

$$\mathbf{s}_i = \mathbf{s} + \boldsymbol{\theta} \times \mathbf{r}_i. \tag{4.21}$$

The components of this vector equation are given by

$$\left.\begin{aligned} s_{ix} &= s_x + \theta_y z_i - \theta_z y_i, \\ s_{iy} &= s_y + \theta_z x_i - \theta_x z_i, \\ s_{iz} &= s_z + \theta_x y_i - \theta_y x_i. \end{aligned}\right\} \tag{4.22}$$

The deflection of the $i-th$ uniaxial element is as follows

$$\begin{aligned} u_i &= s_{ix}\alpha_i + s_{iy}\beta_i + s_{iz}\gamma_i = s_x\alpha_i + s_y\beta_i + s_z\gamma_i + \\ &+\theta_x\left(y_i\gamma_i - z_i\beta_i\right) + \theta_y\left(z_i\alpha_i - x_i\gamma_i\right) + \theta_z\left(x_i\beta_i - y_i\alpha_i\right) \\ &(i = 1, ..., 3N). \end{aligned} \tag{4.23}$$

Inserting eq. (4.23) into the following expression for the potential energy

$$\Pi = \frac{1}{2}\sum_{i=1}^{3N} c_i u_i^2, \tag{4.24}$$

yields

$$\begin{aligned} \Pi &= \frac{1}{2}(c_x s_x^2 + c_y s_y^2 + c_z s_z^2 + 2c_{xy}s_x s_y + 2c_{yz}s_y s_z + 2c_{xz}s_x s_z \\ &+2l_{xx}s_x\theta_x + 2l_{xy}s_x\theta_y + 2l_{xz}s_x\theta_z + 2l_{yx}s_y\theta_x + 2l_{yy}s_y\theta_y \\ &+2l_{yz}s_y\theta_z + 2l_{zx}s_z\theta_x + 2l_{zy}s_z\theta_y + 2l_{zz}s_z\theta_z + M_{xx}\theta_x^2 \\ &+M_{yy}\theta_y^2 + M_{zz}\theta_z^2 + 2M_{xy}\theta_x\theta_y + 2M_{yz}\theta_y\theta_z + 2M_{zx}\theta_x\theta_z). \end{aligned}$$

The following notation is introduced here

$$\left.\begin{aligned} c_x = \sum_{i=1}^{3N} c_i\alpha_i^2,\ c_y = \sum_{i=1}^{3N} c_i\beta_i^2,\ c_z = \sum_{i=1}^{3N} c_i\gamma_i^2, \\ c_{xy} = \sum_{i=1}^{3N} c_i\alpha_i\beta_i,\ c_{yz} = \sum_{i=1}^{3N} c_i\beta_i\gamma_i,\ c_{xz} = \sum_{i=1}^{3N} c_i\alpha_i\gamma_i, \end{aligned}\right\} \tag{4.25}$$

$$\left.\begin{aligned} l_{xx} &= \sum_{i=1}^{3N} c_i\alpha_i\left(y_i\gamma_i - z_i\beta_i\right),\quad l_{xy} = \sum_{i=1}^{3N} c_i\alpha_i\left(z_i\alpha_i - x_i\gamma_i\right), \\ l_{xz} &= \sum_{i=1}^{3N} c_i\alpha_i\left(x_i\beta_i - y_i\alpha_i\right); \\ l_{yx} &= \sum_{i=1}^{3N} c_i\beta_i\left(y_i\gamma_i - z_i\beta_i\right),\quad l_{yy} = \sum_{i=1}^{3N} c_i\beta_i\left(z_i\alpha_i - x_i\gamma_i\right), \\ l_{yz} &= \sum_{i=1}^{3N} c_i\beta_i\left(x_i\beta_i - y_i\alpha_i\right); \\ l_{zx} &= \sum_{i=1}^{3N} c_i\gamma_i\left(y_i\gamma_i - z_i\beta_i\right),\quad l_{zy} = \sum_{i=1}^{3N} c_i\gamma_i\left(z_i\alpha_i - x_i\gamma_i\right), \\ l_{zz} &= \sum_{i=1}^{3N} c_i\gamma_i\left(x_i\beta_i - y_i\alpha_i\right); \end{aligned}\right\} \tag{4.26}$$

$$\left.\begin{aligned}
M_{xx} &= \sum_{i=1}^{3N} c_i \left(y_i\gamma_i - z_i\beta_i\right)^2, \quad M_{yy} = \sum_{i=1}^{3N} c_i \left(z_i\alpha_i - x_i\gamma_i\right)^2, \\
M_{zz} &= \sum_{i=1}^{3N} c_i \left(x_i\beta_i - y_i\alpha_i\right)^2; \\
M_{xy} &= \sum_{i=1}^{3N} c_i \left(y_i\gamma_i - z_i\beta_i\right)\left(z_i\alpha_i - x_i\gamma_i\right), \\
M_{yz} &= \sum_{i=1}^{3N} c_i \left(z_i\alpha_i - x_i\gamma_i\right)\left(x_i\beta_i - y_i\alpha_i\right), \\
M_{xz} &= \sum_{i=1}^{3N} c_i \left(x_i\beta_i - y_i\alpha_i\right)\left(y_i\gamma_i - z_i\beta_i\right).
\end{aligned}\right\} \tag{4.27}$$

The Rayleigh dissipative function is known to be given by

$$\Phi = \frac{1}{2}\sum_{i=1}^{3N} b_i \dot{u}_i^2, \tag{4.28}$$

see [81]. Obtaining $\dot{u}_i$ from eq. (4.23) and substituting the result into eq. (4.28) yields

$$\begin{aligned}
\Phi \;=\; & \frac{1}{2}(b_x\dot{s}_x^2 + b_y\dot{s}_y^2 + b_z\dot{s}_z^2 + 2b_{xy}\dot{s}_x\dot{s}_y + 2b_{yz}\dot{s}_y\dot{s}_z + 2b_{xz}\dot{s}_x\dot{s}_z \\
& +2e_{xx}\dot{s}_x\dot{\theta}_x + 2e_{xy}\dot{s}_x\dot{\theta}_y + 2e_{xz}\dot{s}_x\dot{\theta}_z + 2e_{yx}\dot{s}_y\dot{\theta}_x + 2e_{yy}\dot{s}_y\dot{\theta}_y \\
& +2e_{yz}\dot{s}_y\dot{\theta}_z + 2e_{zx}\dot{s}_z\dot{\theta}_x + 2e_{zy}\dot{s}_z\dot{\theta}_y + 2e_{zz}\dot{s}_z\dot{\theta}_z + G_{xx}\dot{\theta}_x^2 \\
& +G_{yy}\dot{\theta}_y^2 + G_{zz}\dot{\theta}_z^2 + 2G_{xy}\dot{\theta}_x\dot{\theta}_y + 2G_{yz}\dot{\theta}_y\dot{\theta}_z + 2G_{xz}\dot{\theta}_x\dot{\theta}_z).
\end{aligned} \tag{4.29}$$

Equations for b_x, b_y,...b_{xz}; e_{xx},...e_{zz}; G_{xx}...G_{xz} are obtained from eqs. (4.25), (4.26) and (4.27), respectively, by replacing c_i with b_i.

For the sake of compactness we introduce vector and matrix notation. The vector of the generalised coordinates $\mathbf{q}$, which is a column with the following six entries

$$q_1 = s_x,\; q_2 = s_y,\; q_3 = s_z,\; q_4 = \theta_x,\; q_5 = \theta_y,\; q_6 = \theta_z, \tag{4.30}$$

determines the position of the object. Introducing the mass matrix

$$A = \begin{Vmatrix} m & 0 & 0 & 0 & 0 & 0 \\ 0 & m & 0 & 0 & 0 & 0 \\ 0 & 0 & m & 0 & 0 & 0 \\ 0 & 0 & 0 & m\rho_x^2 & 0 & 0 \\ 0 & 0 & 0 & 0 & m\rho_y^2 & 0 \\ 0 & 0 & 0 & 0 & 0 & m\rho_z^2 \end{Vmatrix}, \tag{4.31}$$

the stiffness matrix

$$C = \begin{Vmatrix} c_x & c_{xy} & c_{xz} & l_{xx} & l_{xy} & l_{xz} \\ c_{xy} & c_y & c_{yz} & l_{yx} & l_{yy} & l_{yz} \\ c_{xz} & c_{yz} & c_{zz} & l_{zx} & l_{zy} & l_{zz} \\ l_{xx} & l_{yx} & l_{zx} & M_{xx} & M_{xy} & M_{xz} \\ l_{xy} & l_{yy} & l_{zy} & M_{xy} & M_{yy} & M_{yz} \\ l_{xz} & l_{yz} & l_{zz} & M_{xz} & M_{yz} & M_{zz} \end{Vmatrix} \tag{4.32}$$

and the damping matrix

$$B = \begin{Vmatrix} b_x & b_{xy} & b_{xz} & e_{xx} & e_{xy} & e_{xz} \\ b_{xy} & b_y & b_{yz} & e_{yx} & e_{yy} & e_{yz} \\ b_{xz} & b_{yz} & b_{zz} & e_{zx} & e_{zy} & e_{zz} \\ e_{xx} & e_{yx} & e_{zx} & G_{xx} & G_{xy} & G_{xz} \\ e_{xy} & e_{yy} & e_{zy} & G_{xy} & G_{yy} & G_{yz} \\ e_{xz} & e_{yz} & e_{zz} & G_{xz} & G_{yz} & G_{zz} \end{Vmatrix}, \tag{4.33}$$

we can represent the kinetic T and potential Π energy, and the Rayleigh function Φ as quadratic forms

$$T = \frac{1}{2}\dot{\mathbf{q}}^T A\dot{\mathbf{q}}, \tag{4.34}$$

$$\Pi = \frac{1}{2}\mathbf{q}^T C\mathbf{q}, \tag{4.35}$$

$$\Phi = \frac{1}{2}\dot{\mathbf{q}}^T B\dot{\mathbf{q}}, \tag{4.36}$$

where T denotes the transpose. Small vibration of the system is governed by Lagrange's equations

$$\frac{d}{dt}\frac{\partial T}{\partial \dot{\mathbf{q}}} - \frac{\partial T}{\partial \mathbf{q}} = -\frac{\partial \Pi}{\partial \mathbf{q}} - \frac{\partial \Phi}{\partial \dot{\mathbf{q}}} + \mathbf{Q}, \tag{4.37}$$

where $\mathbf{Q}$ denotes the vector of external forces acting on the object. In the case in which the base must be protected from the vibrating object the components of vector $\mathbf{Q}$ are the projections of the resultant vector and the resultant moment, i.e.

$$\left.\begin{aligned} &Q_1 = \sum_{k=1}^{h} F_{kx}, \quad Q_2 = \sum_{k=1}^{h} F_{ky}, \quad Q_3 = \sum_{k=1}^{h} F_{kz}, \\ &Q_4 = \sum_{k=1}^{h} (F_{kz}Y_k - F_{ky}Z_k), \quad Q_5 = \sum_{k=1}^{h} (F_{kx}Z_k - F_{kz}X_k), \\ &Q_6 = \sum_{k=1}^{h} (F_{ky}X_k - F_{kx}Y_k). \end{aligned}\right\} \tag{4.38}$$

Here F_{kx}, F_{ky} and F_{kz} are components of the resultant vectors along the corresponding coordinate axes while X_k, Y_k and Z_k are the coordinates of the point which this force is applied to. In the case in which the object must be protected from the vibrating base, $\mathbf{Q}$ implies the vector of the force of inertia in the relative motion. Assuming the base to be a rigid body, its displacement can be expressed in terms of the vector of displacement $\boldsymbol{\xi}$ of the coincident point (a point coinciding with the center of inertia of the object in the equilibrium position) and the vector of small angles of rotation $\boldsymbol{\beta}$. Then,

$$\begin{aligned} Q_1 &= -m\ddot{\xi}_x,\ Q_2 = -m\ddot{\xi}_y,\ Q_3 = -m\ddot{\xi}_z, \\ Q_4 &= -m\rho_x^2\ddot{\beta}_x,\ Q_5 = -m\rho_y^2\ddot{\beta}_y,\ Q_6 = -m\rho_z^2\ddot{\beta}_z. \end{aligned} \tag{4.39}$$

Substituting eqs. (4.34)-(4.36) into (4.37) and taking into account that

$$\frac{\partial T}{\partial \dot{\mathbf{q}}} = A\dot{\mathbf{q}},\ \frac{\partial \Pi}{\partial \mathbf{q}} = C\mathbf{q},\ \frac{\partial \Phi}{\partial \dot{\mathbf{q}}} = B\dot{\mathbf{q}},\ \frac{\partial T}{\partial \mathbf{q}} = 0, \tag{4.40}$$

yields

$$A\ddot{\mathbf{q}} + B\dot{\mathbf{q}} + C\mathbf{q} = \mathbf{Q}(t). \tag{4.41}$$

This equation is equivalent to six scalar inhomogeneous linear differential equations with constant coefficients.

We consider first the equation for free vibration. Assuming

$$B = 0,\ \mathbf{Q} = 0,$$

we obtain

$$A\ddot{\mathbf{q}} + C\mathbf{q} = 0. \tag{4.42}$$

In accordance with the general theory of linear differential equations with constant coefficients, a particular solution is sought in the form

$$\mathbf{q} = \mathbf{q}_0 \cos \lambda t. \tag{4.43}$$

This solution implies that all the system coordinates obey the harmonic law with the same frequency and phase angle.

Substitution of eq. (4.43) into (4.42) leads to the following equation

$$\left(C - \lambda^2 A\right)\mathbf{q}_0 = 0, \tag{4.44}$$

which is equivalent to the system of six linear homogeneous equations. For this equation to have a nontrivial solution, it is necessary that the determinant vanishes

$$\Delta\left(\lambda^2\right) = \left|C - \lambda^2 A\right| = 0. \tag{4.45}$$

The latter equation is said to be the frequency equation of the system. By means of expressions for matrices A and C this equation can be recast in the following form

$$\Delta\left(\lambda^2\right) = \tag{4.46}$$

$$\begin{Vmatrix} c_x - m\lambda^2 & c_{xy} & c_{xz} & l_{xx} & l_{xy} & l_{xz} \\ c_{xy} & c_y - m\lambda^2 & c_{yz} & l_{yx} & l_{yy} & l_{yz} \\ c_{xz} & c_{yz} & c_{zz} - m\lambda^2 & l_{zx} & l_{zy} & l_{zz} \\ l_{xx} & l_{yx} & l_{zx} & M_{xx} - m\rho_x^2\lambda^2 & M_{xy} & M_{xz} \\ l_{xy} & l_{yy} & l_{zy} & M_{xy} & M_{yy} - m\rho_y^2\lambda^2 & M_{yz} \\ l_{xz} & l_{yz} & l_{zz} & M_{xz} & M_{yz} & M_{zz} - m\rho_z^2\lambda^2 \end{Vmatrix} = 0.$$

This is an equation of the sixth order with respect to λ^2. As known from the theory of vibration, e.g. [60], this has six positive roots, provided that the system vibrates about the stable equilibrium position. The solutions of eq. (4.46), λ_1, $\lambda_2, ..., \lambda_6$, are called the eigenfrequencies of the system. There exist numerous methods of solving the frequency equation, however they are not addressed here as they are adequately covered in the literature.

If one substitutes one of the eigenfrequencies into eq. (4.44), the resulting equation will have an infinite number of solutions. It is shown, e.g. [60], that in the case of no multiple eigenfrequencies these solutions differ only by an arbitrary scaling factor $a^{(s)}$, i.e.

$$\mathbf{q}_0^{(s)} = a^{(s)}\mathbf{q}_*^{(s)}. \tag{4.47}$$

Each of the system coordinate is seen to obey a harmonic law. These free vibrations are referred to as the principal vibrations. Expression (4.47) points out that the relationships between the coordinate amplitudes are independent of the initial conditions and dependent on the system parameters only. Vector $\mathbf{q}_*^{(s)}$ whose components are ratios of the vibration amplitudes to a value chosen at will is termed the $s - th$ normal mode or mode shape.

Consider now forced vibrations in the undamped system. As the superposition principle is valid for linear systems it is sufficient to consider the case in which all components of vector $\mathbf{Q}(t)$ are harmonic functions of time with the same frequency and phase angle

$$\mathbf{Q}(t) = \mathbf{Q}_0 \cos\omega t, \tag{4.48}$$

where $\mathbf{Q}_0$ is a constant vector.

The solution of the differential equation for the forced vibration

$$A\ddot{\mathbf{q}} + C\mathbf{q} = \mathbf{Q}_0 \cos\omega t, \tag{4.49}$$

is known to consist of the general solution of the homogeneous differential equation (4.42) which describes free vibration in the system and a particular solution of eq. (4.49). Free vibration in the actual systems attenuates because of the dissipative forces, that is, the particular solution describing steady-state vibrations in the systems is of crucial interest. This particular solution is sought in the form

$$\mathbf{q} = \mathbf{a}\cos\omega t, \tag{4.50}$$

where $\mathbf{a}$ is a vector whose components are the amplitudes of vibration of each generalised coordinate.

Substituting eq. (4.50) into eq. (4.49) we obtain the following equation for vector $\mathbf{a}$

$$\left(C - A\omega^2\right)\mathbf{a} = \mathbf{Q}_0, \tag{4.51}$$

which is equivalent to a system of six scalar equations.

The solution of eq. (4.51) can be written in the following form

$$\mathbf{a} = \left(C - A\omega^2\right)^{-1}\mathbf{Q}_0, \tag{4.52}$$

which means that one must invert matrix $\left(C - A\omega^2\right)$. This is possible to do only when the determinant is non-zero

$$\Delta\left(\omega^2\right) = \left|C - A\omega^2\right| \neq 0,$$

i.e. only if $\omega \neq \lambda_s$, $(s = 1, ...6)$.

When the frequency of the external excitation ω coincides with one of the eigenfrequencies of the system, eq. (4.51) has no solution. This case corresponds to *resonance* in the linear system. If ω is close to one of the eigenfrequencies the vibration amplitudes are large. Indeed, the solution of the system of linear algebraic equations corresponding to the vector equation (4.51) is as follows

$$a_k = \frac{\Delta_k\left(\omega^2\right)}{\Delta\left(\omega^2\right)}. \tag{4.53}$$

Here $\Delta\left(\omega^2\right)$ is the determinant given by eq. (4.46) in which λ^2 is replaced by ω^2. Then, $\Delta_k\left(\omega^2\right)$ is obtained by replacement of the $k-th$ column of the determinant $\Delta\left(\omega^2\right)$ by the components of vector $\mathbf{Q}_0$. As $\Delta\left(\lambda_k^2\right) = 0$, a_k increases without bounds when $\omega^2 - \lambda_k^2$ tends to zero.

Hence, for the linear system with six degrees of freedom the frequency band of resonant vibration covers the entire bandwidth of the spectrum of the eigenfrequencies. In order to narrow this frequency band of resonant vibration one can try to bring together the minimum and maximum eigenfrequencies by properly choosing the coordinates of the attachment points and the stiffnesses of the elastic elements.

To study forced vibration in the linear system it is convenient to express the motion of the system in terms of a different set of coordinates, so called principal coordinates. The principal coordinates for a multi-degree-of-freedom system are a set of coordinates related to the generalised coordinates by the following linear transformation

$$\mathbf{q} = R\mathbf{f}, \tag{4.54}$$

where $\mathbf{f}$ is the vector of principal coordinates and R is the matrix whose columns are the vectors $\mathbf{q}_0^{(1)}$, ..., $\mathbf{q}_0^{(6)}$. The transformation is so chosen as to render the vibration problem in terms of the new coordinates easier. In this case, the expressions for kinetic and potential energy contain only quadratic terms in velocities and displacements

$$T = \frac{1}{2}\sum_{s=1}^{6} \mu_s \dot{f}_s^2, \quad \Pi = \frac{1}{2}\sum_{s=1}^{6} \pi_s f_s^2, \tag{4.55}$$

where

$$\mu_s = \left(\mathbf{q}_0^{(s)}\right)^T A\mathbf{q}_0^{(s)}, \quad \pi_s = \left(\mathbf{q}_0^{(s)}\right)^T C\mathbf{q}_0^{(s)} = \mu_s\lambda_s^2.$$

Thus, when the principal coordinates are used as dependent variables, the governing equations are uncoupled

$$\ddot{f}_s + \lambda_s^2 f_s = \frac{1}{\mu_s}\vartheta_s \cos\omega t. \tag{4.56}$$

Here ϑ_s is the component of vector $\boldsymbol{\vartheta}$ given by

$$\boldsymbol{\vartheta} = R^T\mathbf{Q}_0,$$

where R^T denotes the transpose of matrix R. Solving eq. (4.56) we obtain that the dynamic magnification factor in the linear undamped system corresponding to the $s-th$ principal coordinate is

$$K_s = \frac{\lambda_s^2}{\left|\lambda_s^2 - \omega^2\right|}. \tag{4.57}$$

Consequently $K_s < 1$ if $\omega > \lambda_s\sqrt{2}$. Provided that $\omega > \lambda_{s\,\max}\sqrt{2}$, all dynamic magnification factors are less than unity which means that the system of vibration protection is effective.

Consideration of damping complicates the investigation since the principal coordinates do not lead to an uncoupled system of equations in this case. The following can be noted about the influence of damping. Introducing damping has the effect observed in the previous Chapter wherein the single-degree-of-freedom systems are studied. Damping reduces the maximum amplitudes in the resonance domain however it worsens the performance of vibration protection system at high frequencies, i.e. when $\omega > \lambda_{s\ \max}\sqrt{2}$.

4.3 Vibration of an object mounted on nonlinear elastic isolators

To begin with, we derive the differential equations governing vibration of a rigid body suspended on nonlinear elastic isolators having dynamic characteristics (4.2).

We choose the generalised coordinates of Section 4.2 and make use of Lagrange's equations

$$\frac{d}{dt}\left(\frac{\partial T}{\partial \dot{q}_r}\right) - \frac{\partial T}{\partial q_r} = S_r \ \ (r = 1, ..., 6). \tag{4.58}$$

Expression (4.20) for kinetic energy does not change. By means of eq. (4.30) it can be written as follows

$$T = \frac{1}{2} m \left(\dot{q}_1^2 + \dot{q}_2^2 + \dot{q}_3^2 + \rho_x^2 \dot{q}_4^2 + \rho_y^2 \dot{q}_5^2 + \rho_z^2 \dot{q}_6^2\right). \tag{4.59}$$

To obtain the generalised forces S_r we construct an expression for the virtual work of all the forces acting on the object, namely, the reaction forces in the elastic elements $-U_i(u_i, \dot{u}_i)$ and the external forces Q_r. The result is

$$\delta W = \sum_{r=1}^{6} S_r \delta q_r = -\sum_{i=1}^{3N} U_i (u_i, \dot{u}_i)\, \delta u_i + \sum_{r=1}^{6} Q_r \delta q_r. \tag{4.60}$$

As the angles of rotation are assumed to be small, the deflections u_i can be expressed in terms of the generalised coordinates by means of eq. (4.23). Obtaining δu_i from this equation and substituting it into (4.60) results in the following expression for the virtual work

$$\begin{aligned} \delta W \ = \ & -\sum_{i=1}^{3N} U_i \left[u_i (q_1, ..., q_6), \dot{u}_i (\dot{q}_1, ..., \dot{q}_6)\right] \times \\ & \times \left[\alpha_i \delta q_1 + \beta_i \delta q_2 + \gamma_i \delta q_3 + (y_i \gamma_i - z_i \beta_i)\, \delta q_4 + \right. \\ & \left. + (z_i \alpha_i - x_i \gamma_i)\, \delta q_5 + (x_i \beta_i - y_i \alpha_i)\, \delta q_6\right] + \sum_{r=1}^{6} Q_r \delta q_r. \end{aligned} \tag{4.61}$$

This yields

$$\left.\begin{aligned}
S_1 &= Q_1 - \sum_{i=1}^{3N} U_i \left[u_i\left(q_1, ..., q_6\right), \dot{u}_i\left(\dot{q}_1, ..., \dot{q}_6\right)\right] \alpha_i, \\
S_2 &= Q_2 - \sum_{i=1}^{3N} U_i \left[u_i\left(q_1, ..., q_6\right), \dot{u}_i\left(\dot{q}_1, ..., \dot{q}_6\right)\right] \beta_i, \\
S_3 &= Q_3 - \sum_{i=1}^{3N} U_i \left[u_i\left(q_1, ..., q_6\right), \dot{u}_i\left(\dot{q}_1, ..., \dot{q}_6\right)\right] \gamma_i, \\
S_4 &= Q_4 - \sum_{i=1}^{3N} U_i \left[u_i\left(q_1, ..., q_6\right), \dot{u}_i\left(\dot{q}_1, ..., \dot{q}_6\right)\right] \left(y_i\gamma_i - z_i\beta_i\right), \\
S_5 &= Q_5 - \sum_{i=1}^{3N} U_i \left[u_i\left(q_1, ..., q_6\right), \dot{u}_i\left(\dot{q}_1, ..., \dot{q}_6\right)\right] \left(z_i\alpha_i - x_i\gamma_i\right), \\
S_6 &= Q_6 - \sum_{i=1}^{3N} U_i \left[u_i\left(q_1, ..., q_6\right), \dot{u}_i\left(\dot{q}_1, ..., \dot{q}_6\right)\right] \left(x_i\beta_i - y_i\alpha_i\right).
\end{aligned}\right\} \quad (4.62)$$

General analysis of the equations of motion which are obtained by substituting eqs. (4.59) and (4.62) into eq. (4.58) is usually not possible. This can be explained by the fact that functions U_i have very complicated nonlinear dependence on all generalised coordinates and their time derivatives. For this reason we will use another, more simple form of the differential equations of motion.

We choose deflections u_i as a new set of generalised coordinates. In the general case the number of generalised coordinates exceeds the number of degrees of freedom of the system. Thus, the standard form of Lagrange's equations is no longer applicable and we are forced to use a special form of Lagrange's equations, the so called Lagrange's equations with "redundant" coordinates, also called Ferrers's equations, [29]. The old set of the generalised coordinates can be expressed in terms of deflections of any six elastic elements by means of eq. (4.23). Let these six elements be given the ordinal numbers from 1 to 6. For the first six equations in (4.23) to be solved for coordinates $q_1, ..., q_6$ it is necessary and sufficient that the determinant composed of the factors on the right hand side of eqs. (4.23) is not zero. It is easy to show that this is equivalent to requiring that the object is fixed by means of the chosen elements. In other words, it is admissible that the object suspended on the elements 1,..., 6 can move without deforming elastic elements. Indeed, if the determinant of the first six relationships in (4.23) vanishes, then this system of linear algebraic equations will have non-zero solutions q_1,..., q_6 for $u_i = 0$ $(i = 1, ..., 6)$.

Provided that the condition of fixation is satisfied, then coordinates q_1,..., q_6 are uniquely expressed in terms of u_1,..., u_6. The corresponding relations are linear

$$q_r = \sum_{k=1}^{6} \sigma_{rk} u_k \ \ (r = 1, ..., 6). \tag{4.63}$$

Inserting these expressions into the remaining $3N - 6$ relationships (4.23) we can express $u_7,..., u_{3N}$ in terms of $u_1,..., u_6$

$$u_i = \sum_{k=1}^{6} \nu_{ik} u_k \ \ (i = 7, ..., 3N). \tag{4.64}$$

It is convenient to rewrite the latter equation in another form

$$\varphi_e (u_1, ..., u_6, u_l) = u_l - \sum_{k=1}^{6} \nu_{lk} u_k = 0 \ \ (i = 7, ..., 3N). \tag{4.65}$$

Equation (4.63) yields

$$\dot{q}_r = \sum_{k=1}^{6} \sigma_{rk} \dot{u}_k \ \ (r = 1, ..., 6). \tag{4.66}$$

Substituting this into eq. (4.59) gives

$$T = \frac{1}{2} \sum_{k,r=1}^{6} \mu_{kr} \dot{u}_k \dot{u}_r. \tag{4.67}$$

By introducing the notation

$$m_1 = m_2 = m_3 = m, \ m_4 = \rho_x^2 m, \ m_5 = \rho_y^2 m, \ m_6 = \rho_z^2 m \tag{4.68}$$

we can recast expressions for μ_{kr} in the symmetric form

$$\mu_{kr} = \sum_{l=1}^{6} m_l \sigma_{lk} \sigma_{lr} = \mu_{rk}. \tag{4.69}$$

Consider expression (4.60) for the virtual work. The first sum does not change its appearance, the second sum contains some prescribed functions of time, while by virtue of eq. (4.63) variations δ_{qr} are given by

$$\delta_{qr} = \sum_{k=1}^{6} \sigma_{rk} \delta u_k. \tag{4.70}$$

Therefore, the generalised force U_i^* corresponding to coordinate u_i is defined as follows

$$U_i^* = -U_i (u_i, \dot{u}_i) + \sum_{r=1}^{6} Q_r \sigma_r \ \ (i = 1, ..., 6), \tag{4.71}$$

$$U_i^* = -U_i\left(u_i, \dot{u}_i\right) \;\; (i = 7, ..., 3N). \tag{4.72}$$

Lagrange's equations with "redundant" coordinates are written in the following form

$$\frac{d}{dt}\frac{\partial T}{\partial \dot{u}_i} - \frac{\partial T}{\partial u_i} = U_i^* + \sum_{l=7}^{3N} \Lambda_l \frac{\partial \varphi_l}{\partial u_i} \;\; (i = 1, ..., 3N), \tag{4.73}$$

where Λ_l $(i = 7, ..., 3N)$ are new unknown functions of time which are known as Lagrange multipliers. Using eqs. (4.67), (4.65), (4.71) and (4.72) we obtain

$$\sum_{k=1}^{6} \mu_{ki}\ddot{u}_k + U_i\left(u_i, \dot{u}_i\right) + \sum_{l=7}^{3N} \Lambda_l \nu_{li} = \sum_{r=1}^{6} Q_r \sigma_{ri} \;\; (i = 1, ..., 6) \tag{4.74}$$

$$U_i\left(u_i, \dot{u}_i\right) - \Lambda_i = 0 \;\; (i = 7, ..., 3N). \tag{4.75}$$

Thus, there are $6N-6$ equations, eqs. (4.74), (4.75) and (4.65), and $6N-6$ unknown functions of time, u_1, ..., u_{3N}, Λ_7, ..., Λ_{3N}. Now the number of unknowns and the number of equations is larger than that in the original equation (4.58). However, the nonlinear functions in the new equations have quite a simple form, namely, each function is dependent only on one coordinate and its time derivative. This enables us to linearise these functions by one of the methods described in Section 3.1.

In practice, the number of differential equations can be reduced. First of all this is possible in the cases in which some constraints reducing the number of degrees of freedom are imposed on the motion of the object to be protected. For example, the object in a plane motion has three degrees of freedom while the object rotating about an axis possesses a single degree of freedom etc. Often one can assert *a priori* that only a part of the coordinates is active under a certain excitation. This is usually the case in which the vibration protection system is symmetric about some of the coordinate planes and the vibration excitation is directed along a coordinate axis, see Table 4.1. In all these cases the object is conditionally termed as being *partially isolated.*

Let p and q_1, ..., q_p be the number of degrees of freedom of the partially isolated object and its generalised coordinates, respectively. Usually, some of the elastic elements are not deformed due to the constraints imposed on the system. This means that the numbers of elastic elements acting is not necessarily equal to $3N$, with this in mind we denote this number as N'.

The generalised coordinates can be expressed in terms of the element deflections by means of relations of type (4.63), i.e.

Planes of symmetry of vibration protection system	**Vibration excitation direction**		
	axis x	**axis y**	**axis z**
xCy	s_x, s_y, θ_z	s_x, s_y, θ_z	s_z, θ_x, θ_y
yCz	s_x, θ_y, θ_z	s_y, s_z, θ_x	s_y, s_z, θ_x
zCx	s_x, s_z, θ_y	s_y, θ_x, θ_z	s_x, s_z, θ_y
xCy and yCz	s_x, θ_z	s_y	s_z, θ_x
yCz and zCx	s_x, θ_y	s_y, θ_x	s_z
zCx and xCy	s_x	s_y, θ_x	s_z, θ_y
xCy, yCz and zCx	s_x	s_y	s_z

TABLE 4.1.

$$q_r = \sum_{k=1}^{p} \sigma_{rk} u_k \ \ (r = 1, ..., p). \tag{4.76}$$

Furthermore, $N' - p$ constraint conditions analogous to (4.65) must be imposed on deflections u_k $(r = 1, ..., N')$

$$\varphi_l (u_1, ..., u_p, u_l) = u_l - \sum_{k=1}^{p} \nu_{lk} u_k = 0. \tag{4.77}$$

Expressions for kinetic energy and factors μ_{kr} can be obtained from eqs. (4.67) and (4.69), to give

$$T = \frac{1}{2} \sum_{k,r=1}^{p} \mu_{kr} \dot{u}_k \dot{u}_r, \tag{4.78}$$

$$\mu_{kr} = \sum_{l=1}^{p} m_l \sigma_{lk} \sigma_{lr}. \tag{4.79}$$

After a little algebra we obtain the following equations of motion

$$\sum_{k=1}^{p} \mu_{ki} \ddot{u}_k + U_i (u_i, \dot{u}_i) + \sum_{l=p+1}^{N'} \Lambda_l \nu_{li} = \sum_{r=1}^{p} Q_r \sigma_{ri} \ (i = 1, ..., p), \tag{4.80}$$

$$U_i (u_i, \dot{u}_i) - \Lambda_i = 0 \ (i = p + 1, ..., N'). \tag{4.81}$$

The forthcoming Sections of the present Chapter are concerned with the general methods of solution of the differential equations of motion given in the form of eqs. (4.74) and (4.75) or eqs. (4.80) and (4.81). At this stage we consider some particular cases in which the equations of motion are considerably simplified.

Let the number of elastic elements be equal to the number of degrees of freedom. This is not feasible for completely isolated systems when conventional elastic elements are used, because as each elastic element allows the object to move in three directions, only two elastic elements can be utilised. When the system is partially isolated the coincidence of the numbers of elastic elements with the number of degrees of freedom is not only possible but even expedient since this case corresponds to the simplest design scheme of vibration isolation which utilises a minimum number of elastic elements.

Equations (4.80) are simplified if $N' = p$ since all $\Lambda_i = 0$, in this case eq. (4.81) no longer exists. Provided that the attachment points and the axes of the elastic elements are chosen so that the following condition holds

$$\mu_{kr} = 0 \text{ if } k \neq r, \tag{4.82}$$

equations (4.80) become uncoupled. The following uncoupled equation is obtained for the deflection of the $i - th$ element

$$\mu_{ii} \ddot{u}_i + U_i (u_i, \dot{u}_i) = \sum_{r=1}^{p} Q_r \sigma_{ri}, \tag{4.83}$$

which can be investigated by the methods of the previous Chapter.

Uncoupling the differential equations in (4.80) implies that each elastic element can be calculated regardless of the other ones. Each differential equation describes motion of the system consisting of mass m_{ii} mounted on the nonlinear elastic element with characteristic $U_i (u_i, \dot{u}_i)$. In the case in which the object should be protected from the vibrating base, eq. (4.83) can be recast in another, more convenient form. To this aim we introduce the following notation

$$\begin{aligned} &\xi_x = \xi_1, \quad \xi_y = \xi_2, \quad \xi_z = \xi_3, \\ &\beta_x = \xi_4, \quad \beta_x = \xi_5, \quad \beta_x = \xi_6. \end{aligned} \tag{4.84}$$

Expression (4.39) due to notation (4.68) takes the following form

$$Q_r = -m_r \ddot{\xi}_r \quad (r = 1, ..., 6). \tag{4.85}$$

Let $\eta_i \quad (i = 1, ..., N')$ denote the component of displacement of the attachment point of the $i-th$ elastic element for motion relative to the base along the element axis, i.e. u_i. It is apparent that ξ_r can be expressed in terms of $\eta_k \quad (k = 1, ..., p)$ by means of equations of type (4.76)

$$\xi_r = \sum_{k=1}^{p} \sigma_{rk} \eta_k \quad (r = 1, ..., p). \tag{4.86}$$

Then,

$$\sum_{r=1}^{p} Q_r \sigma_{ri} = -\sum_{r=1}^{p} m_r \ddot{\xi}_r \sigma_{ri} = -\sum_{r=1}^{p} m_r \sigma_{ri} \sum_{k=1}^{p} \sigma_{rk} \ddot{\eta}_k = -\sum_{k=1}^{p} \mu_{ik} \ddot{\eta}_k. \tag{4.87}$$

Provided that condition (4.82) holds we have

$$\sum_{r=1}^{p} Q_r \sigma_{ri} = -\mu_{ii} \ddot{\eta}_i. \tag{4.88}$$

Thus, the vibration excitation acting on the $i-th$ element is completely defined by the relative motion of the point of its attachment to the object. This "autonomy" of each elastic element turns out to be useful in many cases.

Condition (4.82) is of merely geometric nature. This does not depend upon the dynamic characteristics of the isolators and is equally suitable both for nonlinear and linear systems. In the latter case, u_i are the principal coordinates. It is worth noting that the presence of the dissipative forces does not affect uncoupling. This is not always the case if the generalised coordinates $q_i, ..., q_p$ are the same as the principal coordinates. There exists no principal difficulty in satisfying conditions (4.82) since the number of conditions in eq. (4.82), $p\,(p-1)\,/2$, is always smaller than the number of factors σ_{rk}. Let us show how these conditions can be met for some particular cases.

a) Let the rigid body of small dimensions (so called "material point") having mass m be attached to the base by means of three elastic elements ("springs"), Fig. 4.5. Clearly, the number of elastic elements is coincident with the number of degrees of freedom.

As

$$\left.\begin{aligned} q_1 &= s_x = u_1\alpha_1 + u_2\alpha_2 + u_3\alpha_3, \\ q_2 &= s_y = u_1\beta_1 + u_2\beta_2 + u_3\beta_3, \\ q_3 &= s_z = u_1\gamma_1 + u_2\gamma_2 + u_3\gamma_3. \end{aligned}\right\} \tag{4.89}$$

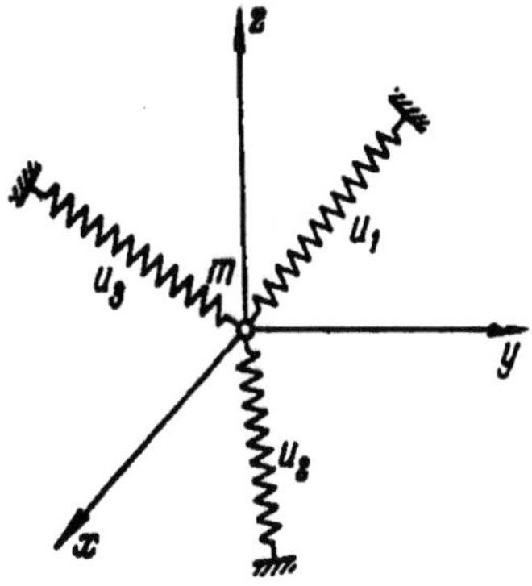

FIGURE 4.5.

we obtain

$$\sigma_{1k} = \alpha_k, \ \sigma_{2k} = \beta_k, \ \sigma_{3k} = \gamma_k, \ (k = 1, 2, 3).$$

Provided that the axes of these three elements are not coplanar the following determinant does not vanish, i.e.

$$|\sigma_{rk}|_1^3 = \begin{vmatrix} \sigma_{11} & \sigma_{12} & \sigma_{13} \\ \sigma_{21} & \sigma_{22} & \sigma_{23} \\ \sigma_{31} & \sigma_{32} & \sigma_{33} \end{vmatrix} \neq 0,$$

Condition (4.82) is met if

$$\left.\begin{aligned} \mu_{12} &= m\,(\alpha_1\alpha_2 + \beta_1\beta_2 + \gamma_1\gamma_2) = 0, \\ \mu_{13} &= m\,(\alpha_1\alpha_3 + \beta_1\beta_3 + \gamma_1\gamma_3) = 0, \\ \mu_{23} &= m\,(\alpha_2\alpha_3 + \beta_2\beta_3 + \gamma_2\gamma_3) = 0, \end{aligned}\right\} \tag{4.90}$$

i.e. the axes of the elastic elements are mutually orthogonal.

b) The rigid body depicted in Fig. 4.6 possesses two degrees of freedom as the motion is assumed to be possible only in the xy plane ($s_z = 0$). The

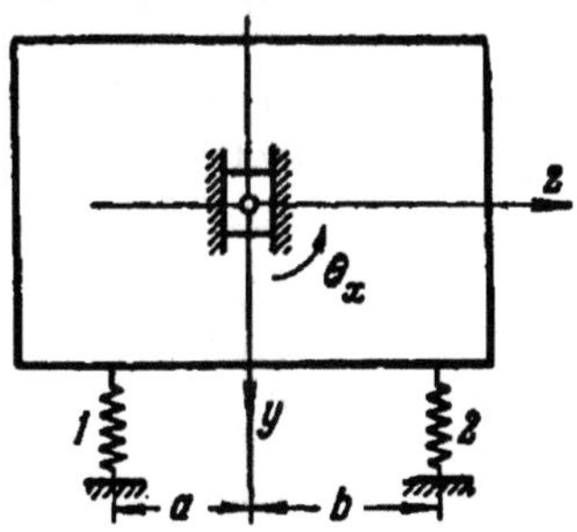

FIGURE 4.6.

generalised coordinates q_1 and q_2 defining the position of the body are y and θ_x which are as follows

$$q_1 = y = \frac{u_1 b + u_2 a}{a+b}, \; q_2 = \theta_x = \frac{u_1 - u_2}{a+b}, \; m_1 = m, \; m_2 = m\rho_x^2. \tag{4.91}$$

Hence

$$\sigma_{11} = \frac{b}{a+b}, \; \sigma_{21} = \frac{1}{a+b}, \; \sigma_{12} = \frac{a}{a+b}, \; \sigma_{22} = -\frac{1}{a+b}.$$

Determinant $|\sigma_{rk}|_1^2$ is not equal to zero

$$|\sigma_{rk}|_1^2 = \begin{vmatrix} \dfrac{b}{a+b} & \dfrac{a}{a+b} \\ \dfrac{1}{a+b} & -\dfrac{1}{a+b} \end{vmatrix} = -\frac{1}{a+b} \neq 0.$$

The condition of the equations uncoupling

$$\mu_{12} = m_1\sigma_{11}\sigma_{12} + m_2\sigma_{21}\sigma_{22} = m\left[\frac{ab}{(a+b)^2} - \frac{\rho_x^2}{(a+b)^2}\right] = 0$$

is satisfied if

$$ab = \rho_x^2. \tag{4.92}$$

In this case

$$\left.\begin{aligned} \mu_{11} &= m_1\sigma_{11}^2 + m_2\sigma_{21}^2 = m\left[\frac{b^2}{(a+b)^2} + \frac{\rho_x^2}{(a+b)^2}\right] = \frac{mb}{a+b}, \\ \mu_{22} &= m_1\sigma_{12}^2 + m_2\sigma_{22}^2 = m\left[\frac{a^2}{(a+b)^2} + \frac{\rho_x^2}{(a+b)^2}\right] = \frac{ma}{a+b}. \end{aligned}\right\} \tag{4.93}$$

We notice in passing that there exists no attachment to the elastic elements which ensures that the equations of motion

$$\left.\begin{aligned} m\ddot{y} + U_1\left(y + a\theta_x, \dot{y} + a\dot{\theta}_x\right) + U_2\left(y - b\theta_x, \dot{y} - b\dot{\theta}_x\right) &= Q_1(t), \\ m\rho_x^2\ddot{\theta}_x + aU_1\left(y + a\theta_x, \dot{y} + a\dot{\theta}_x\right) - bU_2\left(y - b\theta_x, \dot{y} - b\dot{\theta}_x\right) &= Q_2(t) \end{aligned}\right\} \tag{4.94}$$

will be uncoupled for arbitrary nonlinear dependencies $U_1(u_1, \dot{u}_1)$ and $U_2(u_2, \dot{u}_2)$.

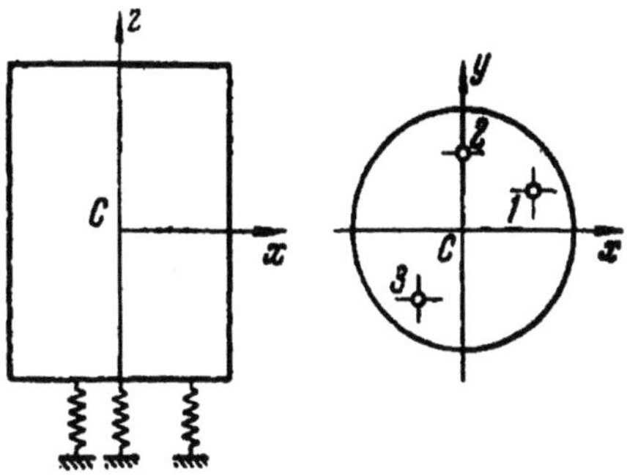

FIGURE 4.7.

c) Let us determine the condition for uncoupling the equations in the case of three elastic elements, Fig 4.7. The following constraints are imposed on the system: no rotation about the z axis and no translation of the center of mass along the x and y axes is allowed. Denoting

$$q_1 = s_z, \; q_2 = \theta_x, \; q_3 = \theta_y,$$

we come by virtue of eq. (4.23) to the following result

$$u_i = q_1 + y_i q_2 - x_i q_3 \;\; (i = 1, 2, 3). \tag{4.95}$$

From the latter equation it follows that

$$\left.\begin{aligned} q_1 &= \left[u_1 \left(x_3 y_2 - x_2 y_3\right) + u_2 \left(x_1 y_3 - x_3 y_1\right) + u_3 \left(x_2 y_1 - x_1 y_2\right)\right]/D, \\ q_2 &= \left[u_1 \left(x_2 - x_3\right) + u_2 \left(x_3 - x_1\right) + u_3 \left(x_1 - x_2\right)\right]/D, \\ q_3 &= \left[u_1 \left(y_2 - y_3\right) + u_2 \left(y_3 - y_1\right) + u_3 \left(y_1 - y_2\right)\right]/D, \end{aligned}\right\} \tag{4.96}$$

where

$$D = x_3 y_2 - x_2 y_3 - x_1 y_3 - x_3 y_1 - x_2 y_1 - x_1 y_2.$$

The condition for uncoupling the equations is

$$\left.\begin{aligned} &\left(x_3 y_2 - x_2 y_3\right)\left(x_1 y_3 - x_3 y_1\right) + \rho_x^2 \left(x_3 - x_2\right)\left(x_1 - x_3\right) \\ &\qquad + \rho_y^2 \left(y_3 - y_2\right)\left(y_1 - y_3\right) = 0, \\ &\left(x_1 y_3 - x_3 y_1\right)\left(x_2 y_1 - x_1 y_2\right) + \rho_x^2 \left(x_1 - x_3\right)\left(x_2 - x_1\right) \\ &\qquad + \rho_y^2 \left(y_1 - y_3\right)\left(y_2 - y_1\right) = 0, \\ &\left(x_2 y_1 - x_1 y_2\right)\left(x_3 y_2 - x_2 y_3\right) + \rho_x^2 \left(x_2 - x_1\right)\left(x_3 - x_2\right) \\ &\qquad + \rho_y^2 \left(y_2 - y_1\right)\left(y_3 - y_2\right) = 0. \end{aligned}\right\} \tag{4.97}$$

For example, if the attachment points are symmetric about the center of mass, namely so that

$$x_1 = 1,\ x_2 = -\frac{1}{2},\ x_3 = \frac{1}{2};\ y_1 = 0,\ y_2 = -\frac{\sqrt{3}}{2},\ y_3 = \frac{\sqrt{3}}{2},$$

condition (4.97) yields

$$\rho_x = \rho_y = \frac{1}{\sqrt{2}}.$$

4.4 Free vibration of a nonlinear vibration protection system

Differential equations for the free vibration of the rigid body mounted on the isolators and possessing six degrees of freedom can be obtained from eqs. (4.74) and (4.75) by adopting the following

$$Q_r = 0\,(r = 1, ..., 6),\ U_i\,(u_i, \dot{u}_i) = U_{ie}\,(u_i)\ \ (i = 1, ..., 3N). \tag{4.98}$$

This implies that the motion is considered without resistance forces and external vibration excitations.

This leads to the following equations of motion

$$\sum_{k=1}^{6} \mu_{ki}\ddot{u}_k + U_{ie}\,(u_i) + \sum_{l=7}^{3N} \nu_{li}\Lambda_l = 0\ \ (i = 1, ..., 6), \tag{4.99}$$

$$U_{ie}\,(u_i) - \Lambda_i = 0\ \ (i = 7, ..., 3N). \tag{4.100}$$

Equation for the constraints, either (4.64) or (4.65), must be added to the above equations.

Obtaining relationships between the eigenfrequencies and the deflection amplitudes is the goal of the analysis of free vibrations. In what follows it is shown that these relationships are very useful for investigating the occurrence of resonant vibrations.

By means of eq. (4.100) one can eliminate the unknown functions Λ_l from eq. (4.99) which reduces the number of unknown variables to $3N$ $(u_1, ..., u_{3N})$. The resulting equations are given by

$$\sum_{k=1}^{6} \mu_{ki}\ddot{u}_k + U_{ie}\,(u_i) + \sum_{l=7}^{3N} U_{le}\,(u_l)\,\nu_{li} = 0\ \ (i = 1, ..., 6). \tag{4.101}$$

An approximate solution of the system of equations, eqs. (4.101) and (4.64), is sought in the form

$$u_i = a_{0i} + a_i \cos \lambda t. \tag{4.102}$$

This assumes that the system exhibits motions in which the elastic elements' deflections are close to being harmonic.

Let us recall that a similar assumption based on the existence of exact periodic solution was made while investigating the free vibration of single-degree-of-freedom systems. The existence of periodic solutions for systems with many degrees of freedom cannot be ensured in general. This can be proved only for a particular class of systems, for example, for Lyapunov's systems, [85]. The conservative system under consideration is a Lyapunov system if all functions $U_{ie}(u_i)$ are analytic.

Existence of periodic vibrations of sufficiently small amplitude can also be proved for conservative systems with nonanalytic elastic characteristics. The existence of periodic solutions of large amplitude is proved by the method of small parameters only for weakly nonlinear systems, i.e. for those systems whose elastic characteristics slightly differ from being linear. In what follows, a periodic solution is assumed to exist which can be treated as substantiation for using approximate methods.

Expression (4.102) has no $\sin \lambda t$ term. This is allowed for a conservative system which is invariant under replacement of the time t by $-t$.

In order to find for solutions of type (4.102) we apply the method of harmonic linearisation, i.e. we replace the nonlinear functions by the following linear function

$$U_{ie}(u_i) \approx U_{0i} + q_i(u_i - a_{0i}) = U_{0i} + q_i u_i^0, \tag{4.103}$$

where the harmonic linearisation factors U_{0i} and q_i depend on parameters a_{0i} and a_i.

Introducing eq. (4.103) into (4.101) we arrive at the following linearised equations

$$\sum_{k=1}^{6} \mu_{ki}\ddot{u}_k^0 + U_{i0} + q_i u_i^0 + \sum_{l=7}^{3N} \left(U_{l0} + q_l u_l^0\right)\nu_{li} = 0 \quad (i = 1, ..., 6). \tag{4.104}$$

Now we substitute eq. (4.102) into eqs. (4.104) and (4.64). Separating the expressions obtained for the constant and harmonic terms we arrive at $6N$ equations relating $6N + 1$ parameters $(a_{10}, ..., a_{3N,0}, a_1, ..., a_{3N}, \lambda)$

$$-\lambda^2 \sum_{k=1}^{6} \mu_{ki}a_i + q_i a_i + \sum_{l=7}^{3N} q_l a_l \nu_{li} = 0 \ (i = 1, ..., 6), \tag{4.105}$$

$$a_i = \sum_{k=1}^{6} \nu_{ik}a_k \ (i = 7, ..., 3N), \tag{4.106}$$

$$U_{i0} + \sum_{l=7}^{3N} U_{l0} \nu_{li} = 0 \ \ (i = 1, ..., 6), \tag{4.107}$$

$$U_{i0} = \sum_{k=1}^{6} \nu_{ik} U_{k0} \ \ (i = 7, ..., 3N). \tag{4.108}$$

We consider first the system of equations containing U_{i0}. Substituting eq. (4.108) into (4.107) we reduce the number of equations up to six

$$U_{i0} + \sum_{k=1}^{6} U_{k0} \sum_{l=7}^{3N} \nu_{li} \nu_{lk} = 0 \quad (i = 1, ..., 6). \tag{4.109}$$

It is assumed in what follows that this system of linear homogeneous equations can have only trivial solutions. Then,

$$U_{i0} (a_{0i}, a_i) = 0 \ \ (i = 1, ..., 3N), \tag{4.110}$$

i.e. the elastic force in each isolator has no constant component. Thus, the dependence $a_{0i} (a_i)$ is the same as in the case of free vibration in the system with one degree of freedom (clearly, for the corresponding values of the static deflections u_{si}). This enables us to express a_{0i} in terms of a_i and obtain the linearisation factors q_i as functions of a_i only. The dependence $q_i (a_i)$ represents the backbone curve which can be calculated in advance for each isolator.

The system of equations (4.105) and (4.106) contain $3N + 1$ unknowns, namely amplitudes $a_1, ..., a_{3N}$ and frequency λ. As the number of unknowns exceeds the number of equations one unknown can be prescribed at will.

Generally speaking, the equations obtained are transcendental equations and their solution encounters computational difficulties. Use of the algorithm developed below allows one to reduce the solution to a sequence of operations, namely, solving a sixth order algebraic equation and solving a system of linear algebraic equations.

Assume that characteristic $U_{ie} (u_i)$ has continuous derivatives at $u_i = 0$, this permits us to adopt the standard assumption used in the theory of small vibrations, i.e.

$$U_{ie} (u_i) \approx U'_{ie} (0) \, u_i .$$

In other words, we let

$$q_l = U'_{le} (0) = c_l \tag{4.111}$$

in eq. (4.105). This leads to the following system of linear differential algebraic equations

$$-\lambda^2 \sum_{k=1}^{6} \mu_{ki} a_i + c_i a_i + \sum_{l=7}^{3N} c_l a_l \nu_{li} = 0 \ \ (i = 1, ..., 6), \tag{4.112}$$

$$a_i = \sum_{k=1}^{6} \nu_{ik} a_k \ \ (i = 7, ..., 3N). \tag{4.113}$$

Inserting a_i $(i > 6)$ from the latter equation into (4.112) yields

$$\sum_{k=1}^{6} \left(c_i \delta_{ki} - \lambda^2 \mu_{ki} + \sum_{l=7}^{3N} c_l \nu_{lk} \nu_{li} \right) a_k = 0 \ \ (i = 1, ..., 6), \tag{4.114}$$

with δ_{ik} being the Kronecker delta

$$\delta_{ik} = \begin{cases} 1 & \text{if } i = k, \\ 0 & \text{if } i \neq k. \end{cases}$$

This system has nontrivial solutions for those values of λ for which the determinant of the system vanishes. The vanishing determinant, which is the system's frequency equation, renders the eigenfrequencies $\lambda_1, ..., \lambda_6$ coincident with those of Section 4.2.

The corresponding harmonic solutions are the principal vibrations of the system. Equation (4.114) allows us to determine the shape of the principal vibrations, i.e. the set of the amplitude values $a_1^{(s)}, ..., a_6^{(s)}$ defined up to an arbitrary constant factor $\gamma^{(s)}$

$$a_i^{(s)} = \gamma^{(s)} \alpha_i^{(s)} \ \ (i = 1, ..., 6). \tag{4.115}$$

Next, by using expression (4.113) we can find the amplitudes of the deflections of the other isolators

$$a_i^{(s)} = \gamma^{(s)} \sum_{k=1}^{6} \nu_{ik} a_k^{(s)} \ \ (i = 7, ..., 3N). \tag{4.116}$$

The linear problem is therefore solved since the eigenfrequencies and the normal modes are determined. This solution differs from that of Section 4.2 only through the choice of generalised coordinates.

Prescribing a value of parameter $\gamma^{(s)}$, e.g. $\gamma^{(s)} = \gamma_1^{(s)}$, we obtain amplitudes $a_{i1}^{(s)}$ by virtue of eqs. (4.115) and (4.116)

$$a_{i1}^{(s)} = \gamma_1^{(s)} \alpha_{i1}^{(s)}.$$

In general, the values of $q_{i1}^{(s)}$ corresponding to $a_{i1}^{(s)}$ do not coincide with those from eq. (4.111). Replacing c_l in eq. (4.114) by q_{i1} and solving the

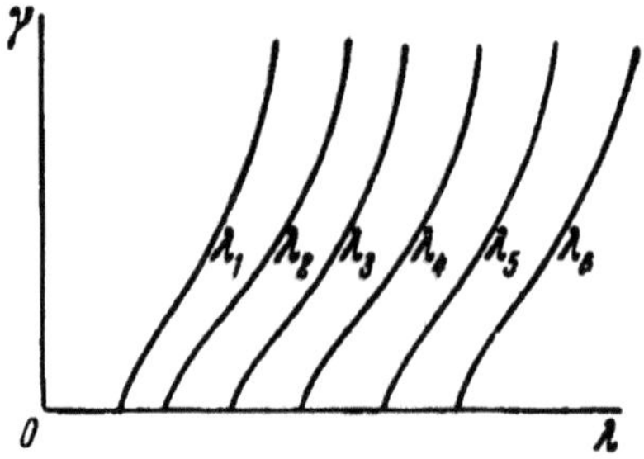

FIGURE 4.8.

equations obtained we find new values of the "eigenfrequencies" and new "mode shapes" $\alpha_{i2}^{(s)}$, the latter not being coincident with the original ones

$$\alpha_{i2}^{(s)} \neq \alpha_{i1}^{(s)}.$$

Inserting these factors into eqs. (4.115) and (4.116) we find new values for amplitudes $\alpha_{i2}^{(s)}$ and then new values of $q_{i2}^{(s)}$. By means of this iterative process one can obtain solution of system (4.105) and (4.106) with any degree of accuracy. It can be shown that this process of successive approximations converges for sufficiently small values of $\gamma^{(s)}$. The smaller the value of $\gamma^{(s)}$ and the smaller the deviation of characteristic $U_{ie}(u_i)$ from the linear one, the faster the convergence.

The design must be carried out for all six normal modes ($s = 1, ..., 6$) and for several values of $\gamma^{(s)}$ from the interval of concern. It is expedient to take the normal mode factors obtained for the previous $\gamma^{(s)}$ as a first approximation to the next value of $\gamma^{(s)}$.

This results in the dependences of all six "eigenfrequencies" of the system on parameter γ, the latter representing the amplitudes of deflection in the isolators up to a certain scaling factor. In a certain sense, these dependences are analogous to the backbone curves, see Fig. 4.8.

The above is equally valid for the partly isolated systems, the only difference being that the number of degrees of freedom is not coincident with the number of unknowns. In particular, if p and N' denote the number of degrees of freedom and the number of elastic elements, respectively, the equations for the vibration amplitudes take the form

$$-\lambda^2 \sum_{k=1}^{6} \mu_{ki} a_i + q_i a_i + \sum_{l=p+1}^{3N} q_l a_l \nu_{li} = 0 \quad (i = 1, ..., p), \tag{4.117}$$

$$a_i = \sum_{k=1}^{p} \nu_{ik} a_k. \tag{4.118}$$

Let us consider the following simple example. Let the object depicted in Fig. 4.9 move in the xy plane. The generalised coordinates of the object are

$$q_1 = s_x, \; q_2 = s_y, \; q_3 = \theta_z.$$

The coordinates of the points of attachment of the elastic elements are given by

$$x_1 = x_3 = -a, \; x_2 = x_4 = a, \; y_1 = y_2 = y_3 = y_4 = -h.$$

We express deflections of the elastic elements in terms of the new generalised coordinates

$$\begin{aligned} u_1 &= s_y - a\theta_z, \\ u_1 &= s_y + a\theta_z, \\ u_3 &= u_4 = s_x + h\theta_z. \end{aligned}$$

This yields

$$\begin{aligned} q_1 &= s_x = u_3 - \frac{h}{2a}(u_2 - u_1), \\ q_2 &= s_y = \frac{u_1 + u_2}{2}, \\ q_3 &= \theta_z = \frac{u_2 - u_1}{2a}. \end{aligned}$$

By virtue of eq. (4.79) we determine μ_{kr}

$$\begin{aligned} \mu_{11} &= \mu_{22} = \frac{m}{4}\left(1 + \frac{h^2}{a^2} + \frac{\rho_z^2}{a^2}\right), \; \mu_{12} = \frac{m}{4}\left(1 - \frac{h^2}{a^2} - \frac{\rho_z^2}{a^2}\right), \\ \mu_{13} &= \frac{mh}{2a}, \; \mu_{23} = -\frac{mh}{2a}, \; \mu_{33} = m. \end{aligned} \tag{4.119}$$

Variable u_4 is related to the new generalised coordinates u_1, u_2 and u_3 by means of the simple equation $u_3 = u_4$. Hence, $\nu_{41} = \nu_{42} = 0$ and $\nu_{43} = 1$. Equations (4.117) and (4.118) take the form

$$\left.\begin{aligned} -\lambda^2(\mu_{11}a_1 + \mu_{12}a_2 + \mu_{13}a_3) + q_1 a_1 &= 0, \\ -\lambda^2(\mu_{12}a_1 + \mu_{22}a_2 + \mu_{23}a_3) + q_2 a_2 &= 0, \\ -\lambda^2(\mu_{13}a_1 + \mu_{23}a_2 + \mu_{33}a_3) + q_3 a_3 + q_4 a_4 &= 0, \\ a_4 &= a_3. \end{aligned}\right\} \tag{4.120}$$

Assume that

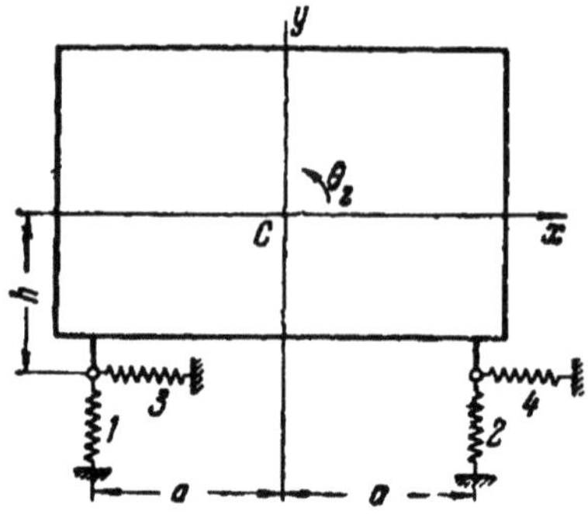

FIGURE 4.9.

$$\frac{h}{a} = \frac{1}{2}, \ \frac{\rho_z}{a} = \frac{\sqrt{3}}{2},$$

then,

$$\begin{aligned} \mu_{11} &= \mu_{22} = \frac{m}{2}, \ \mu_{12} = 0, \ \mu_{13} = \frac{m}{4}, \\ \mu_{13} &= -\frac{m}{4}, \ \mu_{33} = m. \end{aligned}$$

Assume also that all elements have elastic characteristics of type (3.6)

$$\left.\begin{aligned} U_{1e} = \frac{2c\Delta}{\pi}\tan\frac{\pi u_1}{2\Delta}, \ U_{2e} = \frac{2c\Delta}{\pi}\tan\frac{\pi u_2}{2\Delta}, \\ U_{3e} = \frac{2c_1\Delta_1}{\pi}\tan\frac{\pi u_3}{2\Delta_1}, \ U_{4e} = \frac{2c_1\Delta_1}{\pi}\tan\frac{\pi u_4}{2\Delta_1}, \end{aligned}\right\} \tag{4.121}$$

where

$$c_1 = \frac{c}{2}, \ \Delta_1 = \frac{\Delta}{2}. \tag{4.122}$$

The linearisation factors obtained by the method of linearisation based upon the distribution function, Section 3.1, are as follows

$$\begin{aligned} f_1 &= \frac{2}{\sqrt{3}a_1}\frac{2c\Delta}{\pi}\tan\left(\frac{\pi}{2\Delta}\frac{\sqrt{3}}{2}a_1\right) = c\frac{d}{a_1}\tan\left(\frac{a_1}{d}\right), \\ f_2 &= c\frac{d}{a_2}\tan\left(\frac{a_2}{d}\right), \ f_3 = \frac{cd}{4a_3}\tan\left(\frac{2a_3}{d}\right), \ f_4 = \frac{cd}{4a_4}\tan\left(\frac{2a_4}{d}\right), \end{aligned}$$

where

$$d = \frac{4\Delta}{\pi\sqrt{3}}.$$

Introducing these expressions into eq. (4.120) leads to

$$\left.\begin{aligned} -z^2\left(\frac{a_1}{2}+\frac{a_3}{4}\right)+a_1\frac{d}{a_1}\tan\left(\frac{a_1}{d}\right) &= 0,\\ -z^2\left(\frac{a_2}{2}+\frac{a_3}{4}\right)+a_2\frac{d}{a_2}\tan\left(\frac{a_2}{d}\right) &= 0,\\ -z^2\left(\frac{a_1}{4}-\frac{a_2}{4}+a_3\right)+\frac{a_3}{2}\frac{d}{a_3}\tan\left(\frac{a_3}{d}\right) &= 0, \end{aligned}\right\} \tag{4.123}$$

where $z = \sqrt{m/c}$ is the nondimensional frequency.

Provided that amplitudes a_1, a_2 and a_3 are small we can adopt the following

$$\begin{aligned} \frac{d}{a_1}\tan\left(\frac{a_1}{d}\right) &= \frac{d}{a_2}\tan\left(\frac{a_2}{d}\right) \approx 1\\ \frac{d}{a_3} &= \tan\left(\frac{2a_3}{d}\right) \approx 2. \end{aligned}$$

In this case eq. (4.123) is linear.

The vanishing determinant of the system

$$\begin{vmatrix} 1-z^2/2 & 0 & -z^2/4 \\ 0 & 1-z^2/2 & z^2/4 \\ -z^2/4 & z^2/4 & 1-z^2 \end{vmatrix} = 0$$

yields the following eigenfrequencies

$$z_1^2 = 0,845,\ z_2^2 = 2,\ z_3^2 = 3,155.$$

Now we determine the normal modes by substituting the values of the eigenfrequencies into the linearised equations (4.121)

$$\begin{aligned} \left(1-\frac{z_i^2}{2}\right)a_1^{(i)} - \frac{z_i^2}{4}a_3^{(i)} &= 0,\\ \left(1-\frac{z_i^2}{2}\right)a_2^{(i)} - \frac{z_i^2}{4}a_3^{(i)} &= 0,\\ -\frac{z_i^2}{4}a_1^{(i)} + \frac{z_i^2}{4}a_2^{(i)} + \left(1-z_i^2\right)a_3^{(i)} &= 0. \end{aligned}$$

Solving these equations yields

$$\begin{aligned}
a_1^{(1)} &= -a_2^{(1)} = 0,366a_3^{(1)},\\
a_1^{(2)} &= a_2^{(2)},\ a_3^{(2)} = 0,\\
a_1^{(3)} &= -a_2^{(3)} = -1,365a_3^{(3)}.
\end{aligned}$$

It is easy to see that the second mode corresponds to translatory vibration of the object along axis y. In the system under consideration this mode is uncoupled with the other modes at any amplitude, thus, its backbone curve corresponds to that of a single-degree-of-freedom system. For this reason, we study only the first and third modes describing coupled vibrations along the x and z axes.

Taking into account that $a_2 = -a_1$ for these modes, we can simplify the system of equations (4.123) by reducing it to the following two equations in two unknowns

$$\left.\begin{aligned}
\left[\frac{d}{a_1}\tan\left(\frac{a_1}{d}\right) - \frac{z^2}{2}\right]a_1 - \frac{z^2}{4}a_3 = 0,\\
-\frac{z^2}{2}a_1 + \left[\frac{d}{2a_3}\tan\left(\frac{2a_3}{d}\right) - z^2\right]a_3 = 0.
\end{aligned}\right\} \tag{4.124}$$

These equations are used in what follows for obtaining dependences of the frequencies of free vibration on the amplitude of deflections of the elastic elements.

First, we consider vibration with the lower frequency. Prescribing a value for one of the amplitudes, for example

$$a_3^{(1)} = 0,25d,$$

we obtain, due to the expression for the first mode of the linear system,

$$a_1^{(1)} = 0,366 \cdot 0,25d = 0,0915d.$$

One can adopt approximately

$$\begin{aligned}
\frac{d}{a_1^{(1)}}\tan\left(\frac{a_1^{(1)}}{d}\right) &= \frac{\tan 0,0915}{0,0915} \approx 1,\\
\frac{d}{2a_3^{(1)}}\tan\left(\frac{2a_3^{(1)}}{d}\right) &= \frac{\tan 0,5}{0,5} = 1,092.
\end{aligned}$$

Introducing these expressions into (4.124) we arrive at the new frequency equation

$$\begin{vmatrix} 1 - z^2/2 & -z^2/4 \\ -z^2/2 & 1,092 - z^2 \end{vmatrix} = 0. \tag{4.125}$$

Solving this equation yields

$$z_1^2 = 0,904, \; z_2^2 = 3,22.$$

Clearly, we are interested in the value of the lower frequency as we proceeded from the first normal mode while obtaining the more accurate frequency equation (4.125).

The more accurate normal mode is given by

$$a_1^{(1)} = a_3^{(1)} \frac{z_1^2}{2\left[2\dfrac{d}{a_1}\tan\left(\dfrac{a_1}{d}\right) - z_1^2\right]} = 0,412 a_3^{(1)}.$$

Substituting the chosen value of $a_3^{(1)}$ into the latter equation gives

$$a_1^{(1)} = 0,412 \cdot 0,25d = 0,103d.$$

Actually, this does not affect the frequency equation since as before

$$\frac{d}{a_1^{(1)}} \tan\left(\frac{a_1^{(1)}}{d}\right) = \frac{\tan 0,103}{0,103} \approx 1.$$

Thus, we can adopt that for $a_3^{(1)} = 0,25d$ and $z_1^2 = 0,904$

$$a_1^{(1)} = 0,412 a_3^{(1)} = 0,103d.$$

Let us now take

$$a_3^{(1)} = 0,4d.$$

Choosing the normal mode corresponding to $a_3^{(1)} = 0,25d$ as a first approximation yields

$$a_1^{(1)} = 0,412 \cdot 0,4d = 0,165d.$$

Then,

$$\frac{d}{a_1^{(1)}} \tan\left(\frac{a_1^{(1)}}{d}\right) = \frac{\tan 0,165}{0,165} = 1,01,$$

$$\frac{d}{2a_3^{(1)}} \tan\left(\frac{2a_3^{(1)}}{d}\right) = \frac{\tan 0,8}{0,8} = 1,29.$$

Solving the frequency equation

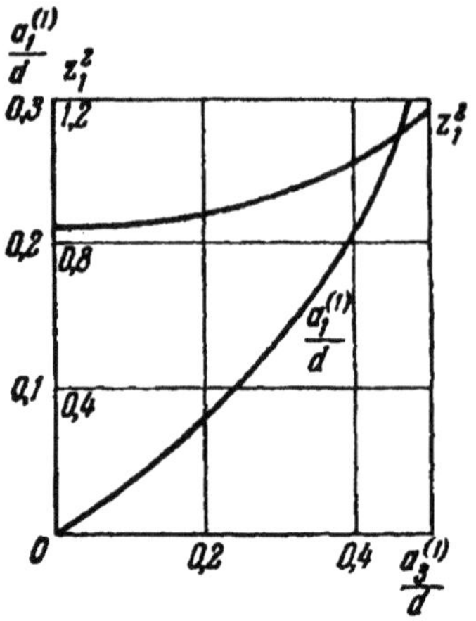

FIGURE 4.10.

$$\begin{vmatrix} 1,01 - z^2/2, & -z^2/4 \\ -z^2/2, & 1,29 - z^2 \end{vmatrix} = 0,$$

yields $z_1^2 = 1,02$. This value leads to the more accurate value of the normal mode

$$a_1^{(1)} = \frac{z_1^2}{2\,(2,02 - z_1^2)} = 0,51 a_3^{(1)}.$$

Thus, $a_1^{(1)} = 0,51 \cdot 0,4d = 0,204d$ and

$$\frac{d}{a_1^{(1)}} \tan\left(\frac{a_1^{(1)}}{d}\right) = \frac{\tan 0,204}{0,204} = 1,014.$$

The frequency shift caused by this correcture is actually of no importance as it is accurate to 3 significant figures, so we can adopt $a_3^{(1)} = 0,4d$, $z_1^2 = 1,02$ and $a_1^{(1)} = 0,204d$.

Values of the lower frequency of free vibration corresponding to another values of $a_3^{(1)}$ can be obtained by analogy. Dependences $z_1^2\left[a_3^{(1)}\right]$ and $a_1^{(1)}\left[a_3^{(1)}\right]$ are displayed in Fig. 4.10.

Analysis of the vibration with the third frequency does not differ from the performed one. The results of the calculations for two values of $a_3^{(1)}$ are collected in Table 4.2.

Number	$a_3^{(3)}$	z_3^2	$a_1^{(3)}$
1	$0,25d$	$3,32$	$-0,335d$
2	$0,4d$	$3,45$	$-0,56d$

TABLE 4.2.

4.5 Resonant vibration

Various resonant phenomena studied in detail in Chapter 3 can take place in nonlinear systems with many degrees of freedom. In contrast to linear systems, resonant vibration may occur in cases in which the frequency of external excitation is considerably higher than the eigenfrequencies. It goes without saying that such resonances are inadmissible. In other words, the very possibility of the resonances must be precluded in a properly designed system of vibration protection.

The physical nature of the resonance phenomena in multi-degree-of-freedom systems is the same as in single-degree-of-freedom systems, see for detail Section 3.7. Any resonance implies motion close to free vibration in the conservative system and can be deemed to be a "free vibration" supported by the external vibrational excitation. This is the main reason for the detailed analysis of free vibrations carried out in Section 4.4.

For example, we consider equations of motion, eqs. (4.80) and (4.81), for the object to be protected. The case of $p = 6$ and $N' = 3N$ corresponds to equations (4.74) and (4.75). Removing Λ_l from eq. (4.80) by means of eq. (4.81) we obtain

$$\sum_{k=1}^{p} \mu_{ki}\ddot{u}_k + U_i\left(u_i, \dot{u}_i\right) + \sum_{l=p+1}^{N'} \nu_{li}U_l\left(u_l, \dot{u}_l\right) = \sum_{r=1}^{p} Q_r\sigma_{ri} \;\; (i = 1, ..., p)\,. \tag{4.126}$$

In what follows, the external vibration excitation is assumed to be periodic with period $T = 2\pi/\omega$.

Separating elastic and dissipative components of forces U_i we rewrite eq. (4.126) in the following form

$$\sum_{k=1}^{p} \mu_{ki}\ddot{u}_k + U_{ie}\left(u_i\right) + \sum_{l=p+1}^{N'} \nu_{li}U_{le}\left(u_l\right) \tag{4.127}$$

$$= -U_{id}\left(u_i, \dot{u}_i\right) - \sum_{l=p+1}^{N'} \nu_{li} U_{ld}\left(u_l, \dot{u}_l\right) + \sum_{r=1}^{p} Q_r \sigma_{ri} \quad (i = 1, ..., p).$$

The following equations for the constraints must be added to these differential equations

$$u_i - \sum_{k=1}^{p} \nu_{ik} u_k = 0 \quad (i = p+1, ..., N'). \tag{4.128}$$

Equating the right hand sides in equations (4.127) to zero yields equations for free vibration. As the resonance vibration is close to free vibration, the periodic solution of system (4.127) and (4.128) may be sought using Galerkin's approach with the family of periodic solutions of equations for free vibration having frequency ω.

Let

$$\widetilde{u}_i = \varphi_i\left(t + \alpha\right) \quad (i = 1, ..., N') \tag{4.129}$$

describe the family of periodic solutions of equations for free vibration having period T. As noted above, the family of solutions depending upon parameter α exists because the governing equations are autonomous. According to the standard procedure of Galerkin's method, e.g. [65], the following equation

$$\int_0^T \sum_{i=1}^{p} \left\{ U_{id}\left[\varphi_i\left(t+\alpha\right), \dot{\varphi}_i\left(t+\alpha\right)\right] + \sum_{l=p+1}^{N'} U_{ld}\left[\varphi_l\left(t+\alpha\right), \dot{\varphi}_l\left(t+\alpha\right)\right] \nu_{li} \right.$$
$$\left. - \sum_{r=1}^{p} \sigma_{ri} Q_r\left(t\right) \right\} \dot{\varphi}_i\left(t+\alpha\right) dt = 0 \tag{4.130}$$

determines the value of α. This value corresponds to the approximate solution of eqs. (4.127) and (4.128) which is close to free vibration.

The resonant vibration of period T occurs in the system if this equation has real roots. Provided that this equation has no real root, the very possibility of the occurrence of a resonant vibration is excluded.

Equation (4.130) means the condition of energy balance, namely work of the dissipative forces in free vibration is equal to work of the external excitation. A similar condition is established in the previous Chapter for single-degree-of-freedom systems.

Let an approximate harmonic solution of the equation for free vibration at frequency ω be given by

$$\widetilde{u}_i = a_{0i} + a_i \cos\omega\left(t + \alpha\right), \quad \dot{\widetilde{u}}_i = -\omega a_i \sin\omega\left(t + \alpha\right). \tag{4.131}$$

All external forces $Q_r(t)$ are assumed to be harmonic ones having frequency ω, i.e.

$$Q_r(t) = A_r \cos \omega t. \tag{4.132}$$

Inserting eqs. (4.131) and (4.132) into (4.130) yields

$$\begin{aligned}
&\int_0^T \sum_{i=1}^{p} \{U_{id}\left[a_{0i} + a_i \cos\omega(t+\alpha), -\omega a_i \sin\omega(t+\alpha)\right] \\
&+ \sum_{l=p+1}^{N'} \nu_{li} U_{ld}\left[a_{0l} + a_l \cos\omega(t+\alpha), -\omega a_l \sin\omega(t+\alpha)\right] \\
&- \sum_{r=1}^{p} A_r \cos\omega t \cdot \sigma_{ri} \Big\} \left[-\omega a_i \sin\omega(t+\alpha)\right] dt = 0.
\end{aligned} \tag{4.133}$$

Taking into account that

$$\begin{aligned}
&-\int_0^T U_{id}\left[a_{0i} + a_i \cos\omega(t+\alpha), -\omega a_i \sin\omega(t+\alpha)\right] \sin\left[\omega(t+\alpha)\right] dt \\
= \; &-\frac{1}{\omega} \int_0^{2\pi} U_{id}\left(a_{0i} + a_i \cos\psi, -\omega a_i \sin\psi\right) \sin\psi d\psi = \pi b_{id} a_i,
\end{aligned}$$

where b_{id} denotes the dynamical resistance factor and

$$\int_0^T \cos\omega t \sin\left[\omega(t+\alpha)\right] dt = \frac{\pi}{\omega} \sin\omega\alpha,$$

allows us to simplify eq. (4.133), to get

$$\sum_{i=1}^{p} \left\{ \pi b_{id} \omega a_i^2 + \sum_{l=p+1}^{N'} \pi b_{ld} a_l a_i \omega \nu_{li} + \sum_{r=1}^{p} \pi A_r a_i \sin\omega\alpha \cdot \sigma_{ri} \right\} = 0. \tag{4.134}$$

It follows from this equation that

$$\sin\omega\alpha = \frac{\omega \sum_{i=1}^{p} \left(b_{id} a_i^2 + \sum_{l=p+1}^{N'} b_{ld} a_l a_i \nu_{li} \right)}{\sum_{i=1}^{p} \sum_{r=1}^{p} A_r a_i \sigma_{ri}}. \tag{4.135}$$

By virtue of eq. (4.128)

$$\sum_{i=1}^{p}\sum_{l=p+1}^{N'} b_{ld}a_l a_i \nu_{li} = \sum_{l=p+1}^{N'} b_{ld}a_l \sum_{i=1}^{p} a_i \nu_{li} = \sum_{l=p+1}^{N'} b_{ld}a_l^2, \tag{4.136}$$

which leads to

$$\sin\omega\alpha = -\frac{\omega \sum\limits_{i=1}^{N'} b_{id}a_i^2}{\sum\limits_{i=1}^{p}\sum\limits_{r=1}^{p} A_r a_i \sigma_{ri}}. \tag{4.137}$$

Resonant vibrations can occur if $|\sin\omega\alpha| \leq 1$, i.e. if

$$\omega \sum_{i=1}^{N'} b_{id}a_i^2 \leq \sum_{i=1}^{p}\sum_{r=1}^{p} A_r a_i \sigma_{ri}. \tag{4.138}$$

Clearly, this condition can be violated by means of sufficiently large values of b_{id}, i.e. by increasing the damping in the isolators. In this case, no resonant vibration of frequency ω is observed in the system.

Conditions for the existence of resonant vibrations in a system undergoing periodic external excitation can be established by analogy. The condition for resonance coincides with that given by eq. (4.138) provided that one ignores higher harmonics of free vibration. An explanation for this coincidence is that the higher harmonics of the external force do no work during harmonic free vibration. Obtaining the higher harmonics of free vibrations, one can establish the existence conditions for the ultrasubharmonic resonances. They are subharmonic resonances in the case of the harmonic external excitation.

Investigating conditions for the occurrence of a resonance we can approximately estimate the resonant amplitude since the fact that the resonant vibration is close to free vibration amplitude means that their amplitudes are close as well. More accurate estimates are usually not needed.

Let us analyse condition (4.138) in more detail because this condition describes resonant vibration at the frequency of the driving. More often than not, these vibrations are the most dangerous.

Consider first the system with dissipative forces proportional to the deflection velocities, i.e. all b_{id} are assumed to be constant factors. We also assume that the amplitudes of the external excitations, A_r, do not change and frequency ω, $\omega_{\min} < \omega < \omega_{\max}$, covers the spectrum of the system's eigenfrequencies.

In order to determine the "dangerous" frequencies at which the resonant vibrations occur, it is convenient to use the system's "backbone curves" which express the dependence of the frequencies of free vibration on the amplitude of deflection of one of the elastic elements, Fig. 4.11.

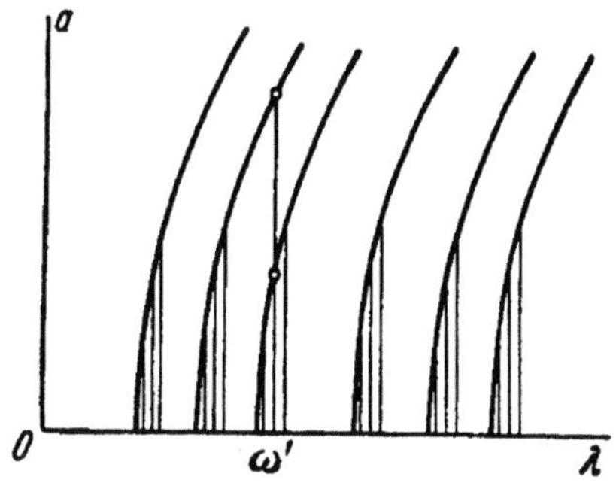

FIGURE 4.11.

The left hand side of inequality (4.138) is proportional to the square of the amplitudes a_i whereas the right hand side is linear in the amplitudes. Hence, the inequality is always satisfied for sufficiently small values of a_i. This means that in systems with resistance forces proportional to velocities resonant vibrations always occur at frequencies sufficiently close to the eigenfrequencies of the system, see Fig. 4.11. The values of b_{id} and A_r as well as the normal modes at the frequency under consideration define the bandwidth of each resonance.

To decrease the total bandwidth of the system, one usually intends to make the eigenfrequencies as close as possible. The frequency domains of the resonances merge, i.e. the resonances can be observed at any frequency ω from the interval $\lambda_{\min} < \omega \le \lambda_{\max}$. Figure 4.11 indicates that a number of resonant vibrations corresponding to different backbone curves, i.e. having different modes, can occur at the same frequency ω'.

The left hand side of inequality (4.138) is proportional to ω, thus, the possibility of the resonance decreases with the growth of ω. This conclusion is valid only in the case of constant amplitudes A_r. If these amplitudes grow as ω increases, for example as ω^2, then the possibility of resonance can increase at higher frequencies. The appearance of additional branches of the resonance curves and another effects may be observed under these circumstances.

We demonstrate now for the particular example how to use the equations obtained for defining the intensity of the dissipative forces which is required to suppress the resonant vibration.

Let the base of the system depicted in Fig. 4.9 vibrate harmonically along the x axis. Let the amplitude and frequency be respectively ξ_0 and

$$\omega = \sqrt{3,45\frac{c}{m}} = 1,86\sqrt{\frac{c}{m}}.$$

As follows from Table 4.2 the amplitudes of free vibration with this frequency are $a_1 = -0,56d$ and $a_3 = 0,4d$. By virtue of eq. (4.88) we obtain

$$\begin{aligned} Q_1\sigma_{11} + Q_3\sigma_{31} &= -\mu_{11}\ddot{\eta}_1 = 0, \\ Q_1\sigma_{13} + Q_3\sigma_{33} &= -\mu_{33}\ddot{\eta}_3 = -m\xi_0\omega^2 \cos\omega t. \end{aligned}$$

From these equations we have

$$A_1\sigma_{11} + A_3\sigma_{31} = 0, \;\; A_1\sigma_{31} + A_3\sigma_{33} = m\xi_0\omega^2,$$

$$\sum_{i=1}^{p}\sum_{r=1}^{p} A_r\sigma_{ri}a_i = m\xi_0\omega^2 a_3 = 1,38cd\xi_0.$$

To suppress the resonant vibration it is necessary to ensure that inequality (4.138) does not hold, i.e.

$$\omega\left(b_{1d}a_1^2 + b_{2d}a_2^2 + b_{3d}a_3^2 + b_{4d}a_4^2\right) > 1,38cd\xi_0.$$

Since $a_1 = a_2$, $a_3 = a_4$, then assuming that $b_{1d} = b_{2d}$, $b_{3d} = b_{4d}$ we obtain

$$1,86\sqrt{\frac{c}{m}}2d^2\left(0,31b_{1d} + 0,16b_{3d}\right) > 1,38cd\xi_0.$$

Let us adopt the following assumptions, see eq. (4.122),

$$\frac{b_{1d}}{\sqrt{c}} = \frac{b_{3d}}{\sqrt{c_1}} \text{ or } b_{1d} = \sqrt{2}b_{3d},$$

which ensure the coincidence of the vibration absorption factors for each elastic element.

In this case

$$2,23d\sqrt{\frac{c}{m}}b_{3d} > 1,38cd\xi_0$$

or

$$b_{3d} > 0,62\sqrt{cm}\frac{\xi_0}{d}.$$

The conditions for occurrence of resonant vibration in a system with dry frictions are somewhat different. In this system

$$b_{id} = \frac{4H_i}{\pi\omega\left|a_i\right|}, \tag{4.139}$$

and condition (4.138) takes the form

$$\sum_{i=1}^{N'}\frac{4}{\pi}H_i\left|a_i\right| \le \sum_{i=1}^{p}\sum_{r=1}^{p} A_r a_i\sigma_{ri}. \tag{4.140}$$

Both sides of this inequality are proportional to amplitudes a_i, hence, for sufficiently large values of H_i no vibration close to free vibration occurs regardless of how small the vibration amplitude is. This means that either the system is "locked" by dry friction forces or some stick-slip motions are observed.

It is worth mentioning that "locking " does not necessarily imply that all degrees of freedom are locked. Dry friction simply reduces the number of degrees of freedom.

Let us consider the latter example under the assumption of dry friction in the isolators. The condition for resonance suppression is given by

$$\frac{4}{\pi}\left(H_1\left|a_1\right|+H_2\left|a_2\right|+H_3\left|a_3\right|+H_4\left|a_4\right|\right)>1,38cd\xi_0.$$

When $H_1 = H_2$ and $H_3 = H_4$ the latter condition reduces to the following one

$$\frac{8}{\pi}\left(0,56H_1+0,4H_3\right)>1,38c\xi_0. \tag{4.141}$$

If this condition is met, no motion close to the coupled free vibrations in coordinates x and θ_z may occur. However this does not mean that the system is locked. Indeed, let

$$H_3=0,\ \ H_1>\frac{1,38\pi}{8\cdot 0,56}c\xi_0=0,97c\xi_0.$$

Then, inequality (4.141) holds. Coordinate θ_z is locked since the amplitude of the external moment which is equal to product of the amplitude of the driving $m\xi_0\omega^2$ and the arm h

$$m\xi_0\omega^2h=m\xi_0\cdot 3,45\frac{c}{m}h=3,45c\xi_0h$$

is smaller than the moment of the dry friction forces

$$2H_1a=4hH_1>3,88c\xi_0h>m\xi_0\omega^2h,$$

see Fig. 4.12. Therefore, translational vibration along the x axis takes place in the system. To determine the amplitude of this vibration it is necessary to consider the single-degree-of-freedom system which is obtained from the original system by assuming infinitely rigid elements u_1 and u_2. The equation for free vibration in this system is given by

$$m\ddot{u}_3+2U_{3e}\left(u_3\right)=0.$$

Linearising nonlinear elastic force leads to the following equation for the backbone curve

$$z^2=\frac{d}{2a_3}\tan\left(\frac{2a_3}{d}\right).$$

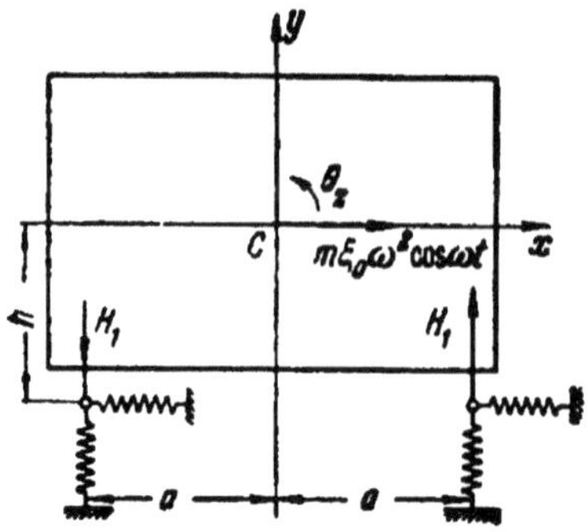

FIGURE 4.12.

Taking, for example, $z^2 = 3,45$ we obtain

$$a_3 = 0,68d = 0,49\Delta.$$

The value of a_3 turns out to be close to the maximum allowed deflection of the elastic elements 3 and 4. This means that free vibration of this amplitude will be accompanied by impacts on stops, i.e. the assumed values of H_1 and H_2 are unacceptable.

Taking

$$H = 2H_3 > 0,72c\xi_0$$

we satisfy condition (4.141) again, however no resonant vibration along the x axis is observed in this case since the rotational motion is no longer locked.

4.6 Forced vibration in systems with Coulomb friction

Provided that no resonant vibration can occur in a vibration protection system subjected to a prescribed external excitation this system can be accepted for use. In order to investigate the vibration protection performance of the system an analysis of forced vibration of non-resonant character is required. The standard linearisation of the nonlinear characteristics suffices to this aim as the isolator deflections are usually small in the case of non-resonant vibration. An exception is for systems with essentially nonlinear characteristics, among them systems with Coulomb's friction are of special interest. The following particular example demonstrates some features of these systems.

Again, we consider the system depicted in Fig. 4.9. In contrast to the previous example we assume that the dry friction forces act in elements 1-4, i.e.

$$U_{i\partial} = H_i \operatorname{sign}\dot{u}_i \;\; (i = 1, ..., 4). \tag{4.142}$$

In what follows it is assumed that

$$H_1 = H_2, \; H_3 = H_4.$$

Let the base vibrate with amplitude ξ_0 and frequency ω in the direction of the x axis. Restricting our analysis to consideration of non-resonant vibration only, we assume the elastic characteristics of all isolators to be linear. Let c denote the stiffness of elements 1 and 2 while c_1 denote that of elements 3 and 4.

Considering eq. (4.87) we can cast the governing equation (4.80) in the form

$$\left.\begin{aligned} \mu_{11}\ddot{u}_1 + \mu_{12}\ddot{u}_2 + \mu_{13}\ddot{u}_3 + cu_1 + H_1\operatorname{sign}\dot{u}_1 &= \mu_{13}\xi_0\omega^2\cos\omega t, \\ \mu_{12}\ddot{u}_1 + \mu_{22}\ddot{u}_2 + \mu_{23}\ddot{u}_3 + cu_2 + H_1\operatorname{sign}\dot{u}_2 &= \mu_{23}\xi_0\omega^2\cos\omega t, \\ \mu_{13}\ddot{u}_1 + \mu_{23}\ddot{u}_2 + \mu_{33}\ddot{u}_3 + 2c_1u_3 + 2H_3\operatorname{sign}\dot{u}_3 &= \mu_{33}\xi_0\omega^2\cos\omega t. \end{aligned}\right\} \tag{4.143}$$

The following constraint equation

$$u_3 = u_4$$

has been taken into account while deriving eq. (4.143). Inserting expressions for μ_{ik}, eq. (4.119), into (4.143) yields

$$\left.\begin{aligned} &\frac{1+r^2+\gamma^2}{4}\ddot{u}_1 + \frac{1-r^2-\gamma^2}{4}\ddot{u}_2 + \frac{\gamma}{2}\ddot{u}_3 + \frac{k^2}{2}u_1 + \frac{H_1}{m}\operatorname{sign}\dot{u}_1 \\ &= \frac{\gamma}{2}\xi_0\omega^2\cos\omega t, \\ &\frac{1-r^2-\gamma^2}{4}\ddot{u}_1 + \frac{1+r^2+\gamma^2}{4}\ddot{u}_2 - \frac{\gamma}{2}\ddot{u}_3 + \frac{k^2}{2}u_2 + \frac{H_1}{m}\operatorname{sign}\dot{u}_2 \\ &= -\frac{\gamma}{2}\xi_0\omega^2\cos\omega t, \\ &\frac{\gamma}{2}\ddot{u}_1 - \frac{\gamma}{2}\ddot{u}_2 + \ddot{u}_3 + k_1^2u_3 + \frac{2H_3}{m}\operatorname{sign}\dot{u}_3 = \xi_0\omega^2\cos\omega t. \end{aligned}\right\} \tag{4.144}$$

Here

$$r = \frac{\rho_z}{a}, \; \gamma = \frac{h}{a}, \; k^2 = \frac{2c}{m}, \; k_1^2 = \frac{2c_1}{m}. \tag{4.145}$$

Forced vibration in coordinates x and θ_z is observed in the system under the prescribed external excitation. This vibration implies the following

$$y_c = 0, \; u_1 = -u_2.$$

The first two equations in (4.144) hold identically, therefore we arrive at the following system of two differential equations in two unknowns

$$\left.\begin{aligned} \left(r^2+\gamma^2\right)\ddot{u}_1+\gamma\ddot{u}_3+k^2u_1+\frac{2H_1}{m}\text{sign}\dot{u}_1 &= \gamma\xi_0\omega^2\cos\omega t, \\ \gamma\ddot{u}_1+\ddot{u}_3+k_1^2u_3+\frac{2H_3}{m}\text{sign}\dot{u}_3 &= \xi_0\omega^2\cos\omega t. \end{aligned}\right\} \tag{4.146}$$

An approximate periodic solution of this system of nonlinear differential equations is sought in the form

$$u_1 = a_{01}+a_1\cos(\omega t+\varphi), \quad u_3 = a_{03}+a_3\cos(\omega t+\psi). \tag{4.147}$$

It can be easily proven by substituting eq. (4.147) into (4.146) that the constant terms in the latter equation vanish.

Harmonic linearisation of the dry friction forces due to eq. (3.111) gives

$$\left.\begin{aligned} \frac{2H_1}{m}\text{sign}\dot{u}_1 &\approx \frac{8H_1}{\pi m a_1\omega}\dot{u}_1, \\ \frac{2H_3}{m}\text{sign}\dot{u}_3 &\approx \frac{8H_3}{\pi m a_3\omega}\dot{u}_3. \end{aligned}\right\} \tag{4.148}$$

Comparing equations (4.146) and (4.147) indicates that the phase angle between external harmonic excitation and $u_1(t)$ is φ. For the sake of further convenience we subtract the phase angle φ on the right hand side of eq. (4.146), recast the linearised equations in the form

$$\left.\begin{aligned} \left(r^2+\gamma^2\right)\ddot{u}_1+\gamma\ddot{u}_3+k^2u_1+\frac{\eta_1}{a_1\omega}\dot{u}_1 &= \gamma\xi_0\omega^2\cos(\omega t-\varphi), \\ \gamma\ddot{u}_1+\ddot{u}_3+k_1^2u_3+\frac{\eta_3}{a_3\omega}\dot{u}_3 &= \xi_0\omega^2\cos(\omega t-\varphi), \end{aligned}\right\} \tag{4.149}$$

where

$$\eta_1 = \frac{8H_1}{\pi m}, \; \eta_3 = \frac{8H_3}{\pi m},$$

and seek the solution in the following form

$$u_1 = a_1\cos\omega t, \quad u_3 = a_3\cos(\omega t+\psi). \tag{4.150}$$

Substituting eq. (4.150) into (4.149) and equating the factors in $\sin\omega t$ and $\cos\omega t$ on both sides to each other leads to the following equations

$$\left.\begin{aligned} \left[k^2-\omega^2\left(r^2+\gamma^2\right)\right]a_1-\omega^2\gamma a_3\cos\psi&=\gamma\xi_0\omega^2\cos\varphi,\\ \omega^2\gamma a_3\sin\varphi-\eta_1&=\gamma\xi_0\omega^2\sin\varphi,\\ -\gamma\omega^2 a_1+\left(k_1^2-\omega^2\right)\cos\psi a_3-\eta_3\sin\psi&=\xi_0\omega^2\cos\varphi,\\ -\left(k^2-\omega^2\right)a_3\sin\psi-\eta_3\cos\psi&=\xi_0\omega^2\sin\varphi. \end{aligned}\right\} \tag{4.151}$$

The unknowns a_1, a_3, φ and ψ can be obtained from these equations, however solving these equations is very laborious. It is expedient to remove the phases, and to this purpose, we multiply the third and fourth equations by γ and subtract the results respectively from the first and second equations. This yields

$$\left.\begin{aligned} \gamma k_1^2 a_3\cos\psi-\gamma\eta_3\sin\psi&=\left(k^2-\omega^2 r^2\right)a_1,\\ \gamma\eta_3\cos\psi+\gamma k_1^2 a_3\sin\psi&=\eta_1. \end{aligned}\right\} \tag{4.152}$$

Raising each equation in the second power and summing leads to the following equation relating a_1 and a_3

$$a_1^2=\frac{\gamma^2 k_1^4 a_3^2+\gamma^2\eta_3^2-\eta_1^2}{\left(k^2-\omega^2 r^2\right)^2}. \tag{4.153}$$

On the other hand, solving the system (4.152) for $\sin\psi$ and $\cos\psi$ we obtain

$$\sin\psi=\frac{\eta_1 k_1^2 a_3-\left(k^2-\omega^2 r^2\right)\eta_3 a_1}{\gamma k_1^4 a_3^2+\gamma\eta_3^2},\quad \cos\psi=\frac{\left(k^2-\omega^2 r^2\right)k_1^2 a_3-\eta_1\eta_3}{\gamma k_1^4 a_3^2+\gamma\eta_3^2}. \tag{4.154}$$

Raising now the first two equations in (4.151) in the second power and adding them yields

$$\begin{aligned} &\left[k^2-\omega^2\left(r^2+\gamma^2\right)\right]^2 a_1^2+\gamma^2\omega^4 a_3^2+\eta_1^2\\ &-2\gamma\omega^2\left[k^2-\omega^2\left(r^2+\gamma^2\right)\right]a_1 a_3\cos\psi-2\gamma\omega^2 a_3\eta_1\sin\psi=\gamma^2\xi_0^2\omega^4. \end{aligned} \tag{4.155}$$

Inserting eq. (4.154) into (4.155) results in a second equation relating a_1 and a_3

$$\begin{aligned} &a_1^2\eta_3^2\omega^2\left[\gamma^2\omega^2-2\left(k^2-\omega^2 r^2\right)\right]+a_3^2\left[\eta_3^2\left(k_1^2-\omega^2\right)^2+k_1^4\left(\eta_3^2-\xi_0^2\omega^4\right)\right]\\ &+a_3^4\left(k_1^2-\omega^2\right)^2 k_1^4+a_1^2 a_3^2\left[k_1^2\gamma^2\omega^4-2\omega^2\left(k_1^2-\omega^2\right)\left(k^2-\omega^2 r^2\right)\right]k_1^2\\ &+2a_1 a_3\eta_1\eta_3\omega^4+\eta_3^2\left(\eta_3^2-\xi_0^2\omega^4\right)=0. \end{aligned} \tag{4.156}$$

Removing a_1 from eq. (4.156) by means of eq. (4.153) we arrive at the following equation of fourth order in a_3^2

$$\begin{aligned}
&\langle a_3^4 k_1^4 \left[\left(k_1^2-\omega^2\right)\left(k^2-\omega^2 r^2\right)-k_1^2\gamma^2\omega^2\right]^2 + a_3^2\{\gamma^2 k_1^4[\gamma^2\omega^4 \\
&-2\omega^2\left(k^2-\omega^2 r^2\right)\xi_0^2] + \left(k^2-\omega^2 r^2\right)^2\left[\eta_3^2\left(k_1^2-\omega^2\right)^2\right. \\
&\left.+\left(\eta_3^2-\xi_0^2\omega^4\right)k_1^4\right] + k_1^2\left(\gamma^2\eta_3^2-\eta_1^2\right)[\gamma^2\omega^4 k_1^2 \\
&-2\omega^2\left(k_1^2-\omega^2\right)\left(k^2-\omega^2 r^2\right)]\} + \left(\gamma^2\eta_3^2-\eta_1^2\right)[\gamma^2\omega^4 \\
&-2\omega^2\left(k^2-\omega^2 r^2\right)] + \eta_3^2\left(\eta_3^2-\xi_0^2\omega^4\right)\left(k^2-\omega^2 r^2\right)^2\rangle^2 \qquad (4.157) \\
= \;& 4a_3^4\eta_1^2\eta_3^2\omega^{12}\left(k^2-\omega^2 r^2\right)^2\gamma^2 + 4a_3^2\eta_1^2\eta_3^2\omega^8\left(k^2-\omega^2 r^2\right)^2\left(\gamma^2\eta_3^2-\eta_1^2\right).
\end{aligned}$$

The absence of positive roots of this equations means that the system under certain excitation is partly or completely locked. The proper value of a_3^2 must also ensure that eq. (4.153) yields a positive value of a_1^2, i.e. the following condition must hold

$$\gamma^2 k_1^4 a_3^2 + \gamma^2\eta_3^2 - \eta_1^2 > 0. \qquad (4.158)$$

Obtaining values of the amplitudes and phases one can estimate the performance of the vibration protections system. To this aim, one calculates the absolute acceleration of a point of the object to be protected, the center of inertia being considered in the cases. By analogy with a single-degree-of-freedom systems one should determine this acceleration directly from the equations of motion rather than from the linearised equations.

In the example under consideration, the absolute acceleration is directed along the x axis and is given by

$$\ddot{x} = \ddot{u}_3 - \xi_0\omega^2\cos\omega t. \qquad (4.159)$$

It follows from the second equation in (4.146) that

$$\ddot{x} = -\gamma\ddot{u}_1 - k_1^2 u_3 - \frac{2H_3}{m}\operatorname{sign}\dot{u}_3. \qquad (4.160)$$

Obtaining $\ddot{u}_1$ from the first equation in (4.146) and substituting this into (4.160) yields

$$\ddot{x} = \frac{\gamma}{r^2+\gamma^2}\left(\gamma\ddot{x} + k^2 u_1 + \frac{2H_1}{m}\operatorname{sign}\dot{u}_1\right) - k_1^2 u_3 - \frac{2H_3}{m}\operatorname{sign}\dot{u}_3$$

or

$$\ddot{x} = \frac{\gamma}{r^2}\left(k^2 u_1 + \frac{2H_1}{m}\operatorname{sign}\dot{u}_1\right) - \frac{r^2+\gamma^2}{r^2}\left(k_1^2 u_3 + \frac{2H_3}{m}\operatorname{sign}\dot{u}_3\right). \qquad (4.161)$$

In the case of high frequency external excitation, amplitudes a_1 and a_2 are usually small. While obtaining the amplitude of $\ddot{x}$ we can neglect terms which are linear in u_1 and u_3 in eq. (4.161) and adopt the following first approximation

$$|\ddot{x}|_{\max} \approx \frac{\gamma}{r^2} \cdot \frac{2H_1}{m} + \frac{\gamma^2 + r^2}{r^2} \cdot \frac{2H_3}{m}. \tag{4.162}$$

We are allowed to ignore signum-functions in eq. (4.161) because there always exists such an instant of time at which both terms have the same sign (provided that $\psi \neq 0$).

Let us return to eq. (4.157). If the factor in the higher order of the unknown, i.e. a_3^8, tends to zero, one of the roots tends to infinity which implies that this root unconditionally satisfies condition (4.158). Thus, at the frequency satisfying the following equation

$$\left(k_1^2 - \omega^2\right)\left(k^2 - \omega^2 r^2\right) - k_1^2 \gamma^2 \omega^2 = 0, \tag{4.163}$$

there occurs a resonance in the system. The vibration amplitude grows without bounds and the vibration ultimately leaves the range of linearity of the elastic characteristics. Provided that such vibrations are inadmissible, the system must remain locked at all frequencies which satisfy eq. (4.163), the latter coinciding with the frequency equation for the linear system. It is easy to show that any system with linear elastic characteristics and dry friction dampers possesses a similar property, namely, if the system is locked at the frequency of external excitation coinciding with one of the eigenfrequencies, then the vibration amplitude in the system increases (theoretically) without bounds. Considering this property one usually chooses the dry friction forces in such a way that the system is locked at all eigenfrequencies.

Assuming that both conditions hold in the example under consideration we obtain from eq. (4.146)

$$\frac{2H_1}{m} > \gamma \xi_0 \lambda_{\max}^2, \quad \frac{2H_3}{m} > \xi_0 \lambda_{\max}^2. \tag{4.164}$$

Solving frequency equation (4.163) yields

$$\lambda_{\max}^2 = \frac{k^2 + k_1^2 r^2 + k_1^2 \gamma^2 + \sqrt{\left(k^2 + k_1^2 r^2 + k_1^2 \gamma^2\right)^2 - 4 k_1^2 k^2 r^2}}{2r^2}. \tag{4.165}$$

Condition (4.162) now renders

$$|\ddot{x}|_{\max} > \frac{r^2 + 2\gamma^2}{r^2} \xi_0 \lambda_{\max}^2 = \left(1 + \frac{2\gamma^2}{r^2}\right) \xi_0 \lambda_{\max}^2. \tag{4.166}$$

Therefore, similar to single-degree-of-freedom systems, Coulomb friction essentially worsens vibration protection properties of the system at high

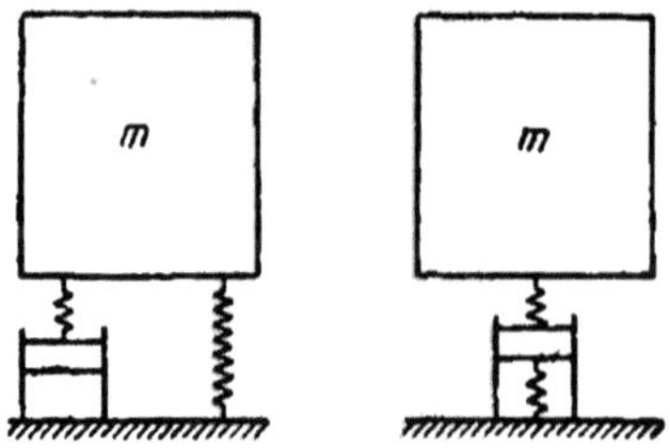

FIGURE 4.13.

frequency external vibration excitation. The vibration protection performance can be achieved by making changes in the design of an isolator with dry friction. For example, the dry friction damper can be attached to the object by means of additional elastic elements, as schematically shown in Fig. 4.13. A detailed analysis of such isolators was carried out in [60].

4.7 Forced vibration of elastic bodies

Up until now, the object to be protected has been considered to be a rigid body. However actual systems of vibration protection are elastic objects. This Section is devoted to the methods of analysis of vibration protection systems containing protected objects with linear elastic characteristics. The assumption of linearity of the elastic properties turns out to be acceptable in the majority of cases since the deflections of the bodies to be protected usually remain small within the range of linearity of elastic characteristics.

We restrict our consideration to the analysis of the simplest systems consisting of an elastic body mounted on elastic isolator, see Fig. 4.14. The point of attachment of the isolator to the object is assumed to move along the x axis only. Also, some small element in the vicinity of this attachment point is assumed to move in the same direction. Another assumption is that the base motion obeys the harmonic law, i.e.

$$\xi(t) = \xi_0 \cos \omega t.$$

If the object to be protected were a rigid body, we would have a single-degree-of-freedom system considered at great length in Chapter 3. The number of degrees of freedom of the elastic system is determined by the choice of the model which is either a lumped-parameter model, i.e. an assemblage of rigid bodies connected by elastic elements, or a distributed-parameters model. The methods of choice of the model are adequately covered in the literature and will be not addressed here. We note only that the choice of the model is essentially dependent on the frequency spectrum of the external excitation. Provided that these frequencies are considerably higher than the eigenfrequencies of a solid, one can consider this solid

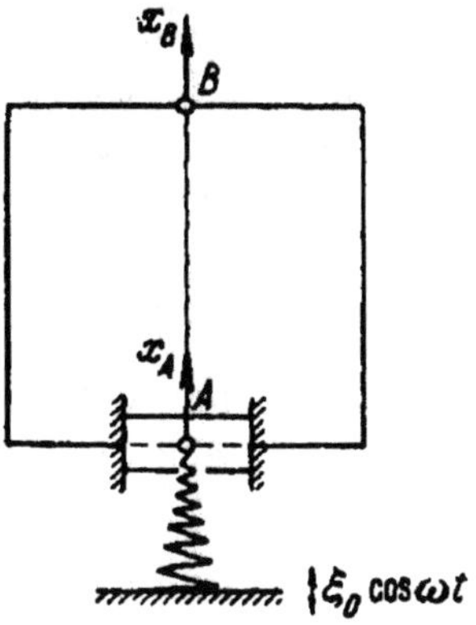

FIGURE 4.14.

as a rigid body. Furthermore, the higher the frequencies of the external excitation, the more complex the model of the object to be protected.

Equations of motion of the vibration protection system contain the single nonlinearity, namely, the characteristics of the elastic isolator, so that they may be solved by the above methods. Very often, this way proves to be too complicated. This is especially true with respect to such elastic objects which enable no simple model but permit the possibility of experimental analysis of the frequency properties. Under these circumstances, it is reasonable to apply another method based of description of the frequency properties of the linear system by means of the transfer matrices. A very short description needed for solving the problem stated above is given here. The details of the transfer matrices method can be found for instance in [129] and [94].

Consider an arbitrary complex linear system which is connected to another systems at points A and B, see Fig. 4.15. Assume that the displacement of each of the points A and B is subjected to constraints and determined by the single coordinate, x_A or x_B, respectively. Let the system undergo harmonic vibration of frequency ω which implies in particular the following

$$x_A = \xi_A \cos(\omega t + \varphi_A), \quad x_B = \xi_B \cos(\omega t + \varphi_B). \tag{4.167}$$

FIGURE 4.15.

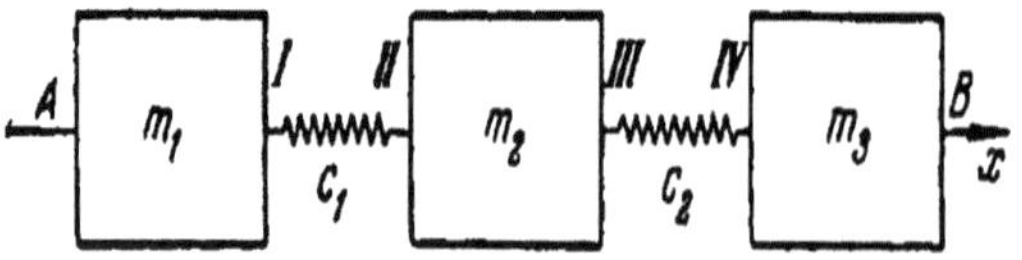

FIGURE 4.16.

The forces at points A and B which represent the forces of interaction of the system under consideration with the adjacent systems obey the harmonic law

$$X_A = P_A \cos(\omega t + \psi_A), \quad X_B = P_B \cos(\omega t + \psi_B). \tag{4.168}$$

As frequency ω is prescribed, the unknowns x_A, x_B, X_A and X_B are completely defined by the complex amplitudes $\bar{\xi}_A$, $\bar{\xi}_B$, $\bar{P}_A$ and $\bar{P}_B$ given by

$$\left|\bar{\xi}_A\right| = \xi_A, \ \arg \bar{\xi}_A = \varphi_A; \ \left|\bar{\xi}_B\right| = \xi_B, \ \arg \bar{\xi}_B = \varphi_B;$$
$$\left|\bar{P}_A\right| = P_A, \ \arg \bar{P}_A = \psi_A; \ \left|\bar{P}_B\right| = P_B, \ \arg \bar{P}_B = \psi_B.$$

The transfer matrix method relies on the following equation

$$\left.\begin{aligned} \bar{\xi}_B &= s_{11}(\omega)\,\bar{\xi}_A + s_{12}(\omega)\,\bar{P}_A, \\ \bar{P}_B &= s_{21}(\omega)\,\bar{\xi}_A + s_{22}(\omega)\,\bar{P}_A, \end{aligned}\right\} \tag{4.169}$$

which says that there always exist linear dependences between the complex amplitudes. In the latter equation $s_{11}(\omega)$, $s_{12}(\omega)$, $s_{21}(\omega)$ and $s_{22}(\omega)$ are some complex-valued functions of ω. The matrix composed of these functions is called the *transfer matrix* for the linear system

$$S_{AB}(\omega) = \left\| \begin{matrix} s_{11}(\omega) & s_{12}(\omega) \\ s_{21}(\omega) & s_{22}(\omega) \end{matrix} \right\|. \tag{4.170}$$

The transfer matrices can be introduced in the more general case in which the "input" displacement x_A and the "output" displacement x_B are determined by a few parameters, i.e. x_A and x_B may be viewed as vectors, however this case is not addressed in what follows. In order to obtain the transfer matrix, the system is split into a number of parts in series. The transfer matrix must be determined for each part, the transfer matrix of the entire system being obtained by multiplying together the transfer matrices of the separate parts.

Consider, for example, the system which consists of a number of masses which are linked by elastic elements. Figure 4.16 displays a system of three masses m_1, m_2 and m_3 connected by the elastic elements of stiffnesses c_1 and c_2. In what follows we use the spring compliances $E_1 = c_1^{-1}$ and $E_2 = c_2^{-1}$. In order to apply free-body diagrams of the vibrating masses, we separate the system in the cross-sections I-IV into the different parts. The first part is the mass m_1. When this is subjected to harmonic vibration of frequency ω and amplitude $\bar{\xi}_A$, then, clearly

$$\left.\begin{aligned}\bar{\xi}_I &= \bar{\xi}_A,\\ \bar{P}_I &= \bar{P}_A - m_1\omega^2\bar{\xi}_A.\end{aligned}\right\} \tag{4.171}$$

The transfer matrix for mass m_1 is thus obtained

$$S_{A-I} = \left\| \begin{matrix} 1 & 0 \\ -m_1\omega^2 & 1 \end{matrix} \right\|. \tag{4.172}$$

The governing equations for part I-II are given by

$$\left.\begin{aligned}\bar{\xi}_{II} &= \bar{\xi}_I + E\bar{P}_I\\ \bar{P}_{II} &= \bar{P}_I,\end{aligned}\right\} \tag{4.173}$$

therefore

$$S_{I-II} = \left\| \begin{matrix} 1 & E_1 \\ 0 & 1 \end{matrix} \right\|. \tag{4.174}$$

We obtain by analogy

$$\left.\begin{aligned} S_{II-III} = \left\| \begin{matrix} 1 & 0 \\ -m_2\omega^2 & 1 \end{matrix} \right\|, \quad S_{III-IV} = \left\| \begin{matrix} 1 & E_2 \\ 0 & 1 \end{matrix} \right\|,\\ S_{IV-B} = \left\| \begin{matrix} 1 & 0 \\ -m_3\omega^2 & 1 \end{matrix} \right\|.\end{aligned}\right\} \tag{4.175}$$

The relationships between $\bar{\xi}_B$, $\bar{P}_B$ and $\bar{\xi}_A$, $\bar{P}_A$ can be written down in matrix form as follows

$$\left\| \begin{matrix} \bar{\xi}_B \\ \bar{P}_B \end{matrix} \right\| = S_{AB} \left\| \begin{matrix} \bar{\xi}_A \\ \bar{P}_A \end{matrix} \right\|. \tag{4.176}$$

Matrix S_{AB} is defined as the product of transfer matrices for each part

$$S_{AB} = S_{IV-B} \cdot S_{III-IV} \cdot S_{II-III} \cdot S_{I-II} \cdot S_{A-I} = \left\| \begin{matrix} s_{11} & s_{12} \\ s_{21} & s_{22} \end{matrix} \right\|.$$

Multiplying the matrices in the latter equation yields

$$\begin{aligned}
s_{11}(\omega) &= \left(1 - m_1E_1\omega^2\right)\left(1 - m_2E_2\omega^2\right) - E_2m_1\omega^2,\\
s_{12}(\omega) &= E_1 + E_2 - E_1E_2m_2\omega^2,\\
s_{21}(\omega) &= -\left(m_1 + m_2 + m_3\right)\omega^2 + \left(m_1m_2E_1 + m_1m_3E_2 + m_1m_3E_1\right.\\
&\quad \left.+ m_3m_2E_2\right)\omega^4 - m_1m_2m_3E_1E_2\omega^6,\\
s_{22}(\omega) &= \left(1 - m_2E_1\omega^2\right)\left(1 - m_3E_2\omega^2\right) - m_3E_1\omega^2.
\end{aligned}$$

The entries are all real-valued which is explained by the fact that no system damping is taken into account.

Let us construct the transfer matrix for the elastic element with linear friction. The element deflection u is related to the forces X_A and X_B acting on the element by means of the following equation

$$X_A = X_B = b\dot{u} + cu. \tag{4.177}$$

If $X_A = \bar{P}_A e^{i\omega t}$, eq. (4.177) yields

$$u = \frac{\bar{P}_A}{bi\omega + c} e^{i\omega t}. \tag{4.178}$$

On the other hand, $x_B = x_A + u$, i.e.

$$\begin{aligned} \bar{\xi}_B &= \bar{\xi}_A + \frac{\bar{P}_A}{bi\omega + c}, \\ \bar{P}_B &= \bar{P}_A, \end{aligned}$$

which means that the transfer matrix is

$$S_{AB} = \left\| \begin{array}{cc} 1 & \bar{E} \\ 0 & 1 \end{array} \right\|, \tag{4.179}$$

where $\bar{E} = (bi\omega + c)^{-1}$ is referred to as the complex compliance of a linear element with friction.

The method of dynamic compliances is one of the modifications of the transfer matrix method. In the simplest case under consideration, the dynamic compliance of the linear system at point A is the ratio of the amplitude of displacement of this point to the amplitude of the acting force

$$e_A(\omega) = \frac{\bar{\xi}_A}{\bar{P}_A}. \tag{4.180}$$

Dividing the first of the equations in (4.169) by the second one yields

$$\bar{e}_B(\omega) = \frac{\bar{\xi}_B}{\bar{P}_B} = \frac{s_{11}(\omega)\,\bar{e}_A(\omega) + s_{12}(\omega)}{s_{21}(\omega)\,\bar{e}_A(\omega) + s_{22}(\omega)}. \tag{4.181}$$

Thus, the dynamic compliances of the input and output of the linear system are related by a certain rational relationship.

Governing equation (4.171) for the lump mass m_1 yields

$$\bar{e}_B(\omega) = \frac{\bar{e}_A}{1 - m_1\omega^2\bar{e}_A}. \tag{4.182}$$

For the elastic element with friction one obtains

$$\bar{e}_B = \bar{e}_A + \bar{E}. \tag{4.183}$$

By virtue of the latter two equations one can express the dynamic compliance at point B for the three mass system under consideration. Expression for the dynamic compliance is the following continued fraction

$$\begin{aligned}\bar{e}_B &= \frac{\bar{e}_{IV}}{1-m_3\omega^2\bar{e}_{IV}} = \frac{1}{-m_3\omega^2+\dfrac{1}{\bar{e}_{IV}}} \\ &= \frac{1}{-m_3\omega^2+\dfrac{1}{E_2+\bar{e}_{III}}} = \frac{1}{-m_3\omega^2+\dfrac{1}{E_2+\dfrac{1}{-m_2\omega^2+\bar{e}_{II}^{-1}}}} \\ &= \frac{1}{-m_3\omega^2+\dfrac{1}{E_2+\dfrac{1}{-m_2\omega^2+\dfrac{1}{E_1+\dfrac{1}{-m_1\omega^2+\bar{e}_A^{-1}}}}}}. \end{aligned} \tag{4.184}$$

Now we consider the question as to how can the transfer matrix method be used for investigating vibration protection systems.

Assume that the forced vibration in the system "linear elastic body plus nonlinear isolator" subjected to harmonic driving is close to harmonic vibration. Let the vibration amplitude of a certain point B in a chosen direction be sought, Fig. 4.14. In this case, eq. (4.169) relates the system output x_B to the system input x_A.

Provided that displacement x_B causes no force F_B, i.e. the system has a "free" output, see. Fig. 4.14, then

$$\bar{P}_B = s_{21}(\omega)\,\bar{\xi}_A(\omega) + s_{22}(\omega)\,\bar{P}_A = 0. \tag{4.185}$$

The harmonic linearisation of the dynamic characteristic of the vibration isolator yields

$$U(u,\dot{u}) \approx U_0 + c_d u^0 + b_d \dot{u}^0, \tag{4.186}$$

where U_0, c_d and b_d are factors of the harmonic linearisation for nonlinear force $U(u,\dot{u})$ and functions of the constant component a_0, amplitude a and frequency ω.

If no time-independent external force acts on the protected object (this is always the case provided that the object is completely isolated and has no motionless points), then

$$U_0(a,a_0) = 0,$$

that is a_0 is determined in terms of a which in turn implies that factors c_d and b_d can be considered as known functions of amplitude a.

We can construct now the transfer matrix for the linearised element. The following relations hold

$$\left.\begin{aligned}\bar{\xi}_A &= \xi_0 + \frac{\bar{P}_0}{c_d + b_d i\omega} = \xi_0 + \bar{E}_d(a,\omega)\,\bar{P}_0,\\ \bar{P}_A &= \bar{P}_0,\end{aligned}\right\} \tag{4.187}$$

where

$$\bar{E}_d(a,\omega) = \frac{1}{c_d(a) + b_d(a,\omega)\,i\omega}.$$

Inserting eq. (4.187) into condition (4.185) we obtain

$$s_{21}(\omega)\,\xi_0(\omega) + \left[s_{21}(\omega)\,\bar{E}_d(a,\omega) + s_{22}(\omega)\right]\bar{P}_0 = 0. \tag{4.188}$$

It follows from this equation that

$$\bar{P}_0 = -\xi_0\frac{s_{21}(\omega)}{\bar{E}_d(a,\omega)\,s_{21}(\omega) + s_{22}(\omega)} = -\xi_0\frac{1}{\bar{E}_d(a,\omega) + \dfrac{s_{22}(\omega)}{s_{21}(\omega)}}. \tag{4.189}$$

On the other hand, eq. (4.178) yields

$$\begin{aligned}a &= \left|\frac{\bar{P}_0}{c_d + b_d i\omega}\right| = \left|\bar{E}_d(a,\omega)\,\bar{P}_0\right| = \xi_0\left|\frac{\bar{E}_d}{\bar{E}_d + \dfrac{s_{22}(\omega)}{s_{21}(\omega)}}\right| = \\ &= \xi_0\left|\frac{1}{1 + \dfrac{s_{22}(\omega)}{s_{11}(\omega)}\left[c_d(a) + b_d(a,\omega)\,i\omega\right]}\right|.\end{aligned} \tag{4.190}$$

The amplitude of the isolator deflection can be obtained from this equation. By virtue of eq. (4.185)

$$\frac{s_{22}(\omega)}{s_{21}(\omega)} = -\frac{\bar{\xi}_A}{\bar{P}_A} = -\bar{e}_A(\omega), \tag{4.191}$$

we are able to cast eqs. (4.189) and (4.190) in the following form

$$\bar{P}_0 = -\frac{\xi_0}{\bar{E}_d(a,\omega) - \bar{e}_A(\omega)}, \tag{4.192}$$

$$a = \frac{\xi_0}{\left|1 - \bar{e}_A(\omega)\left[c_d(a) + b_d(a,\omega)\,i\omega\right]\right|}. \tag{4.193}$$

In all these equations ξ_0 is a real value, i.e. the phase angle of the external excitation is assumed to be equal to zero.

Taking into account that $\xi_0 = 0$ and $\bar{P}_0 = 0$ for free vibration one can obtain from eq. (4.192) the following equation for the system eigenfrequencies

$$\bar{E}_d(a, \omega) - \bar{e}_A(\omega) = 0. \tag{4.194}$$

Letting $b_d = 0$ in the latter equation leads to the conventional frequency equation

$$\bar{e}_A(\omega)\, c_d(a) = 1. \tag{4.195}$$

Solving this equation allows one to determine the dependence of the system eigenfrequency on the amplitude of deflection of the isolator.

Obtaining amplitude a from eq. (4.193) and substituting the result into the expression for $\bar{E}_d$ one can find $\bar{P}_0$, $\bar{\xi}_A$ and $\bar{P}_A$ from eqs. (4.189) and (4.187). The stated problem will be solved when one inserts the expressions for $\bar{\xi}_A$ and $\bar{P}_A$ into (4.169) and obtains $\bar{\xi}_B$.

The displacement amplitude of any "free" point of the object to be protected can be obtained by analogy. In what follows we consider some peculiarities of the approach in the case of "non-free" output for two particular examples. The above analysis shows that while analysing the forced vibration of the system it is necessary to dispose the dynamic compliance of the elastic object $\bar{e}_A(\omega)$ at the point of the isolator attachment. This characteristic can be obtained experimentally with relative ease. To this end, it suffices to mount the object to be protected without isolator on a shaker at point A, and then record the amplitude of the interaction force between the shaker and the object. Provided that the amplitude and the frequency of the harmonic vibration of the shaker are ω and ξ_A, respectively, one obtains due to eq. (4.185)

$$\bar{e}_A(\omega) = \frac{\bar{P}_A}{\xi_A}.$$

If amplitude $\bar{\xi}_B$ is also recorded within this test, one can easily determine the value of this amplitude after the object has been mounted on the isolator.

Indeed,

$$\begin{aligned} \bar{\xi}_B &= s_{11}(\omega)\,\bar{\xi}_A + s_{12}(\omega)\,\bar{P}_A \\ &= \left[s_{11}(\omega)\,\bar{e}_A(\omega) + s_{12}(\omega)\right]\bar{P}_A = \bar{e}'(\omega)\,\bar{P}_A. \end{aligned} \tag{4.196}$$

The value of $\bar{e}'(\omega)$ does not depend upon the isolator's properties, that is this can be determined experimentally. Considering eq. (4.192) one can obtain $\bar{\xi}_B$ for the isolated system in the following form

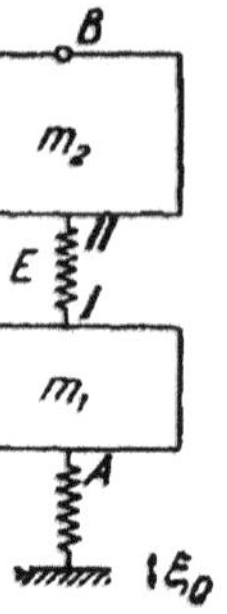

FIGURE 4.17.

$$\bar{\xi}_B = \bar{e}'(\omega)\bar{P}_A = \bar{e}'(\omega)\bar{P}_0 = -\xi_0 \frac{\bar{e}'(\omega)}{\bar{E}_d(a,\omega) - \bar{e}_A(\omega)}, \tag{4.197}$$

the deflection amplitude a being determined from eq. (4.193).

The system "output" is not necessarily the displacement of one of the points of the isolated object. This can be any parameter of interest, for example, stress in one of the structural elements or even voltage provided that the isolated object is an electromechanical system. Clearly, it is necessary to ensure the linear dependence of the chosen output parameter on the input, i.e. vibration at point A. Experimental proof of the linearity of the system causes ordinarily no difficulties.

We proceed now to some particular examples.

1. Let us determine the transfer matrix for the two-degrees-of-freedom system depicted in Fig. 4.17. Splitting the system into two parts and applying eqs. (4.172) and (4.174) yields

$$\begin{aligned} S_{AB} &= \left\| \begin{matrix} 1 & 0 \\ -m_2\omega^2 & 1 \end{matrix} \right\| \left\| \begin{matrix} 1 & E \\ 0 & 1 \end{matrix} \right\| \left\| \begin{matrix} 1 & E \\ -m_1\omega^2 & 1 - Em_1\omega^2 \end{matrix} \right\| = \\ &= \left\| \begin{matrix} 1 - Em_1\omega^2 & E \\ -m_1\omega^2 - m_2\omega^2 + Em_1m_2\omega^4 & 1 - Em_2\omega^2 \end{matrix} \right\|. \end{aligned}$$

The dynamic compliance in point A is therefore as follows

$$e_A(\omega) = \frac{1 - Em_2\omega^2}{m_1\omega^2 + m_2\omega^2 - Em_1m_2\omega^4} = -\frac{s_{22}(\omega)}{s_{21}(\omega)}. \tag{4.198}$$

The same result can be obtained by means of another method, namely, using eqs. (4.182) and (4.183) we arrive at the following chain of equations

$$\bar{e}_B(\omega) = \frac{1}{-m_2\omega^2 + \dfrac{1}{\bar{e}_{II}}} = \frac{1}{-m_2\omega^2 + \dfrac{1}{E + e_1}}$$

$$= \frac{1}{-m_2\omega^2 + \cfrac{1}{E + \cfrac{1}{-m_1\omega^2 + \cfrac{1}{e_A}}}}. \tag{4.199}$$

On the other hand, $\bar{e}_B(\omega) = \infty$ because $P_B = 0$. Thus, the denominator in the latter equation must vanish identically. This leads to the expression (4.198) for e_A.

Let us assume that the isolator consists of a linear elastic element and a dry friction damper. Then,

$$\frac{1}{\bar{E}_d(a,\omega)} = c + \frac{4H}{\pi a\omega} i\omega = c + \frac{\eta}{a} i, \tag{4.200}$$

where

$$\eta = \frac{4H}{\pi}.$$

In this case, eq. (4.193) takes the following form

$$a = \frac{\xi_0}{\left|1 - \bar{e}_A(\omega)\left(c + \frac{\eta}{a} i\right)\right|} = \frac{\xi_0}{\sqrt{[1 - ce_A(\omega)]^2 + \frac{e_A^2(\omega)\eta^2}{a^2}}}. \tag{4.201}$$

Solving this equation yields

$$a = \frac{\sqrt{\xi_0^2 - e_A^2(\omega)\eta^2}}{|1 - e_A c|}. \tag{4.202}$$

The amplitude-frequency characteristic for the system is thus determined. From the latter equation is it easy to establish the condition under which the isolator is locked

$$\xi_0 < |e_A(\omega)|\,\eta. \tag{4.203}$$

When $E = 0$, i.e. the masses are rigidly attached to one another, eq. (4.198) gives

$$e_A(\omega) = \frac{1}{(m_1 + m_2)\omega^2}.$$

Inserting this expression into eqs. (4.202) and (4.203) we arrive at the expressions corresponding to the isolation of the rigid body having mass $m_1 + m_2$.

We are now in a position to define the vibration amplitudes at various points of the isolated object. We first determine $\bar{P}_0$ by means of eq. (4.192), to get

$$\bar{P}_0 = -\frac{\xi_0}{\dfrac{1}{c+\dfrac{\eta}{a}i} - e_A(\omega)} = -\frac{\xi_0\left(c+\dfrac{\eta}{a}i\right)}{1-e_A(\omega)\left(c+\dfrac{\eta}{a}i\right)} = \bar{P}_A.$$

Amplitude $\bar{\xi}_A$ is obtained from eq. (4.187)

$$\bar{\xi}_A = \xi_0 + \bar{E}_d P_0 = \xi_0 - \frac{\xi_0}{1-e_A(\omega)\left(c+\dfrac{\eta}{a}i\right)} = -\xi_0\frac{e_A\left(c+\dfrac{\eta}{a}i\right)}{1-e_A\left(c+\dfrac{\eta}{a}i\right)}.$$

Now we can determine $\bar{\xi}_B$ as follows

$$\begin{aligned}
\bar{\xi}_B &= s_{11}(\omega)\bar{\xi}_A + s_{12}(\omega)\bar{P}_A \\
&= \left(1-Em_1\omega^2\right)\left[-\xi_0\frac{e_A\left(c+\dfrac{\eta}{a}i\right)}{1-e_A\left(c+\dfrac{\eta}{a}i\right)}\right] - \frac{E\xi_0\left(c+\dfrac{\eta}{a}i\right)}{1-e_A\left(c+\dfrac{\eta}{a}i\right)} \\
&= \frac{\xi_0\left(c+\dfrac{\eta}{a}i\right)}{1-e_A\left(c+\dfrac{\eta}{a}i\right)}\left(Ee_Am_1\omega^2 - e_A - E\right), \qquad (4.204)
\end{aligned}$$

$$\left|\bar{\xi}_B\right| = \xi_0\left|Ee_Am_1\omega^2 - e_A - E\right|\frac{\sqrt{c^2+\dfrac{\eta^2}{a^2}}}{\sqrt{\left(1-ce_A\right)^2+\dfrac{\eta^2}{a^2}}}. \qquad (4.205)$$

It is easy to find the amplitude of any system parameter. For example, let us determine the amplitude of the force in the elastic element E

$$\begin{aligned}
\bar{P}_I &= \bar{P}_A - m_1\omega^2\bar{\xi}_A = \\
&= \xi_0\left[-\frac{c+\dfrac{\eta}{a}i}{1-e_A\left(c+\dfrac{\eta}{a}i\right)} + \frac{m_1\omega^2 e_A\left(c+\dfrac{\eta}{a}i\right)}{1-e_A\left(c+\dfrac{\eta}{a}i\right)}\right] \\
&= \xi_0\frac{c+\dfrac{\eta}{a}i}{1-e_A\left(c+\dfrac{\eta}{a}i\right)}\left(m_1\omega^2 e_A - 1\right), \qquad (4.206)
\end{aligned}$$

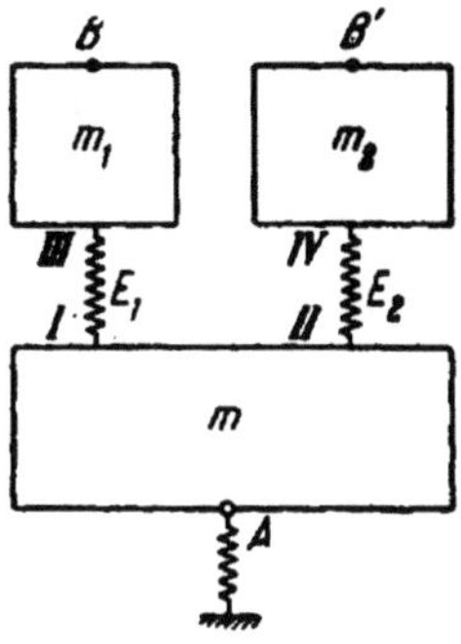

FIGURE 4.18.

$$\left|\bar{P}_I\right| = \xi_0 \left|m_1\omega^2 e_A - 1\right| \frac{\sqrt{c^2 + \dfrac{\eta^2}{a^2}}}{\sqrt{\left(1 - ce_A\right)^2 + \dfrac{\eta^2}{a^2}}}. \tag{4.207}$$

2. Consider now the three-degree-of-freedom object shown in Fig. 4.18. The forthcoming analysis is restricted by obtaining the dynamic compliance at point A. The other dynamic parameters can be determined by analogy.

In order to obtain e_A, we use the method of dynamic compliances. We take into account that the total dynamic compliance of two elastic elements in parallel is given by

$$\frac{1}{e_t} = \frac{1}{e_I} + \frac{1}{e_{II}}. \tag{4.208}$$

The dynamic compliance at points B and B' is equal to ∞. Hence,

$$\begin{aligned}
-e_A &= \frac{1}{-m\omega^2 + \dfrac{1}{e_t}} = \frac{1}{-m\omega^2 + \dfrac{1}{e_I} + \dfrac{1}{e_{II}}} \\
&= \frac{1}{-m\omega^2 + \dfrac{1}{e_{III} + E_1} + \dfrac{1}{e_{IV} + E_2}} \\
&= \frac{1}{-m\omega^2 + \dfrac{1}{E_1 + \dfrac{1}{-m_1\omega^2 + \dfrac{1}{e_B}}} + \dfrac{1}{E_2 + \dfrac{1}{-m_2\omega^2 + \dfrac{1}{e_B}}}} \\
&= \left(1 - E_1 m_1\omega^2\right)\left(1 - E_2 m_2\omega^2\right)\left[m_1\omega^2\left(E_2 m_2\omega^2 - 1\right)\right. \qquad (4.209) \\
&\quad \left. + m_2\omega^2\left(E_1 m_1\omega^2 - 1\right) - m\omega^2\left(E_1 m_1\omega^2 - 1\right)\left(E_2 m_2\omega^2 - 1\right)\right]^{-1}.
\end{aligned}$$

The minus in front of e_A is needed because we "move" from points B and B' to point A while deriving the equation for e_A. In the previous example an opposite direction was chosen.

It goes without saying that the conventional methods of deriving the equations of motion and their integration may be applied for solving the simple systems considered above. The merits of the approach of this Section manifest themselves while dealing with complex systems.

5

Nonlinear active systems

The influence of nonlinearity in passive elastic and dissipative elements on the dynamics of vibration protection systems has been studied in detail in Chapters 3 and 4, see also [65]. It was shown that the nonlinearity of elastic restoring characteristics and especially the presence of elastic and rigid stops lead to undesirable resonances. To suppress them, one should increase either the clearance or the system dissipation. In the present Chapter we study the influence of active feedbacks on the resonance phenomena. In addition to this, the influence of the nonlinearity in the active systems on the efficiency of the vibration protection system will be considered.

5.1 Resonant vibrations in nonlinear systems under harmonic excitation

Consider the simplest one-dimensional active system depicted in Fig. 5.1. Let us assume that the object is linear. We also assume that without the vibration isolator the dynamic excitations $F_1, ..., F_k$ result in a given displacement $\xi(t)$ of point A. According to the superposition principle, mounting the vibration isolator leads to the following equation for displacement $x(t)$

$$x(t) = \xi(t) + e_A(p)R(t), \tag{5.1}$$

where $e_A(p)$ is the dynamic compliance operator at point A and $R(t)$ stands for the total reaction of the vibration isolator's elements. This reaction is assumed to consist of the elastic restoring force

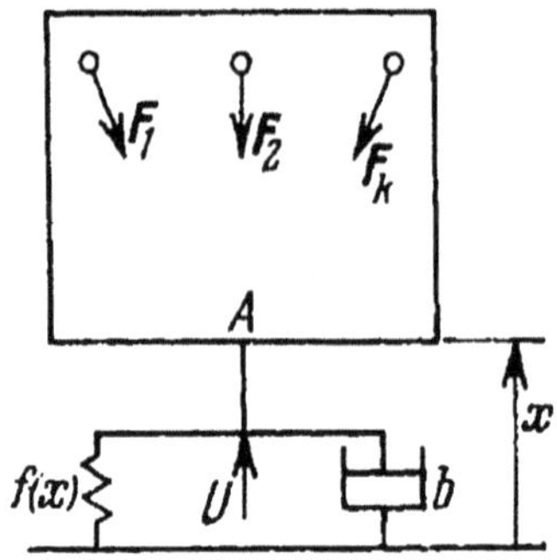

FIGURE 5.1.

$$R_e = -cx - f(x), \tag{5.2}$$

with $f(x)$ being some nonlinear characteristic, a linear dissipative force

$$R_d = -b\dot{x}, \tag{5.3}$$

and a control force produced by linearly feeding back the absolute displacement of point A

$$U = -w(p)x. \tag{5.4}$$

Substituting eqs. (5.2)-(5.4) into (5.1) yields

$$[d(p) + bp + c + w(p)]x + f(x) = d(p)\xi(t). \tag{5.5}$$

Here $d(p) = e_A^{-1}(p)$ denotes the dynamic rigidity operator at point A. Provided that all dynamic excitations are harmonic with frequency ω, the process $\xi(t)$ is harmonic, too. In this case the method of harmonic linearisation can be applied to find an approximate solution. Assuming that the solution is in the form

$$x(t) = a_0 + a\cos(\omega t + \varphi) = a_0 + x^0(t) \tag{5.6}$$

and replacing the nonlinear function $f(x)$ in (5.5) by the linear one

$$f^*(x) = f_0 + qx^0, \tag{5.7}$$

where

$$\left.\begin{aligned} f_0(a_0, a) &= \frac{1}{2\pi}\int_0^{2\pi} f(a_0 + a\cos\psi)d\psi, \\ q(a_0, a) &= \frac{1}{\pi a}\int_0^{2\pi} f(a_0 + a\cos\psi)\cos\psi d\psi, \end{aligned}\right\} \tag{5.8}$$

leads to the linearised equation

$$[d(p) + bp + c + w(p)](a_0 + x^0) + f_0 + qx^0 = d(p)\xi\,(t)\,. \tag{5.9}$$

Separating the constant and harmonic terms yields

$$[d(p) + bp + c + w(p) + q]x^0 = d(p)\{\xi_0 \cos\omega t\}, \tag{5.10}$$

$$[d\,(0) + c + \omega\,(0) + q]a_0 + f_0(a_0, a) = 0. \tag{5.11}$$

Using eq. (5.11) to express a_0 in terms of a and substituting the result into the linearisation factor $q\,(a_0, a)$ we obtain $q\,(a)$ which is only a function of a. Equation (5.10) renders the following expressions for the vibration amplitude a and the phase φ

$$a = \xi_0\,|d\,(i\omega)| \cdot |d\,(i\omega) + bi\omega + c + q(a) + w\,(i\omega)|^{-1}, \tag{5.12}$$

$$\varphi = \arg\{d\,(i\omega)\,[d\,(i\omega) + bi\omega + c + q\,(a) + w\,(i\omega)]^{-1}\}. \tag{5.13}$$

To establish the conditions under which the vibration amplitude a is much greater than ξ_0, i.e. resonance occurs, we write down eq. (5.12) in the form

$$\frac{a}{\xi_0} = \frac{d\,(i\omega)}{\sqrt{[\operatorname{Re} d\,(i\omega) + \operatorname{Re} w\,(i\omega) + c + q]^2 + [\operatorname{Im} d\,(i\omega) + b\omega + \operatorname{Im} w\,(i\omega)]^2}}\,. \tag{5.14}$$

Assuming that the following condition holds

$$|\operatorname{Im} d\,(i\omega) + b\omega + \operatorname{Im} w\,(i\omega)| \ll |d\,(i\omega)| \tag{5.15}$$

in a certain frequency domain implies that the resonances occur at frequencies which satisfy the equation

$$\operatorname{Re} d\,(i\omega) + \operatorname{Re} w\,(i\omega) + c + q\,(a) = 0. \tag{5.16}$$

Equation (5.14) gives the following expression for the vibration amplitude

$$a = \xi_0\,|d\,(i\omega)| \cdot |\operatorname{Im} d\,(i\omega) + b\omega + \operatorname{Im} w\,(i\omega)|^{-1}\,. \tag{5.17}$$

Thus, the resonance amplitude and the resonance frequency can be determined from the system of equations (5.16) and (5.17).

Let us consider the free vibration of an object mounted to the nonlinear elastic element under the assumption that the object and the vibration isolator have no dissipation and the active feedback is absent. In this case, eq. (5.5) is written in the form

$$[d_0(p) + c]x + f\,(x) = 0, \tag{5.18}$$

where $d_0(p)$ is the operator of dynamic rigidity for the object without dissipation. Looking for the solution in the form

$$x = a_0 + a\cos\lambda t \tag{5.19}$$

and applying the method of harmonic linearisation, it is easy to obtain the expression

$$d_0(i\lambda) + c + q\,(a) = 0, \tag{5.20}$$

which relates the amplitude and the frequency of the free vibration. The graph of dependence $a\,(\lambda)$ is known as the *backbone curve*, see Chapter 3. When the object is a rigid body of mass m, the equation for the backbone curve is written in the form

$$q\,(a) + c = m\lambda^2, \tag{5.21}$$

because in this case $d_0(i\lambda) = -m\lambda^2$.

Equation (5.16) may be viewed as an analogy of eq. (5.20) for the corresponding active system. For this reason, the graphs of the dependence $a\,(\lambda)$ corresponding to this equation are referred to as the *modified backbone curves.* To clarify the physical meaning of eq. (5.17) we consider some energetic relationships. Let us calculate the power of all forces in eq. (5.5) under the harmonic vibration (5.6) and average the result over the period of vibration. Multiplying both sides of eq. (5.5) by

$$\dot{x} = -a\omega\sin\left(\omega t + \varphi\right)$$

and integrating over time yields

$$\frac{1}{2}[\operatorname{Im} d\,(i\omega) + b\omega + \operatorname{Im} w\,(i\omega)]a^2\omega = \frac{1}{2}\left|d\,(i\omega)\right|\xi_0 a\omega\sin\psi, \tag{5.22}$$

where ψ denotes the phase angle between the response $x\,(t)$ and the "excitation force" $d\,(p)\,\xi\,(t)$. As $|\sin\psi| < 1$ eq. (5.22) gives

$$a \leq \left|d\,(i\omega)\right|\xi_0\left|\operatorname{Im} d\,(i\omega) + b\omega + \operatorname{Im} w\,(i\omega)\right|^{-1}. \tag{5.23}$$

Therefore, eq. (5.17) determines the maximal amplitude of vibration which can be obtained from the condition of power balance. With this in view, we refer to the graph $a\,(\omega)$ due to eq. (5.17) as the *curve of maximal amplitudes.*

The points of intersection of the modified backbone curves with the curve of maximal amplitudes determine the values of the resonant frequencies and amplitudes. Evidently, for vibrations to be resonant, their frequencies must satisfy eq. (5.15). Let us assume that the nonlinearity $f\,(x)$ is due to the rigid stops which are set symmetrically with the gap Δ from the equilibrium point. Then,

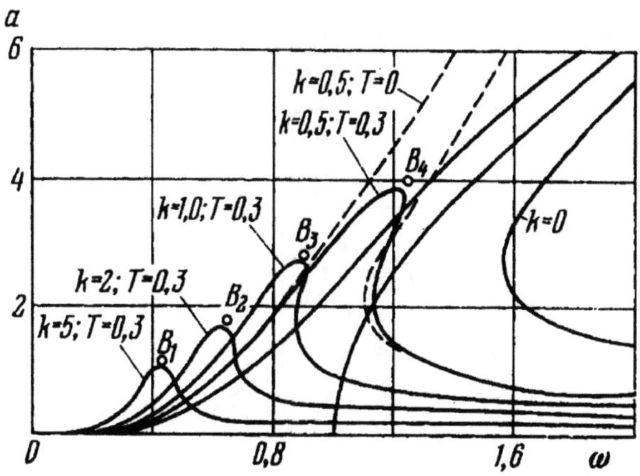

FIGURE 5.2.

$$q(a) \equiv 0 \text{ if } a < \Delta,\ a_0 = 0.$$

In this case, the resonant vibration can not be accompanied by impacts with the stops if

$$\Delta > \xi_0 \left| d(i\omega) \right| \left| \operatorname{Im} d(i\omega) + b\omega + \operatorname{Im} w(i\omega) \right|^{-1}. \tag{5.24}$$

Indeed, if the latter condition holds, the resonance amplitudes due to eq. (5.17) is smaller than Δ. So that, condition (5.24) determines a "safety" distance between the equilibrium and the rigid stop which ensures no periodic regimes with impacts upon the stops. As follows from this condition the safety distance can be reduced by means of such an active feedback which has Im $w(i\omega) > 0$. The substantiation is that Im $d(i\omega) > 0$, see Section 1.2.

In general, the feedback increases the safety distance at those frequencies at which Im $w(i\omega) < 0$ and, correspondingly, decreases this distance when Im $w(i\omega) > 0$. If the frequency can vary in a certain band, the distance Δ must be taken from the following condition

$$\Delta \geq \max_{\omega}\{\xi_0 \left| d(i\omega) \right| \left| \operatorname{Im} d(i\omega) + b\omega + \operatorname{Im} w(i\omega) \right|^{-1}\}. \tag{5.25}$$

Figure 5.2 displays the resonance curves (5.12) which correspond to

$$d(p) = mp^2,\ w(p) = \frac{kp^2}{Tp+1},\ f(x) = \gamma x^3.$$

The nondimensional values

$$\gamma = 0,1,\ \xi_0 = 1,\ c = 1,\ m = 1,\ b = 0,1.$$

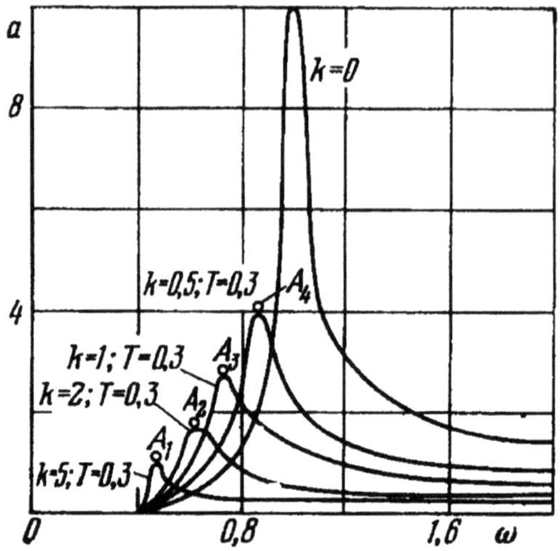

FIGURE 5.3.

were taken and the values of k and T are shown in Fig. 5.2. Figure 5.3 displays the curves for the same transfer functions $d(p)$ and $w(p)$, but for

$$f(x) = \gamma \text{sign} x.$$

The same numerical values of the parameters are taken. Figure 5.4 shows the points of intersection of the curves (5.16) and (5.17) plotted in Figs. 5.2 and 5.3. These points, depicted in Figs. 5.2 and 5.3, lie sufficiently close to the maxima of the resonance curves.

2. We consider now the resonant vibrations of a linear object mounted on N nonlinear vibration isolators, Fig. 5.5. Let $\boldsymbol{\xi}$ denote the vector of the kinematic excitations, i.e. the absolute displacements of the points of attachment of the vibration isolators to the base. We assume that $\boldsymbol{\xi}$ is a periodic function of time. Also, let $\mathbf{x}$ denote the displacement vector of the points of attachment of the vibration isolators to the object and $\mathbf{u}$ denote the relative displacement vector. Then,

$$\mathbf{x} = \mathbf{u} + \boldsymbol{\xi}. \tag{5.26}$$

Let us assume that the action of $k-th$ vibration isolator on the object consists of the nonlinear elastic force

$$F_k = -c_k u_k - f_k(u_k),$$

the dissipative force

$$G_k = -b_k \dot{u}_k$$

and the control force U_k produced by an active feedback. Evidently,

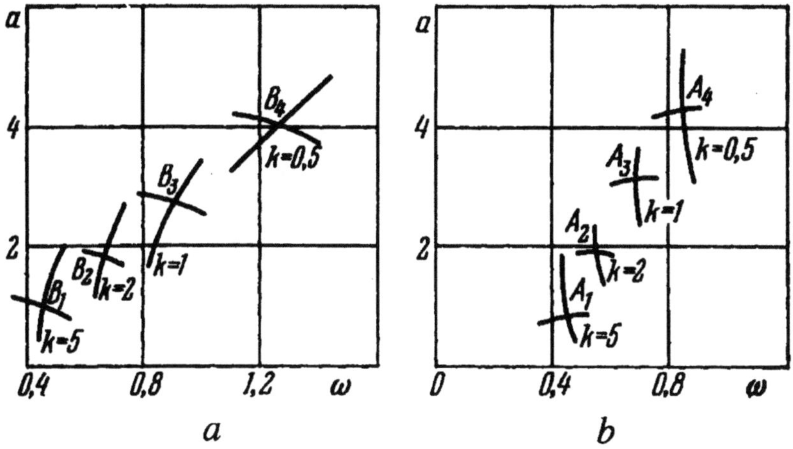

FIGURE 5.4.

$$\mathbf{x} = E(p)\{\mathbf{U} + \mathbf{G} + \mathbf{F}\}, \tag{5.27}$$

$E(p)$ being the matrix of dynamic compliance of the object at the attachment points; $\mathbf{U}$, $\mathbf{G}$ and $\mathbf{F}$ denoting the vectors of the corresponding forces. Assuming that

$$\mathbf{U} = -W(p)\,\mathbf{x},$$

where $W(p)$ is the matrix of transfer functions of the active feedbacks, we obtain from eqs. (5.27) and (5.26)

$$\mathbf{u} + \boldsymbol{\xi} = E(p)\{-W(p)(\mathbf{u} + \boldsymbol{\xi}) - Bp\mathbf{u} - C\mathbf{u} - \mathbf{f}(\mathbf{u})\}. \tag{5.28}$$

Here B and C are the diagonal matrices

$$B = diag\{b_1, ..., b_N\}, \ C = diag\{c_1, ..., c_N\}$$

and $\mathbf{f}(\mathbf{u})$ is the vector with components $f_k(u_k)$. Equation (5.28) yields

$$\{I_N + E(p)[W(p) + Bp + C]\}\mathbf{u} + E(p)\mathbf{f}(\mathbf{u}) = -\{I_N + E(p)W(p)\}\boldsymbol{\xi}(t). \tag{5.29}$$

Resonant vibrations are periodic solutions of the latter equation. The period of the resonant vibration either coincides with that of the dynamic excitation T or is equal to nT where n is a natural number. In this Section our analysis will be limited to vibration with period T. The method of small parameters whose modifications are studied in detail in [10] and [85] is suitable for such an analysis.

Assume that the excitations and the components of matrix B are proportional to a small parameter μ, i.e.

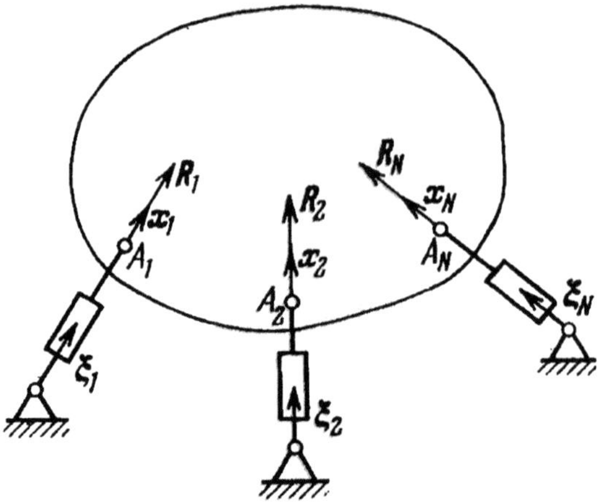

FIGURE 5.5.

$$\boldsymbol{\xi} = \mu \boldsymbol{\xi}_1, \ B = \mu B_1, \tag{5.30}$$

whereas operators $D(p) = E^{-1}(p)$ and $W(p)$ may be cast in the form

$$D(p) = D_0(p) + \mu D_1(p), \ W(p) = W_0(p) + \mu W_1(p), \tag{5.31}$$

with operators $D_0(p)$ and $W_0(p)$ satisfying the following conditions

$$\operatorname{Im} D_0(i\omega) \equiv 0, \ \operatorname{Im} W_0(i\omega) \equiv 0. \tag{5.32}$$

Equation (5.29) then becomes

$$\begin{gathered}[D_0(p) + W_0(p) + C]\mathbf{u} + \mathbf{f}(\mathbf{u}) = \\ -\mu \left\{[D(p) + W(p)]\boldsymbol{\xi}_1(t) - [B_1 p + D_1(p) + W_1(p)]\mathbf{u}\right\}.\end{gathered} \tag{5.33}$$

By introducing the small parameter into the equation of motion we state the problem of finding a periodic solution which is close to the solution of the generating equation, the latter being obtained from eq. (5.33) by setting $\mu = 0$ and having the form

$$[D_0(p) + W_0(p) + C]\mathbf{u}_0 + \mathbf{f}(\mathbf{u}_0) = 0. \tag{5.34}$$

The latter equation does not contain the dynamic excitation, hence, the assumptions (5.30)-(5.32) correspond to the assumption that the resonant vibration governed by eq. (5.29) is close to the "free" vibration due to eq. (5.34).

Let $\mathbf{u}_0(t)$ denote the T-periodic solution of eq. (5.34). As this equation is autonomous, i.e. it does not contain time t explicitly, there is a family of solutions of the sort $\mathbf{u}_0(t+\tau)$ where τ is a parameter. For the solution of eq. (5.34) which is close to a function from the above family to exist, it is necessary that the equation for τ

$$\Phi(\tau) = \int_0^T \{[D(p) + W(p)]\xi_1(t) + [B_1 p + D_1(p) + W_1(p)]\mathbf{u}_0(t+\tau)\}\,\dot{\mathbf{u}}_0(t+\tau)dt = 0, \tag{5.35}$$

possesses the real roots. A rigorous derivation can be found in the papers devoted to the substantiation of the method of small parameter, see e.g. [85]. This condition has a simple physical meaning and implies that the work done by all forces in (5.29) on the generating solution $\mathbf{u}_0(t+\tau)$ over the period T must vanish.

Let us study the case of harmonic excitation

$$\boldsymbol{\xi}_1 = \boldsymbol{\xi}_0 \cos\omega t \tag{5.36}$$

at greater length. By using the method of harmonic balance, an approximate solution of the generating equation is sought in the form

$$\mathbf{u}_0 = \mathbf{a}_0 + \mathbf{a}\cos\omega t. \tag{5.37}$$

Inserting eq. (5.37) into (5.34) we replace the nonlinear function $\mathbf{f}(\mathbf{u}_0)$ by the following linearised one

$$\mathbf{f}(\mathbf{u}_0) \approx \mathbf{f}_0 + G\mathbf{a}\cos\omega t, \tag{5.38}$$

where $G(\mathbf{a}_0, \mathbf{a})$ is the diagonal matrix whose entries are the linearisation factors for nonlinear functions $f_k(\omega_k)$

$$g_k(a_{0k}, a_k) = \frac{1}{\pi a_k}\int_0^{2\pi} f_k(a_{0k} + a_k\cos\psi)\cos\psi d\psi,$$

while $\mathbf{f}_0(\mathbf{a}_0, \mathbf{a})$ is the vector composed of the factors $f_{0k}(a_{0k}, a_k)$. Equating the constant and the harmonic components of the linearised equation

$$[D_0(p) + W_0(p) + C](\mathbf{a}_0 + \mathbf{a}\cos\omega t) + \mathbf{f}_0 + G\mathbf{a}\cos\omega t = 0 \tag{5.39}$$

to zero results in the equation for vectors $\mathbf{a}$ and $\mathbf{a}_0$

$$\mathbf{f}_0(\mathbf{a}_0, \mathbf{a}) + C\mathbf{a}_0 = 0, \tag{5.40}$$

$$[D_0(i\omega) + W_0(i\omega) + C + G(\mathbf{a}_0 + \mathbf{a})]\mathbf{a} = 0. \tag{5.41}$$

The first of these vector equations is actually a set of N independent scalar equations

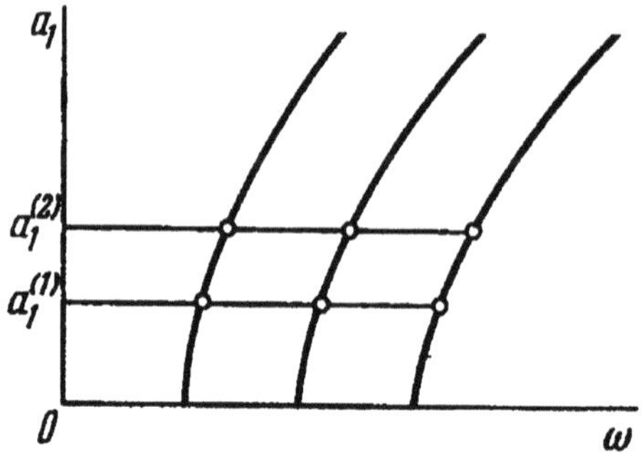

FIGURE 5.6.

$$f_{0k}(a_{0k}, a_k) + c_k a_{0k} = 0 \ (k = 1, ..., N), \tag{5.42}$$

which allows a_{0k} to be expressed in terms of a_k. Substituting the result into matrix G in (5.41) allows one to define the amplitude vector $\mathbf{a}$ which will be a function of ω.

We now demonstrate one of the possible ways of solving this equation which can be cast in the form

$$\{D_0(i\omega) + W_0(i\omega) + C + G[\mathbf{a}_0(\mathbf{a}), \ \mathbf{a})]\}\mathbf{a} = 0. \tag{5.43}$$

To begin with, we assume that amplitudes a_k are small, so that $g_k = 0$ in the first approximation. Equation (5.43) is then the linear vector equation

$$[D_0 \, (i\omega) + W \, (i\omega) + C] \, \mathbf{a} = 0.$$

It has nontrivial solution only for those values of ω which satisfy the following equation

$$\det [D_0 \, (i\omega) + W \, (i\omega) + C] = 0.$$

Vector $\mathbf{a}$ is defined up to an arbitrary scalar factor, for example, one of the components of vector $\mathbf{a}$, say a_1, can be taken to this aim. Taking a sufficiently small value of $a_1^{(1)}$, Fig. 5.6, we can determine the points in the plane (a_1, ω) which correspond to the periodic solutions of the linear equation

$$[D_0(p) + W_0(p) + C]\mathbf{u}_0 = 0. \tag{5.44}$$

This means that the frequencies and the modes of free vibration are determined.

We increase now the value of a_1 up to $a_1^{(2)}$ and assume the same vibration mode. This allows us to find another values for the components of vector $\mathbf{a}^{(1)}$ and thus another components of matrix G. Solving the equation

$$\det[D_0(i\omega) + W_0(i\omega) + C + G] = 0,$$

yields other frequencies ω. This set of points is shown in Fig. 5.6 as well. Solving the equation

$$\det\left[D_0(i\omega) + W_0(i\omega) + C + G\left(\mathbf{a}^{(1)}\right)\right]\mathbf{a} = 0,$$

yields the form $\mathbf{a}^{(2)}$ which will be used for the next step. Repeating the procedure we can plot the curves shown schematically in Fig. 5.6 which means that the approximate harmonic solution (5.37) of eq. (5.34) is found.

Let us consider now eq. (5.35). Inserting

$$\mathbf{u}_0(t+\tau) = \mathbf{a}_0 + \mathbf{a}\cos\omega(t+\tau),$$

into it yields

$$\begin{aligned}
&\int_0^T [D(p) + W(p)]\boldsymbol{\xi}_1(t)\,\dot{\mathbf{u}}_0(t+\tau)dt \\
&= -\int_0^T [D_0(p) + W_0(p)](\boldsymbol{\xi}_0\cos\omega t)\mathbf{a}\omega\sin\omega(t+\tau)dt \\
&= -[D_0(i\omega) + W_0(i\omega)]\boldsymbol{\xi}_0\cdot\mathbf{a}\omega\int_0^T \cos\omega t\sin\omega(t+\tau)dt \\
&= -\pi[D_0(i\omega) + W_0(i\omega)]\boldsymbol{\xi}_0\cdot\mathbf{a}\sin\omega\tau,
\end{aligned} \tag{5.45}$$

$$\begin{aligned}
&\int_0^T [B_1 p + D_1(p) + W_1(p)]\mathbf{u}_0(t+\tau)\dot{\mathbf{u}}_0(t+\tau)dt \\
&= \pi[\operatorname{Im} D_1(i\omega) + \operatorname{Im} W_1(i\omega) + B_1\omega]\mathbf{a}\cdot\mathbf{a}.
\end{aligned} \tag{5.46}$$

We obtain from these equations

$$\sin\omega\tau = \frac{[\operatorname{Im} D_1(i\omega) + \operatorname{Im} W_1(i\omega) + B_1\omega]\,\mathbf{a}\cdot\mathbf{a}}{[D_0(i\omega) + W_0(i\omega)]\,\boldsymbol{\xi}_0\cdot\mathbf{a}}.$$

As $|\sin\omega\tau| < 1$ the condition for the existence of resonant vibrations takes the form

$$[\operatorname{Im} D_1(i\omega) + \operatorname{Im} W_1(i\omega) + B_1\omega]\mathbf{a}\cdot\mathbf{a} < [D_0(i\omega) + W_0(i\omega)]\boldsymbol{\xi}_0\cdot\mathbf{a}. \tag{5.47}$$

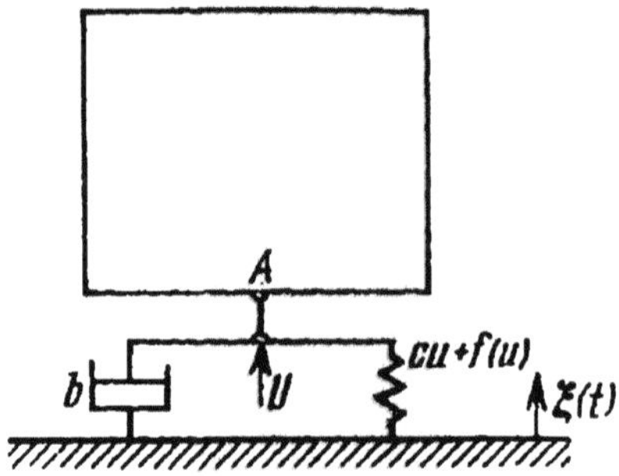

FIGURE 5.7.

Let us apply the results obtained to a single-degree-of-freedom system depicted in Fig. 5.7. The equation of motion

$$[d_A(p) + w(p) + c + bp]u + f\,(u) = -[d_A(p) + w(p)]\xi\,(t) \tag{5.48}$$

is written as follows

$$[d_{A0}(p)+w_0(p)+c]u+f\,(u) = -\mu\{[d_A(p)+w(p)]\xi_1\,(t)+[d_1(p)+w(p)+bp]u\} \tag{5.49}$$

after the small parameter μ has been introduced. One can take the dynamic rigidity operator for the undamped object as $d_{A0}\,(p)$. The solution of the generating equation is sought in the form

$$u_0(t+\tau) = a_0 + a\cos\omega(t+\tau). \tag{5.50}$$

Introducing the harmonic linearisation factors for the nonlinear function $f\,(x)$

$$f_0\,(a_0, a) = \frac{1}{2\pi}\int_0^{2\pi} f\,(a_0 + a\cos\psi)\,d\psi,$$

$$q\,(a_0, a) = \frac{1}{\pi a}\int_0^{2\pi} f\,(a_0 + a\cos\psi)\cos\psi d\psi,$$

yields the system of equations for a_0 and a

$$f_0(a_0, a) + ca_0 = 0, \tag{5.51}$$

$$d_{A0}(i\omega) + w_0(i\omega) + c + q\,(a_0, a) = 0. \tag{5.52}$$

Obtaining a_0 from the first equation and substituting the result into the second equation results in the dependence $a(\omega)$. In other words, we obtain the modified backbone curve which differs from the backbone curve (5.16) in as much as $d_{A0}(i\omega)$ and $w_0(i\omega)$ differ from $\operatorname{Re} d_{A0}(i\omega)$ and $\operatorname{Re} w_0(i\omega)$. The condition for the existence of resonant vibrations is analogous to eq. (5.47) and is given by

$$\left|[\operatorname{Im} d_{A1}(i\omega) + \operatorname{Im} w_1(i\omega) + b\omega]a^2\right| < \left|[d_{A0}(i\omega) + w_0(i\omega)]a\xi_0\right|. \qquad (5.53)$$

Provided that the linear system has rigid stops at distance Δ from the equilibrium point, the resonant vibration accompanied by impacts upon the stops is impossible if

$$\Delta > \xi_0 \frac{|d_{A0}(i\omega) + w_0(i\omega)|}{|\operatorname{Im} d_{A1}(i\omega) + \operatorname{Im} w_1(i\omega) + b\omega|}. \qquad (5.54)$$

As an example we consider the system having a rigid object

$$d(p) = mp^2$$

and an active facility which produces the following control force

$$U = -\frac{kp^2}{T_1 p + 1} x = -w(p)\, x.$$

Assuming that T_1 is a fast time constant, we cast operator $w(p)$ in the form

$$w(p) = kp^2 - \frac{\mu T_1 k p^3}{\mu T_1 p + 1} = w_0(p) + \mu w_1(p).$$

Let $f(u) = \gamma u^3$, then $f_0(a,a) = 0$, $a_0 = 0$, $q(a) = 0{,}75\gamma a^2$. Equation (5.52) renders the equation for the modified backbone curve

$$-(m+k)\,\omega^2 + c + \frac{3}{4}\gamma a^2 = 0. \qquad (5.55)$$

We obtain up to terms of order μ^2

$$\operatorname{Im} d_{A1}(i\omega) \equiv 0, \ \operatorname{Im} w_1(i\omega) = kT_1\omega^3. \qquad (5.56)$$

Expressing a^2 from (5.55) and inserting into (5.53) leads to the condition for the existence of resonant vibrations

$$2(kT_1\omega^2 + b)\omega \frac{|c - (m+k)\,\omega^2|}{3\gamma} < \frac{(m+k)\,\omega^2 \sqrt{c - (m+k)\,\omega^2}}{\sqrt{3\gamma}}. \qquad (5.57)$$

As seen from eq. (5.53) the influence of the control on the condition for the existence of resonant vibrations is twofold: (i) the system backbone curve is changed, the eigenfrequencies being reduced in general, and (ii) the imaginary part of the frequency characteristic (if it is positive) increases dissipation and thus decreases the danger of the occurrence of resonances. When the imaginary part of the frequency characteristic is negative the feedback plays the role of "negative" damping, i.e. increases the probability of occurrence of resonant vibrations.

5.2 Subharmonic vibrations

A nonlinear system subjected to a single frequency harmonic excitation can possess a wide variety of periodic motions in addition to those which have the period of the external excitation; for example, what is called *subharmonic vibration* of order s can occur in which the smallest period of motion is s times the period of the external force. The subharmonic resonances in passive systems of vibration protection are analysed in detail in Chapter 3, see also [65]. These resonances were shown to be vibrations which were close to the free vibrations of the conservative system. In nonlinear conservative systems free vibration of the period sT contains the higher harmonics. The harmonic of order s has the period T which coincides with the period of the external harmonic excitation. For this reason, the free vibrations are "sustained" by the external force.

We consider now the influence of the control on the occurrence of the subharmonic resonances. To this end, we investigate the simplest active system governed by the following equation

$$[mp^2 + bp + c + w(p)]x + f(x) = A_1 \cos s\omega t. \tag{5.58}$$

Here x is the absolute displacement of the object which is a rigid body of mass m, b and c are the factors of the damping and elastic passive elements, respectively, $w(p)$ denotes the transfer function of the control, and $f(x)$ stands for the nonlinear elastic force. The subharmonic vibrations of order s are the solutions of eq. (5.58) having period $T = 2\pi/\omega$. There are a number of approximate methods of solution, which are considered in what follows.

1. The basic assumption of the first method implies that the subharmonic vibration in the active system is close to the free vibration in the nonlinear conservative system. With this in view, we introduce the small parameter μ into eq. (5.58) as follows

$$(mp^2 + c)x + f(x) = -\mu\{[w(p) + bp]x - A_1 \cos s\omega t\}. \tag{5.59}$$

An approximate periodic solution $x_0(t)$ of the generating equation

$$(mp^2 + c)x_0 + f(x_0) = 0 \tag{5.60}$$

having frequency ω is sought in the form

$$x_0 = a_1 \cos \omega t + a_s \cos s\omega t + a_0, \tag{5.61}$$

which implies that we keep the constant term, the first harmonic and the harmonic of order s, the latter having the frequency of the driving force. In general, eq. (5.60) has a family of periodic solutions (5.61) depending upon the parameter τ

$$x_0 = a_1 \cos \omega(t+\tau) + a_s \cos s\omega(t+\tau) + a_0. \tag{5.62}$$

For eq. (5.59) to have the solution close to that of the generating equation, the following equation

$$\Phi(\tau) = \int_0^{2\pi/\omega} \{[w(p) + bp]x_0(t+\tau) - A_1 \cos s\omega t\}\dot{x}(t+\tau)dt = 0 \tag{5.63}$$

must have a real root. The physical meaning of this statement is explained in the previous Section.

Substituting eq. (5.61) into (5.63) and evaluating the integral yields

$$\Phi(\tau) = \left\{-A_1 \sin s\omega\tau \cdot s \cdot a_s\omega + b\omega^2 a_1^2 + s^2 b\omega^2 a_s^2 \right.$$
$$\left. + \operatorname{Im} w(i\omega)\,\omega a_1^2 + \operatorname{Im} w(is\omega)\, s\omega a_s^2 \right\} \pi/\omega = 0.$$

As $|\sin s\omega\tau| < 1$ this equation has the real root only if

$$A_1 s a_s > b\omega(a_1^2 + s^2 a_s^2) + \operatorname{Im} w(i\omega) a_1^2 + \operatorname{Im} w(is\omega) s a_s^2. \tag{5.64}$$

For example, let us find the conditions for existence of the subharmonic vibrations of third order ($s = 3$) in the system with a cubic nonlinearity. The solution of the generating equation

$$\ddot{x}_0 + k^2 x_0 + \gamma x_0^3 = 0 \quad \left(k^2 = \frac{c}{m}\right) \tag{5.65}$$

is sought in the form

$$x_0 = a_1 \cos \omega t + a_3 \cos 3\omega t, \tag{5.66}$$

wherein the oddness of the nonlinear term has been taken into account. Applying the method of harmonic balance yields

$$a_1^2 = \frac{4}{3\gamma}(\omega^2 - k^2), \quad a_3 = \frac{\gamma a_1^3}{4(9\omega^2 - k^2)}. \tag{5.67}$$

Condition (5.64) gives

$$3A_1a_3 > b\omega(a_1^2 + 9a_3^2) + \operatorname{Im} w(i\omega)a_1^2 + \operatorname{Im} w(3i\omega)3a_3^2. \tag{5.68}$$

By inserting the values of a_1 and a_2 from eq. (5.67) into the latter equation one can directly prove whether this condition holds at the given frequency ω for the prescribed values of A_1, b, $\operatorname{Im} w\,(i\omega)$ and $\operatorname{Im} w\,(3i\omega)$.

2. Having inserted the transfer function $w\,(p)$ into the right hand side of eq. (5.59) which contains the small parameter, we assumed that the influence of the transfer function on the system was weak. In particular, we neglected the influence of the transfer function on the values of amplitudes a_1 and a_s. In order to consider this influence we make use of representation (5.31) putting

$$w(p) = w_0(p) + \mu w_1(p), \ \operatorname{Im} w_0(i\omega) \equiv 0. \tag{5.69}$$

Substituting eq. (5.61) into the generating equation

$$[mp^2 + c + w_0(p)]x_0 + f(x_0) = 0 \tag{5.70}$$

and applying the method of harmonic balance leads to the following relationships

$$\left.\begin{aligned}
&[c - m\omega^2 + w_0(i\omega)]a_1 + \frac{1}{\pi}\int_0^{2\pi} f(a_0 + a_1\cos\psi + \\
&\qquad a_s\cos s\psi)\cos\psi d\psi = 0, \\
&[c - ms^2\omega^2 + w_0(is\omega)]a_s + \frac{1}{\pi}\int_0^{2\pi} f(a_0 + a_1\cos\psi + \\
&\qquad a_s\cos s\psi)\cos s\psi d\psi = A_1, \\
&[c + w_0(0)]a_0 + \frac{1}{2\pi}\int_0^{2\pi} f(a_0 + a_1\cos\psi + a_s\cos s\psi)d\psi = 0,
\end{aligned}\right\} \tag{5.71}$$

which allow a_0, a_1 and a_3 to be determined. Condition (5.64) holds true provided that

$$\operatorname{Im} w\,(i\omega) = \operatorname{Im} w_1\,(i\omega)\,, \ \operatorname{Im} w\,(is\omega) = \operatorname{Im} w_1\,(is\omega)\,.$$

If

$$s = 3, \ f\,(x) = m\gamma x^3,$$

equation (5.71) can be rewritten in the form

$$\left.\begin{aligned} a_0 = 0, \\ k^2 - \omega^2 + \frac{3}{4}\gamma a_1^2 + \frac{3}{4}\gamma a_1 a_3 + \frac{3}{2}\gamma a_3^2 + \frac{1}{m} w_0(i\omega) = 0, \\ (k^2 - 9\omega^2) a_3 + \frac{1}{4}\gamma a_1^3 + \frac{3}{2}\gamma a_1^2 a_3 + \frac{3}{4}\gamma a_3^3 + \tfrac{1}{m} w_0(3i\omega) a_3 = A_1. \end{aligned}\right\} \quad (5.72)$$

In practice, one can respectively replace $w_0\,(i\omega)$ and $w_0\,(si\omega)$ by $\operatorname{Re} w\,(i\omega)$ and $\operatorname{Re} w\,(si\omega)$ in eq. (5.71). In this case, representation of the transfer function (5.69) is not used at all.

3. Hayashi, [46], suggested the following method for determination of the subresonance regimes. The solution of eq. (5.58) is sought in the form

$$x = \alpha \cos \omega t + \beta \sin \omega t + \alpha_0 + a_s \cos s\omega t, \quad (5.73)$$

whereas the amplitude of the $s-th$ harmonic is determined from the linear equation

$$[mp^2 + bp + c + w(p)] x_0 = A_1 \cos s\omega t. \quad (5.74)$$

In other words, the nonlinear term $f\,(x)$ is assumed to have a weak influence on the amplitude value a_s. Although the substantiation of this assumption will be not discussed here, we notice at this place that this is acceptable provided that the amplitude $a_1 = \sqrt{\alpha^2 + \beta^2}$ of the harmonic of frequency ω is much greater than a_s, i.e. $a_1 \gg a_s$. Equation (5.74) yields the amplitude of the harmonic solution

$$a_s = \frac{A_1}{\left|c - ms^2\omega^2 + bs\omega i + w(is\omega)\right|}. \quad (5.75)$$

Inserting then eq. (5.73) into eq. (5.58) and equating the constant term and the factors of $\sin \omega t$ and $\cos s\omega t$ to zero results in a system of equation for α_0, α and β. In the particular case in which

$$s = 3, \ \ f\,(x) = m\gamma x^3$$

these equations take the form

$$a_0 = 0,$$

$$\left[k^2 - \omega^2 + \frac{1}{m} \operatorname{Re} w\,(i\omega) + \frac{3}{4}\gamma\left(\alpha^2 + \beta^2\right) + \frac{3}{2}\gamma a^2\right] \alpha$$

$$+ \left[2n\omega + \frac{1}{m} \operatorname{Im} w\,(i\omega)\right] \beta = -\frac{3}{4}\gamma a \left(\alpha^2 - \beta^2\right),$$

$$
-\left[2n\omega + \frac{1}{m}\operatorname{Im} w\,(i\omega)\right]\alpha + \left[k^2 - \omega^2 + \frac{1}{m}\operatorname{Re} w\,(i\omega) \right.
$$
$$
\left. + \frac{3}{4}\gamma\left(\alpha^2+\beta^2\right) + \frac{3}{2}\gamma a^2\right]\beta = \frac{3}{2}\gamma\alpha\beta a,
$$

where $2n = b/m$. Raising the left and right hand sides of these equations to the second power and adding them gives

$$
\begin{aligned}
&\left[k^2 - \omega^2 + \frac{1}{m}\operatorname{Re} w(i\omega) + \frac{3}{4}\gamma z^2 + \frac{3}{2}\gamma a^2\right]^2 \\
&+ \left[2n\omega + \frac{1}{m}\operatorname{Im} w(i\omega)\right]^2 = \frac{9}{16}\gamma^2 a^2 z^2,
\end{aligned}
\tag{5.76}
$$

where $z^2 = \alpha^2 + \beta^2$. For a subharmonic solution to exist, eq. (5.76) must have a root $z^2 > 0$. In order to establish the values of A_1 which ensure such a root, one can use the fact that eq. (5.76) defines a second order curve in the plane of parameters $\left(z^2, a^2\right)$. Denoting

$$
\begin{aligned}
k^2 - \omega^2 + \frac{1}{m}\operatorname{Re} w\,(i\omega) &= M\,(\omega)\,,\ 2n\omega + \frac{1}{m}\operatorname{Im} w\,(i\omega) = N\,(\omega)\,, \\
z^2 &= u,\ a^2 = v
\end{aligned}
$$

we write down eq. (5.76) as follows

$$
F(u,v) = \left(M + \frac{3}{4}\gamma u + \frac{3}{2}\gamma v\right)^2 + N^2 - \frac{9}{16}\gamma^2 uv = 0. \tag{5.77}
$$

It is easy to see that this equation has no real solutions for v if $u = 0$ and no real solutions for u if $v = 0$. The curve (5.77) thus does not intersect the coordinate axes. Let us find the points at which the tangent to the curve is horizontal. The condition

$$
\frac{\partial F}{\partial u} = \frac{3}{2}M\gamma + \frac{9}{8}\gamma^2 u + \frac{27}{16}\gamma^2 v = 0
$$

renders

$$
u = \frac{4}{3\gamma}M - \frac{3}{2}v. \tag{5.78}
$$

Substituting (5.78) into eq. (5.77) yields

$$
\frac{63}{64}\gamma^2 v^2 + \frac{3}{4}\gamma M v + N^2 = 0.
$$

Solving this equation we obtain

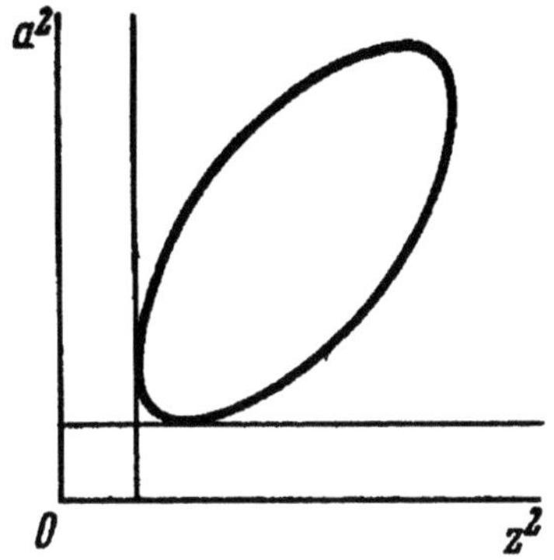

FIGURE 5.8.

$$v = \frac{8}{21\gamma}(-M \pm \sqrt{M^2 - 7N^2}). \tag{5.79}$$

Positive solutions exist when

$$M(\omega) < 0, \ \ M^2(\omega) > 7N^2(\omega). \tag{5.80}$$

In this case, curve (5.77) has two horizontal tangents, and, therefore, is an ellipse, see Fig. 5.8. From eq. (5.79) it follows that

$$\left(a^2\right)_{\min} = \frac{8}{21\gamma}\left(-M\left(\omega\right) \pm \sqrt{M^2\left(\omega\right) - 7N^2\left(\omega\right)}\right).$$

Substituting $a(\omega)$ from eq. (5.75) we obtain the values of A_1 for which the subharmonic vibrations with frequency ω are possible

$$\begin{aligned} A_1^2 \ > \ & \left|k^2 - 9\omega^2 + 2ni\omega + \frac{1}{m}w(3i\omega)\right|^2 \frac{8}{21\gamma}\left\{\omega^2 - k^2 - \frac{1}{m}\operatorname{Re} w(i\omega) - \right. \\ & \left. \sqrt{\left[\omega^2 - k^2 - \frac{1}{m}\operatorname{Re} w(i\omega)\right]^2 - 7\left[\frac{1}{m}\operatorname{Im} w(i\omega) + 2n\omega\right]^2}\right\}. \end{aligned} \tag{5.81}$$

4. Finally, one can seek the solution in the general form

$$x = \alpha_0 + \alpha_1 \cos\omega t + \beta_1 \sin\omega t + \alpha_s \cos s\omega t + \beta_s \sin s\omega t. \tag{5.82}$$

This is achieved by substituting the latter equation into eq. (5.58), expanding the result as a Fourier series and equating the constant term and the factors of the first and $s-th$ harmonic to zero. This method leads usually to rather bulky calculations, however in some cases closed form results can be obtained. For instance, for the above system with a cubic nonlinearity the equations of harmonic balance are given by

$$\left[k^2-\omega^2+\frac{1}{m}\operatorname{Re}w(i\omega)+\frac{3}{4}\gamma(\alpha_1^2+\beta_1^2)+\frac{3}{2}\gamma(\alpha_3^2+\beta_3^2)\right]\alpha_1 \\ -\left[2n\omega+\frac{1}{m}\operatorname{Im}w(i\omega)\right]\beta_1=\frac{3}{4}\gamma(\alpha_1^2-\beta_1^2)\alpha_3+\frac{3}{2}\gamma\alpha_1\beta_1\beta_3, \tag{5.83}$$

$$\left[k^2-\omega^2+\frac{1}{m}\operatorname{Re}w(i\omega)+\frac{3}{4}\gamma(\alpha_1^2+\beta_1^2)+\frac{3}{2}\gamma(\alpha_3^2+\beta_3^2)\right]\beta_1 \\ +\left[2n\omega+\frac{1}{m}\operatorname{Im}w(i\omega)\right]\alpha_1=\frac{3}{4}\gamma(\alpha_1^2-\beta_1^2)\beta_3-\frac{3}{2}\gamma\alpha_1\beta_1\alpha_3, \tag{5.84}$$

$$\left[k^2-9\omega^2+\frac{1}{m}\operatorname{Re}w(3i\omega)+\frac{3}{2}\gamma(\alpha_1^2+\beta_1^2)+\frac{3}{4}\gamma(\alpha_3^2+\beta_3^2)\right]\alpha_3 \\ -\left[6n\omega+\frac{1}{m}\operatorname{Im}w(3i\omega)\right]\beta_3=\frac{1}{4}\gamma\alpha_1^3-\frac{3}{4}\gamma\alpha_1\beta_1^2, \tag{5.85}$$

$$\left[k^2-9\omega^2+\frac{1}{m}\operatorname{Re}w(3i\omega)+\frac{3}{2}\gamma(\alpha_1^2+\beta_1^2)+\frac{3}{4}\gamma(\alpha_3^2+\beta_3^2)\right]\beta_3 \\ +\left[6n\omega+\frac{1}{m}\operatorname{Im}w(3i\omega)\right]\alpha_3=-\frac{1}{4}\gamma\beta_1^3+\frac{3}{4}\gamma\alpha_1^2\beta_1+\frac{A_1}{m}. \tag{5.86}$$

Denoting

$$\begin{aligned} k^2-\omega^2+\frac{1}{m}\operatorname{Re}w(i\omega) &= M_1(\omega), \\ 2n\omega+\frac{1}{m}\operatorname{Im}w(i\omega) &= N_1(\omega), \\ k^2-9\omega^2+\frac{1}{m}\operatorname{Re}w(3i\omega) &= M_3(\omega), \\ 6n\omega+\frac{1}{m}\operatorname{Im}w(3i\omega) &= N_3(\omega), \end{aligned}$$

$$\alpha_1^2+\beta_1^2=u,\ \alpha_3^2+\beta_3^2=v$$

and rearranging eqs. (5.83)-(5.86) yields

$$\left.\begin{aligned} \left(M_1+\frac{3}{4}\gamma u+\frac{3}{2}\gamma v\right)^2+N_1^2=\frac{9}{16}\gamma^2uv, \\ \left[\left(M_3+\frac{3}{2}\gamma v+\frac{3}{4}\gamma u\right)^2+N_3^2\right]v=\frac{1}{16}\gamma^2u^3+\frac{1}{m^2}A_1^2. \end{aligned}\right\} \tag{5.87}$$

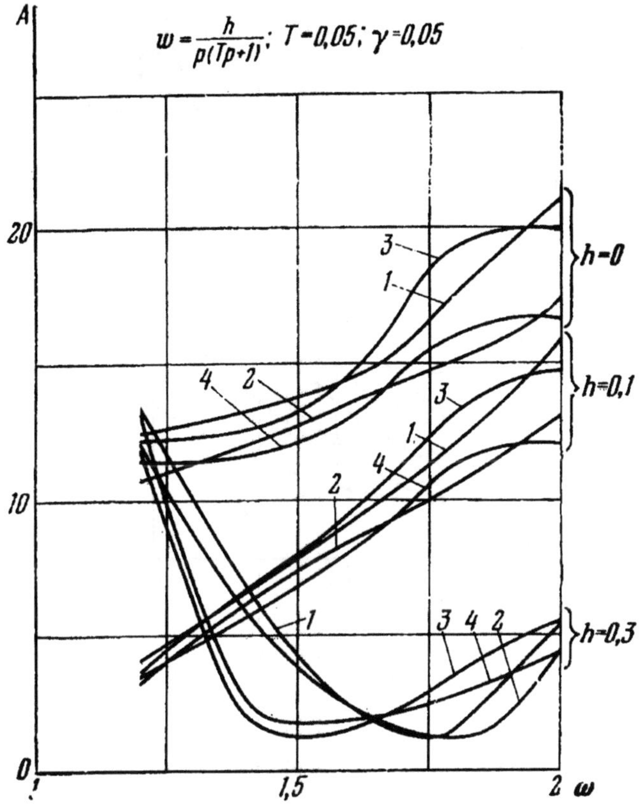

FIGURE 5.9.

A subharmonic regime can occur only for those values of A_1 for which this system of equations has positive solutions $u > 0$ and $v > 0$.

Figures 5.9-5.12 display the dependencies of the minimal amplitude A_1 enabling the subharmonic resonance of third order in the system with the cubic characteristic versus frequency of harmonic excitation. Figures 5.9 and 5.11 show the results of the numerical work performed for the following transfer function

$$w(p) = \frac{h}{p(Tp+1)},$$

whereas Figs. 5.10 and 5.12 were computed for

$$w(p) = \frac{hp^2}{Tp+1}.$$

The following values were taken: $T = 0,05$, $k = 1$ and $n = 0,05$. The numbers indicate the ordinal number of the method utilised to plot the curve. The graphs indicate that the results obtained by the above methods

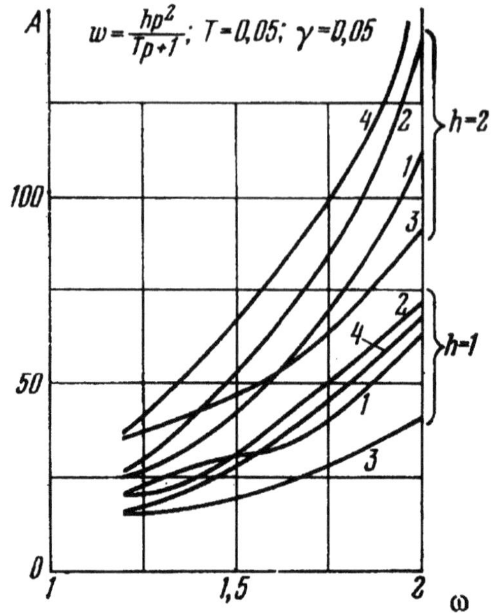

FIGURE 5.10.

coincide qualitatively, the quantitative discrepancy being not very considerable.

Introducing the control due to integration of the displacement leads to an increase in likelihood of subharmonic resonances. This is quite understandable because $\operatorname{Im} w(i\omega) < 0$, hence, introducing this feedback is equivalent to decreasing the damping. Introducing the control due to acceleration decreases danger of the subharmonic resonances.

The methods considered can be generalised to more complex equations. So, for instance, the subharmonic solutions of eq. (5.29) can be sought by the method of small parameters. In accordance with this, the solution of the generating equation (5.34) must be sought in the form

$$\mathbf{u}_0 = \mathbf{a}_0 + \mathbf{a}_1 \cos \omega t + \mathbf{a}_s \cos \omega t, \tag{5.88}$$

where $s\omega$ is the frequency of the harmonic driving force

$$\boldsymbol{\xi} = \mu \boldsymbol{\xi}_0 \cos s\omega t.$$

Inserting eq. (5.88) into eq. (5.35) yields the following equation for τ

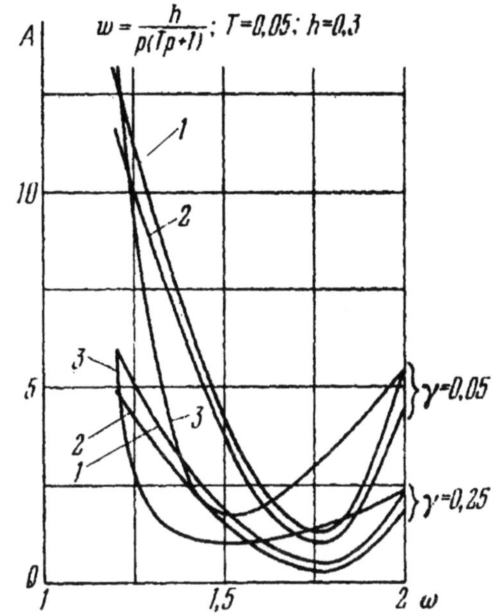

FIGURE 5.11.

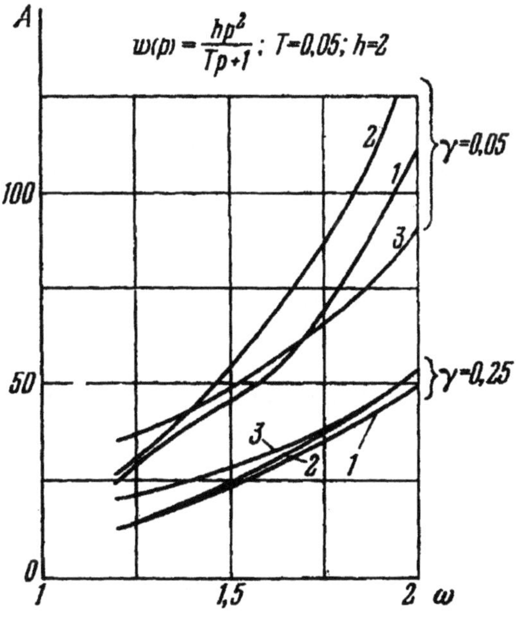

FIGURE 5.12.

$$\begin{aligned}\Phi(\tau) = & \int_0^{2\pi/\omega} \{[D(p)+W(p)]\boldsymbol{\xi}_0 \cos s\omega t + [B_1 p + D_1(p) + W_1(p)] \\ & [\mathbf{a}_0 + \mathbf{a}_1 \cos\omega(t+\tau) + \mathbf{a}_s \cos s\omega(t+\tau)]\} \\ & \cdot[-\mathbf{a}_1\omega \sin\omega(t+\tau) - \mathbf{a}_s \sin\omega s(t+\tau)]dt = 0.\end{aligned} \tag{5.89}$$

Evaluating the integral we obtain

$$\begin{aligned}\Phi(\tau) &= -\pi\,[D_0(is\omega) + W_0(is\omega)]\,\boldsymbol{\xi}_0 \cdot \mathbf{a}_s s \sin s\omega\tau \\ &+\pi\,[\operatorname{Im} D_1(i\omega) + \operatorname{Im} W_1(i\omega) + B_1\omega]\,\mathbf{a}_1 \cdot \mathbf{a}_1 \\ &+\pi s\,[\operatorname{Im} D_1(is\omega) + \operatorname{Im} W_1(is\omega) + B_1 s\omega]\,\mathbf{a}_s \cdot \mathbf{a}_s = 0\end{aligned}$$

up to terms of order μ^2.

Thus, the condition for existence of the subharmonic vibration is written as follows

$$\begin{aligned}&[D_0(is\omega) + W_0(is\omega)]\boldsymbol{\xi}_0 \cdot \mathbf{a}_s > \\ &[\operatorname{Im} D_1(i\omega) + \operatorname{Im} W_1(i\omega) + B_1\omega]\mathbf{a}_1 \cdot \mathbf{a}_1 \\ &+[\operatorname{Im} D_1(is\omega) + \operatorname{Im} W_1(is\omega) + B_1 s\omega]\mathbf{a}_s \cdot \mathbf{a}_s.\end{aligned} \tag{5.90}$$

5.3 Influence of nonlinearities in feedback

As mentioned in Section 2.1 the sensors, compensators and actuators may have nonlinear characteristics. If the active feedback of the vibration protection system contains a nonlinear element the analysis of the efficiency and stability of the system cannot be carried out by the methods of Chapter 2.

Let one of the elements of the control have the nonlinear characteristic shown in Fig. 5.13. This characteristic consists of a dead zone $(|x| < \Delta)$, a linear part with gain k and a saturation zone $(|x| > d)$. For instance, this characteristic is representative of a hydraulic system, see Fig. 2.11, and some sensors.

Making use of the method of equivalent linearisation we can replace the nonlinear characteristic $f(x)$ by a linear element

$$f_l(x) = qx + f_0, \tag{5.91}$$

whose parameters q and f_0 depend on the input parameters. We cast the characteristic in the form

$$\begin{aligned}f(x) \;=\; & -k(d-\Delta) + k(x+d)\eta(x+d) \\ & -k(x+\Delta)\eta(x+\Delta) + k(x-\Delta)\eta(x-\Delta) - k(x-d)\eta(x-d).\end{aligned} \tag{5.92}$$

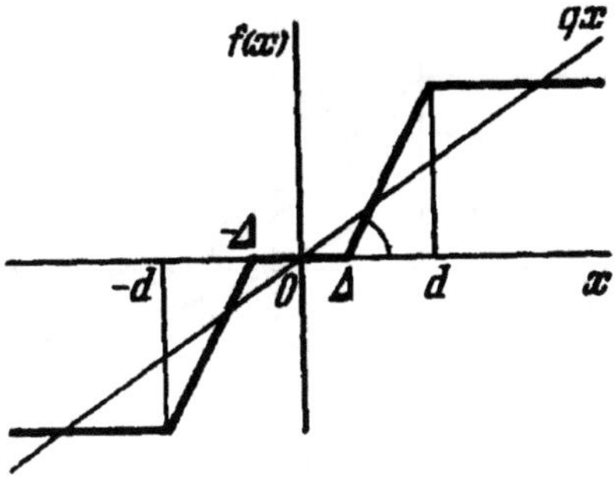

FIGURE 5.13.

Let the input of the nonlinear element be

$$x_* = a\cos\omega t. \tag{5.93}$$

Using the expressions for the factors of harmonic linearisation collected in Table 3.1, see also [102], we obtain

$$\left.\begin{aligned}
&q = 0 && \text{if } a < \Delta,\\
&q = k\left(1 - \frac{2}{\pi}\arcsin\frac{\Delta}{a} - \frac{2\Delta}{\pi a}\sqrt{1-\frac{\Delta^2}{a^2}}\right) && \text{if } \Delta < a < d,\\
&q \equiv \frac{2k}{\pi}\left(\arcsin\frac{d}{a} - \arcsin\frac{\Delta}{a}\right. && \\
&\left. +\frac{d}{a}\sqrt{1-\frac{d^2}{a^2}} - \frac{\Delta}{a}\sqrt{1-\frac{\Delta^2}{a^2}}\right) && \text{if } a > d,\\
&f_0 \equiv 0.
\end{aligned}\right\} \tag{5.94}$$

Figure 5.14 shows q/k versus a/d for some values of Δ/d.

Thus, introducing the nonlinear element results in multiplication of the amplitude of the harmonic signal by the "gain" q which depends on the amplitude. In the case of concern, $q/k < 1$ for any amplitude smaller than unity. This means that both the dead zone and the saturation reduce the feedback efficiency by reducing the equivalent gain.

As a first approximation one can use the methods of linear system analysis explained in Section 2.2, the dependence of the total gain on the amplitude of the nonlinear element's input being however taken into account.

Evidently, the equivalent gain is equal to zero if $a < \Delta$ which implies that vibrations within the dead zone are not controlled by the feedback. Increase in the vibration amplitude increases the gain which approaches k and then when $a > d$ decreases up to zero. Thus, the feedback turns out to be ineffective both for very small ($a < \Delta$) and very large amplitudes.

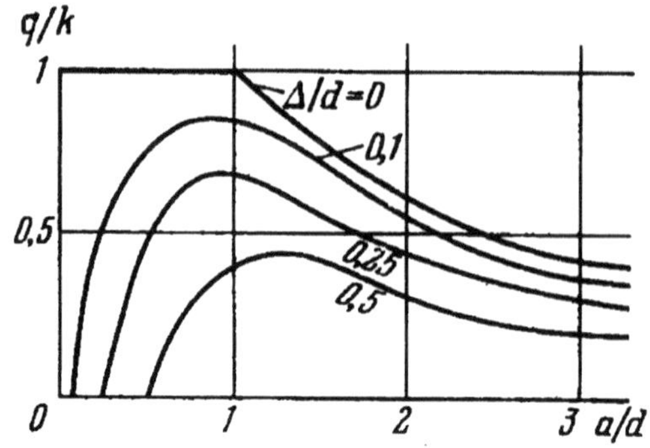

FIGURE 5.14.

Introducing the nonlinear element leads not only to change in the gain for the harmonic of frequency ω, but also to the appearance of higher harmonics of the frequencies being an integral multiple of ω. In other words, the nonlinear element acts as a "generator" of high frequency vibrations. The amplitudes of the higher harmonics can be estimated by means of the expressions for the Fourier coefficients of the periodic process $f(a\sin\omega t)$. Since the function (5.92) is odd the Fourier coefficients of the even harmonics vanish. For the odd harmonics these are

$$\begin{aligned} a_m &= \frac{1}{\pi}\int_0^{2\pi} f(a\sin\psi)\sin m\psi d\psi = \frac{4}{\pi}\int_0^{\pi/2} f(a\sin\psi)\sin m\psi d\psi \\ &= \frac{4}{\pi}\left[\int_{\psi_1}^{\psi_2} ka\sin\psi\sin m\psi d\psi + \int_{\psi_2}^{\pi/2} k(d-\Delta)\sin m\psi d\psi\right], \end{aligned}$$

where

$$\psi_1 = \arcsin\frac{\Delta}{a},\quad \psi_2 = \arcsin\frac{d}{a}.$$

Evaluating the integrals yields

$$\begin{aligned} a_m &= \frac{4k}{\pi}\left\{a\left[\frac{1}{m-1}\cos\frac{m-1}{2}(\psi_1+\psi_2)\sin\frac{m-1}{2}(\psi_2-\psi_1)\right.\right. \\ &\quad \left. -\frac{1}{m+1}\cos\frac{m+1}{2}(\psi_1+\psi_2)\sin\frac{m+1}{2}(\psi_2-\psi_1)\right] \\ &\quad \left. +\frac{d-\Delta}{m}\cos m\psi_2\right\}. \end{aligned} \qquad (5.95)$$

The occurrence of the higher harmonics in the output of the nonlinear element leads to their appearance in $x(t)$ as the system is a closed-loop one. Equation (5.95) does not account for this fact and therefore is an approximate one. Moreover, the amplitude of the first harmonic obtained by an approximate method is inserted into this equation. All these inaccuracies, however, are not of principal importance if the frequencies of the higher harmonics exceed the cutoff frequency of the open-loop system. In this case they are filtered by the system itself and do not influence its isolation properties considerably. To estimate the amplitudes of the higher harmonics one can apply the method suggested in [67].

For the sake of certainty let us assume that the measured value representing the vibration of one of the object points is the input of the nonlinear element. Let us derive the differential equation for this variable. By virtue of eq. (1.90) we obtain

$$x\,(t) = \xi\,(t) + e(p)U\,(t)\,, \tag{5.96}$$

where $\xi\,(t)$ denotes vibration of the point in the system without feedback while $e\,(p)$ denotes the dynamic compliance operator relating the displacement of the point to the control $U\,(t)$ produced by the feedback. Then,

$$U = -w(p)f\,(x)\,, \tag{5.97}$$

with $w(p)$ being the transfer function of the linear elements of the feedback. We obtain from eqs. (5.96) and (5.97)

$$x = \xi - e(p)w(p)f\,(x)\,. \tag{5.98}$$

Letting $\xi\,(t) = \xi_0 \cos\omega t$ and applying the method of harmonic linearisation we seek an approximate solution of eq. (5.98) in the form (5.93). This solution satisfies the linearised equation

$$x_* = \xi_0 \cos\omega t - e(p)w(p)qx_*. \tag{5.99}$$

We rewrite eq. (5.98) in the following form

$$x = \xi - e(p)w(p)qx + e(p)w(p)[qx - f\,(x)] \tag{5.100}$$

and assume that $x_1\,(t)$ is an exact periodic solution of this equation. Subtracting eq. (5.99) from eq. (5.100) gives the equation for the difference between the exact solution and the approximate one

$$\Delta\,(t) = x_1\,(t) - x_*\,(t) = -e\,(p)\,w\,(p)\,q\Delta\,(t) + e\,(p)\,w\,(p)\,[qx_1 - f\,(x_1)]\,,$$

or

$$\Delta\,(t) = v(p)[qx_1 - f\,(x_1)], \tag{5.101}$$

where

$$v(p) = \frac{e(p)w(p)}{1 + e(p)w(p)q}. \tag{5.102}$$

The expression obtained allows the error of the approximate solution to be estimated. We expand the periodic function

$$\varphi[x_1(t)] = qx_1(t) - f[x_1(t)] \tag{5.103}$$

in Fourier series

$$\varphi[x_1(t)] = \sum_{k=1}^{\infty} c_k \cos(\omega kt + \psi_k). \tag{5.104}$$

This expansion has no constant term since $\varphi(x_1)$ is an odd function. Inserting eq. (5.104) into (5.101) yields

$$\Delta(t) = \sum_{k=1}^{\infty} |v(ik\omega)| c_k \cos[k\omega t + \psi_k + \arg v(ik\omega)]. \tag{5.105}$$

According to Parceval's formula the mean square value of the periodic function is equal to the half-sum of squares of the harmonic amplitudes. In the light of this formula, we obtain the following estimation for the mean-squared value of the error

$$\begin{aligned} \sigma_\Delta^2 &= \frac{1}{2}\sum_{k=1}^{\infty} |v(ik\omega)|^2 c_k^2 \leq \frac{1}{2}\max_k \left[|v(ik\omega)|^2\right] \sum_{k=1}^{\infty} c_k^2 = \\ &= \max_k \left[|v(ik\omega)|^2\right] \sigma_\varphi^2, \end{aligned} \tag{5.106}$$

where σ_φ is the mean-squared value of the function (5.104). Apparently,

$$\sigma_\varphi^2 < \left\{\max_t |\varphi[x_1(t)]|\right\}^2.$$

On the other hand, Fig. 5.13 indicates that for this particular nonlinearity the maximum absolute value of $\varphi(x)$ is achieved either for $x = d$ or for $x = \Delta$ or for $x = x_{\max}$. In other words,

$$\max_t |\varphi[x_1(t)]| = \max\{q\Delta,\ k(d - \Delta) - qd, qx_{\max} - k(d - \Delta)\}. \tag{5.107}$$

In order to determine this maximum one can take in the first approximation that $x_{\max} = a$. Inserting (5.107) into eq. (5.106) yields

$$\sigma_\Delta^2 < \max_k \left[|v(ik\omega)|^2\right] \left\{\max_t |\varphi[x_1(t)]|\right\}^2. \tag{5.108}$$

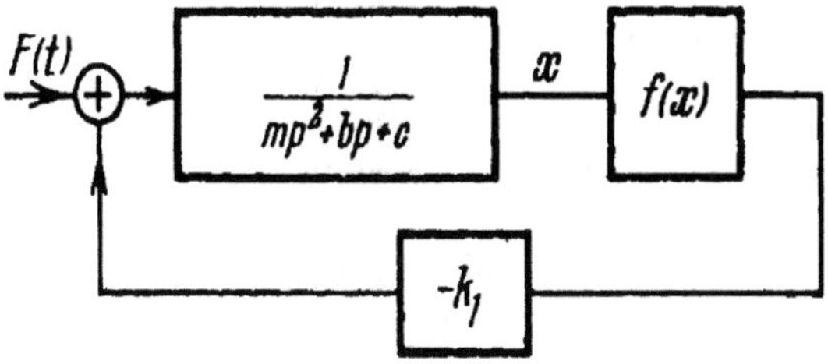

FIGURE 5.15.

This expression indicates that the mean-squared value of the error depends essentially upon the properties of operator $v(p)$. It follows from eq. (5.102) that

$$|v(ik\omega)| = \left|\frac{1}{1+e(ik\omega)w(ik\omega)q}\right| |e(ik\omega)w(ik\omega)|.$$

The first factor is the efficiency factor of the linearised feedback at frequency $k\omega$ which is much smaller than unity in the performance zone and close to unity at frequencies higher than the cutoff frequency. The second factor is equal to the ratio of the amplitude-frequency characteristic of the linearised open-looped system to the linearisation factor q. This second factor does not exceed q^{-1} at frequencies higher than the cutoff frequency. This permits us to affirm that the value of $|v(ik\omega)|$ in the entire frequency domain does not exceed q^{-1} and thus

$$\sigma_\Delta^2 \leq \frac{1}{q^2}\left\{\max_t |\varphi[x_1(t)]|\right\}^2. \tag{5.109}$$

The periodic process (5.105) contains the first harmonic as well as the high frequency components. We take the latter, i.e. we consider the process

$$\Delta_1(t) = \sum_{k=3}^{\infty} |v(ik\omega)| c_k \cos[k\omega t + \psi_k + \arg v(ik\omega)], \tag{5.110}$$

then the mean-squared value of $\Delta_1(t)$ describes the level of high frequency vibrations in the nonlinear system. By analogy with (5.106) we obtain

$$\sigma_{\Delta_1}^2 \leq \max_{k\geq 3}\left\{|v(ik\omega)|^2\right\}\sigma_\varphi^2. \tag{5.111}$$

As an example, let us study now the system shown in the block diagram in Fig. 5.15. This system consists of the mass m mounted on the elastic element and the damper and involves the feedback with the nonlinear element $f(x)$ whose characteristic is depicted in Fig. 5.13. In the system under consideration

$$e(p) = \left(mp^2 + bp + c\right)^{-1}, \quad w(p) = -k_1.$$

Variable $x(t)$ is governed by the following equation

$$x = (mp^2 + bp + c)^{-1} F(t) - k_1 (mp^2 + bp + c)^{-1} f(x). \qquad (5.112)$$

We determine now the amplitude of the approximate harmonic solution assuming $F(t) = F_0 \cos \omega t$. Applying the method of harmonic linearisation we form the linearised equation

$$x = (mp^2 + bp + c)^{-1} [F_0 \cos \omega t - k_1 q(a) x]. \qquad (5.113)$$

Solving this equation one obtains

$$x = a \cos(\omega t + \psi),$$

where a is determined by the equation

$$a = F_0 \left\{ \left[c + k_1 q(a) - m\omega^2\right]^2 + b^2 \omega^2 \right\}^{-1/2}. \qquad (5.114)$$

This equation can be solved for example by the geometrical method described in detail in Section 5.1. This method allows one to find the vibration amplitude a and the linearisation factor q. Equation (5.114) may have several solutions and each of these solutions corresponds to an approximate solution of eq. (5.112).

Let estimate now the error of the approximate solution. Inserting expressions for $e(p)$ and $w(p)$ into eq. (5.102) yields

$$v(p) = \frac{k_1}{mp^2 + bp + c + qk_1}. \qquad (5.115)$$

The maximal value of $|v(p)|$ is achieved at the resonance frequency

$$\omega_r \approx \sqrt{\frac{c + k_1 q(a)}{m}},$$

which depends on a. Comparing eqs. (5.114) and (5.115) we find that

$$|v(i\omega)| = k_1 \left\{ \left[c + k_1 q(a) - m\omega^2\right]^2 + b^2 \omega^2 \right\}^{-1/2} = \frac{k_1 a}{F_0}. \qquad (5.116)$$

Therefore, the dependencies $|v(i\omega)|$ and $a(\omega)$ coincide qualitatively. Figure 5.16 displays the amplitude a versus frequency ω. Let us find the error of the solution at frequency ω_1 at which $a = a_1$ and $q = m\omega_0^2$, see Fig. 5.16. By virtue of eq. (5.108)

$$\sigma_\Delta^2 \le \frac{k_1^2 a_1^2}{F_0^2} \left\{ \max_t |\varphi[x_1(t)]| \right\}^2. \qquad (5.117)$$

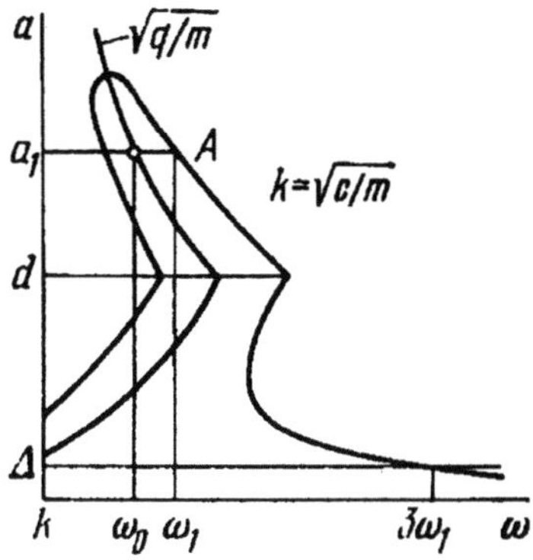

FIGURE 5.16.

Equation (5.107) can be used to estimate the second factor.

We estimate now the level of high frequency vibrations corresponding to vibration of amplitude a_1. By means of eq. (5.111) we obtain

$$\sigma_{\Delta_1}^2 < \frac{k_1^2 a_3^2}{F_0^2} \left\{ \max_t |\varphi [x_1 (t)]| \right\}^2, \tag{5.118}$$

where a_3 is the vibration amplitude at frequency $3\omega_1$, see Fig. 5.16.

Let us now proceed to analysis of the nonlinear system under polyharmonic excitation. Depending on the relation between the frequencies of harmonic components different approaches can be used. Assume first that the excitation has two harmonics with essentially distinct frequencies, i.e.

$$F(t) = A \cos \nu t + B \cos \Omega t, \tag{5.119}$$

wherein $\Omega \gg \nu$. This case is typical for active systems of vibration protection which control low frequency vibrations. As a rule, the system is ineffective at high frequencies. Because of the nonlinearity the presence of the high frequency component may affect the performance of the control due to the low frequency signal.

To investigate this influence we utilize the approach suggested in [63]. Assume that the system motion is governed by eq. (5.98) which is now convenient to write in the form

$$d(p)x = F(t) - w(p)f(x), \tag{5.120}$$

where $d(p) = e^{-1}(p)$. The solution of this equation is sought as the sum of a slowly varying function $x_0(t)$ and a harmonic process of frequency Ω with slowly varying amplitude, i.e.

$$x(t) = x_0(t) + b(t) \cos(\Omega t + \varphi). \tag{5.121}$$

We substitute eqs. (5.121) and (5.119) into (5.120) and select the slowly varying terms (also the low frequency component of the excitation) and the first harmonics of frequency Ω. The harmonic linearisation of the nonlinear function $f(x)$ is carried out as follows

$$f\left[x_0 + b\cos\left(\Omega t + \varphi\right)\right] = f_0\left(x_0, b\right) + q\left(x_0, b\right) b \cos\left(\Omega t + \varphi\right),$$

where

$$f_0(x_0, b) = \frac{1}{2\pi}\int_0^{2\pi} f(x_0 + b\cos\psi)d\psi, \tag{5.122}$$

$$q(x_0, b) = \frac{1}{\pi b}\int_0^{2\pi} f(x_0 + b\cos\psi)\cos\psi d\psi. \tag{5.123}$$

This leads to the following equations

$$d(p)x_0 = A\cos\nu t - w(p) f_0(x_0, b), \tag{5.124}$$

$$\begin{aligned} &[\operatorname{Re} d\,(i\Omega)\cos(\Omega t + \varphi) - \operatorname{Im} d\,(i\Omega)\sin(\Omega t + \varphi)]b = \\ &B\cos\Omega t - [\operatorname{Re} w\,(i\Omega)\cos(\Omega t + \varphi) \\ &- \operatorname{Im} w\,(i\Omega)\sin(\Omega t + \varphi)]\, q\,(x_0, b)\, b. \end{aligned} \tag{5.125}$$

While constructing eq. (5.125) we neglected the terms with derivatives of slowly varying amplitude $b\,(t)$.

By using eq. (5.125) one can find the phase angle φ and express b in terms of the linearisation factor $q\,(x_0, b)$. For example, the amplitude b is given by

$$b = \frac{B}{\sqrt{[\operatorname{Re} d\,(i\Omega) + q\operatorname{Re} w\,(i\Omega)]^2 + [\operatorname{Im} d\,(i\Omega) + q\operatorname{Im} w\,(i\Omega)]^2}}. \tag{5.126}$$

Replacing b in eq. (5.124) by this expression we obtain a nonlinear differential equation for the "slow" component $x_0\,(t)$. We are interested in the periodic solution of this equation having frequency ν. As a first approximation, we limit our analysis to the harmonic solution obtained by the method of harmonic linearisation. Evidently, amplitude $x_0\,(t)$ depends essentially upon amplitude b of the high frequency component of the solution. This is the dependence which seems to present great interest.

It is easy to notice that the difference between eqs. (5.124) and (5.120) in which $F\,(t) = A\cos\nu t$ is that function $f\,(x)$ is replaced by function $f_0\,(x_0, b)$. Thus, the influence of the high frequency component manifests itself in changing the nonlinear function's characteristic which now depends on b.

As an example, we consider the system shown in Fig. 5.15. Equation (5.120) takes the form

$$(mp^2 + bp + c)x = A\cos\nu t + B\cos\Omega t - k_1 f(x). \tag{5.127}$$

The equation for the low frequency component is written in the form (5.124) as follows

$$(mp^2 + bp + c)x_0 = A\cos\nu t - k_1 f_0(x_0, b). \tag{5.128}$$

Expression $f_0(x_0, b)$ for the nonlinearity (5.92) is obtained by means of Table 3.1. Assuming that $x_{0\,\max} + b < d$, i.e. the vibration does not reach the saturation zone, we get

$$\left.\begin{array}{r}
f_0 = 0 \text{ if } x_0 + b < \Delta,\; x_0 - b > -\Delta, \\[2ex]
f_0 = \dfrac{kb}{\pi}\left[\sqrt{1-\left(\dfrac{\Delta - x_0}{b}\right)^2} - \dfrac{\Delta - x_0}{b}\arccos\dfrac{\Delta - x_0}{b}\right] \\
\text{if } \Delta > x_0 - b > -\Delta,\; x_0 + b > \Delta, \\[2ex]
f_0 = k\Delta - \dfrac{kb}{\pi}\left[\sqrt{1-\left(\dfrac{\Delta - x_0}{b}\right)^2} + \dfrac{\Delta + x_0}{b}\arccos\left(-\dfrac{\Delta + x_0}{b}\right)\right] \\
\text{if } x_0 - b < -\Delta,\; -\Delta < x_0 + b < \Delta, \\[2ex]
f_0 = k\Delta + \dfrac{kb}{\pi}\left[\sqrt{1-\left(\dfrac{\Delta - x_0}{b}\right)^2} - \dfrac{\Delta - x_0}{b}\arccos\left(\dfrac{\Delta - x_0}{b}\right) - \right. \\
\left. -\sqrt{1-\left(\dfrac{\Delta + x_0}{b}\right)^2} + \dfrac{\Delta + x_0}{b}\arccos\left(-\dfrac{\Delta + x_0}{b}\right)\right] \\
\text{if } x_0 - b < -\Delta,\; x_0 + b > \Delta.
\end{array}\right\} \tag{5.129}$$

The dependencies $f_0(x_0, b)$ are plotted in Fig. 5.17 where $\Delta = 1$ and $k = 1$. Function $f_0(x_0, b)$ is seen to coincide with $f(x)$ for $b = 0$. A smoothing of the nonlinearity is observed for $b > 0$. The insensitivity zone seems to narrow and the system becomes more effectively controlled due to the low frequency component.

A similar smoothing occurs in the saturation zone, too. Let us construct function $f_0(x_0, b)$ for the nonlinearity (5.92) in which $\Delta = 0$. By virtue of Table 3.1 we obtain

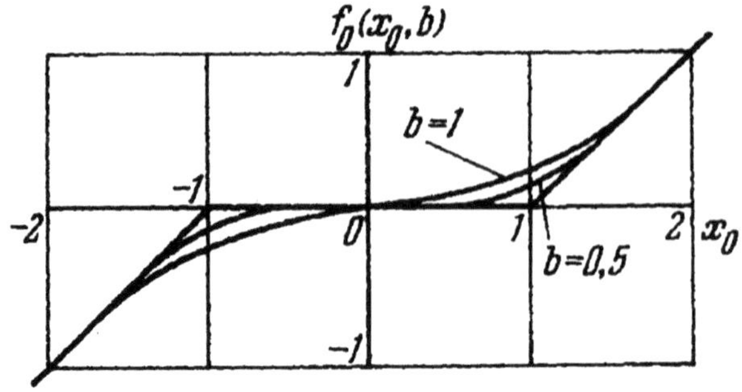

FIGURE 5.17.

$$\left.\begin{aligned}
f_0 &= kx_0 \ \text{ if } x_0 - b < -d,\ x_0 + b < d,\\
f_0 &= -kd + \frac{kb}{\pi}\left[\sqrt{1-\left(\frac{d+x_0}{b}\right)^2} + \frac{d+x_0}{b}\arccos\left(-\frac{d+x_0}{b}\right)\right]\\
&\qquad \text{if } x_0 - b < -d,\ x_0 + b < d,\\
f_0 &= kx_0 - \frac{kb}{\pi}\left[\sqrt{1-\left(\frac{d-x_0}{b}\right)^2} - \frac{d-x_0}{b}\arccos\left(\frac{d-x_0}{b}\right)\right]\\
&\qquad \text{if } x_0 - b > -d,\ x_0 + b > d,\\
f_0 &= -kd + \frac{kb}{\pi}\left[\sqrt{1-\left(\frac{d+x_0}{b}\right)^2} + \frac{d+x_0}{b}\arccos\left(-\frac{d+x_0}{b}\right) -\right.\\
&\left. -\sqrt{1-\left(\frac{d-x_0}{b}\right)^2} + \frac{d-x_0}{b}\arccos\left(\frac{d-x_0}{b}\right)\right]\\
&\qquad \text{if } x_0 - b < -d,\ x_0 + b > d.
\end{aligned}\right\} \tag{5.130}$$

The graphs of function $f_0(x_0, b)$ are plotted in Fig. 5.18 for $k = 1$ and $d = 1$ and a few values of b.

It is worth noting that the nonlinearities in the passive elements are also smoothed out by high frequency processes. This is proved for the cubic nonlinearity and the system with some initial compressive force and Coulomb friction in [63]. As proven there, the high frequency vibrations influence essentially the condition of the occurrence of the resonant vibrations with frequency ν and the subharmonic ones.

The method considered turns out to be inapplicable if the frequencies of the harmonic components of the excitations are not quite distinct. The

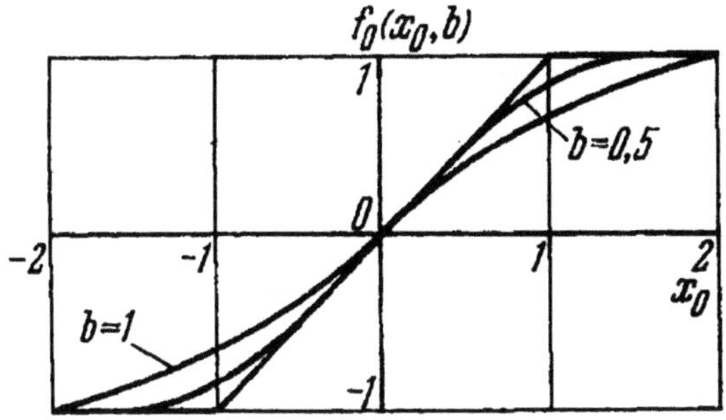

FIGURE 5.18.

method of harmonic linearisation can be applied in this case. Provided that the number of harmonics is sufficiently large and they are excited by independent sources, the Gaussian distribution law can be taken both for the distribution function for the sought-for solution and the distribution density (3.44), see [64]

$$w(u) = \frac{1}{2\pi\sigma_x} \exp\left[-\frac{(u-a_0)^2}{2\sigma_x^2}\right]. \tag{5.131}$$

In this case the linearisation factors for the piecewise linear function can be chosen according to Table 3.1. For the nonlinearity (5.92) without the constant term we have

$$f_0 = 0, \; q = 2k\left[\Phi\left(\frac{d}{\sigma_x}\right) - \Phi\left(\frac{\Delta}{\sigma_x}\right)\right]. \tag{5.132}$$

Dependencies q/k versus σ_x/d are plotted in Fig. 5.19 for some values of Δ/d. Similar to the case of harmonic linearisation, the equivalent gain first increases with increasing σ_x and then decreases. Recall that relation (3.53) holds for σ_x. Thus, a decrease in the feedback efficiency may be conditioned by a single resonant harmonic of large amplitude as well as a large number of harmonics of small amplitude.

Expression (5.132) for the linearisation factor holds true also in the case in which $x(t)$ is a normal stationary random process. In this case, σ_x means the mean-squared value of the solution obtained by averaging over realisations.

The accuracy of the solution obtained by the method of linearisation using the distribution function or by the method of statistical linearisation may be carried out by means of the same approach which has been applied in the case of the harmonic linearisation. Let us consider the case of statistical linearisation in greater depth.

Casting the equation of motion in the form of eq. (5.120) we obtain eq. (5.101) for the difference between the exact stationary stochastic solution $x_1(t)$ and an approximate one. Let $S_\varphi(\omega)$ denote the spectral density of

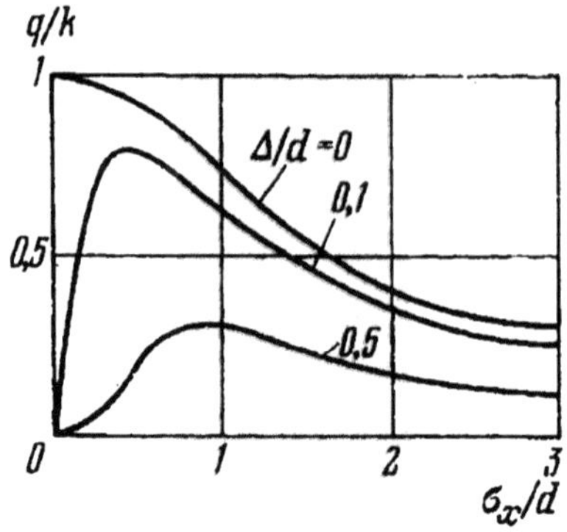

FIGURE 5.19.

function (5.103), then by virtue of the known Wiener-Khinchin relationships, Ref. [128] , we obtain

$$\sigma_{\Delta}^{2}=\frac{1}{2\pi}\int_{-\infty}^{\infty}|v(i\omega)|^{2}\,S_{\varphi}\,(\omega)\,d\omega, \tag{5.133}$$

where $v\,(p)$ is due to eq. (5.102). Equation (5.133) yields the following formula for the mean-squared value of the error

$$\sigma_{\Delta}^{2}\le\max_{\omega}\left[|v(i\omega)|^{2}\right]\sigma_{\varphi}^{2}. \tag{5.134}$$

The estimate of the type (5.107) may be used for σ_{φ}^{2}.

If the random process is not centered, one can estimate mathematic expectation of the error stipulated by the linearisation, too, see [69] and [71].

5.4 Stability of vibration in nonlinear systems

We proceed now to investigation of the influence of both passive and active elements on the stability of the vibrational process.

The equations studied up until now can be cast in the following general form

$$Q(p)x+R(p)f(x)=F\,(t)\,, \tag{5.135}$$

where $Q(p)$ and $R(p)$ are some polynomials in the differentiation operator p, the degree of $Q(p)$ being higher than that of $R(p)$. For example, in eq. (5.112)

$$Q\,(p)=mp^{2}+bp+c,\;\;R\,(p)=k_{1},$$

while in eq. (5.98) $Q(p)$ and $R(p)$ are the numerator and the denominator of the transfer function $e(p)\,w(p)$, respectively.

Assuming that $F(t)$ is a stationary process, no matter if it is harmonic, polyharmonic or random one, we can study the stability of the approximate stationary solutions of eq. (5.135) obtained by the method of equivalent linearisation. As a precursor, we make some remarks.

Provided that an exact stationary solution of a system of nonlinear differential equations is obtained, a study of the stability of the solution is achieved by analysing the variational equations which are the linear equations with stationary coefficients. Solving the variational equations with periodic coefficients presents a problem. When the coefficients are aperiodic the problem becomes very challenging and permits only an approximate solution. On the other hand, if the exact stationary solution is not known *a priori* and only an approximate solution is obtained the very system of equations in variation for the approximate solution is not exact, and the investigation of this system cannot render the rigorous results.

With this in view, we consider here another approach for analysing the stability which is similar in accuracy to the methods used to obtain the approximate solutions. The stability conditions established by this approach usually have a simple physical meaning. With these conditions applied to the periodic solutions, one can obtain known stability criteria in the theory of nonlinear vibration.

The above arguments point out that the results obtained by the approach of this Section are plausible. Nonetheless, by applying these results it is necessary to bear in mind that they are not mathematically rigorous and, for this reason, they cannot be treated as final.

Consider the equation

$$Q(p)x + R(p)f(x) = F(t+\alpha), \tag{5.136}$$

which differs from eq. (5.135) only in parameter α which is assumed to be a random parameter whose values are distributed uniformly on an infinite interval. If $\widetilde{x}(t)$ is a stationary solution of eq. (5.135), then $\widetilde{x}(t+\alpha)$ is a stationary solution of eq. (5.136). From the physical perspective, transition from eq. (5.135) to eq. (5.136) is equivalent to the random choice of the "time datum point", thus, equations (5.136) corresponding to different values of α are physically equivalent. Evidently, if $\widetilde{x}(t)$ is a stable solution of eq. (5.135), then $\widetilde{x}(t+\alpha)$ is a stable solution of eq. (5.136) for any α. If $x(t+\alpha) = \widetilde{x}(t+\alpha) + y(t,\alpha)$ is a perturbed solution corresponding to the unperturbed solution $\widetilde{x}(t+\alpha)$ and the initial condition $y(0,\alpha) = \eta$, then in the case of asymptotic stability the perturbation $y(t,\alpha)$ must tend to zero for any α and for any sufficiently small value of η which is independent of α. Therefore, the probabilistic properties of the process, in particular, the mean value of $y(t,\alpha)$ and that of $\widetilde{x}(t+\alpha)\,y(t,\alpha)$, must tend to zero.

Let

$$\widetilde{x}\,(t+\alpha) = m_x + \widetilde{x}^0\,(t+\alpha) \tag{5.137}$$

be the solution of the linearised equation

$$Q(p)x + F(p)(f_0 + qx^0) = F\,(t+\alpha)\,, \tag{5.138}$$

where m_x denotes the constant term in the solution (or mathematic expectation if $F\,(t)$ is a random process), f_0 and q are the linearisation factors. In general, they can be deemed as given functions of m_x and σ_x^2 (the mean-square value of $\tilde{x}^0\,(t)$)

$$f_0 = f_0(m_x, \sigma_x^2), \;\; q = q(m_x, \sigma_x^2). \tag{5.139}$$

Let us form the equation in variations for the solution $\tilde{x}\,(t+\alpha)$. Let

$$x = \widetilde{x} + y = m_x + \widetilde{x}^0 + m_y + y^0. \tag{5.140}$$

Substituting the perturbed solution (5.140) into eq. (5.138) and introducing the variation of the linearisation factors yields

$$Q(p)(m_x + \widetilde{x}^0 + m_y + y^0) + R(p)[f_0 + \delta f_0 + (q+\delta q)(\widetilde{x}^0 + y^0)] = F(t+\alpha). \tag{5.141}$$

As $\widetilde{x}$ is the solution of eq. (5.138) we obtain

$$Q(p)(m_x + \widetilde{x}^0) + R(p)(f_0 + q\widetilde{x}^0) = F(t+\alpha). \tag{5.142}$$

Subtracting (5.142) from (5.141) gives

$$Q(p)(m_y + y^0) + R(p)(\delta f_0 + qy^0 + \delta q\widetilde{x}^0) = 0. \tag{5.143}$$

In this equation we separate the slowly varying and "vibrational" terms

$$Q(p)m_y + R(p)\delta f_0 = 0, \tag{5.144}$$

$$[Q(p) + qR(p)]y^0 + R(p)(\delta q\widetilde{x}^0) = 0. \tag{5.145}$$

Assuming that eq. (5.139) holds for the varied solution, too, we obtain expressions for δf_0 and δq

$$\left.\begin{aligned} \delta f_0 &= \frac{\partial f_0}{\partial m_x}\delta m_x + \frac{\partial f_0}{\partial \sigma_x^2}\delta\sigma_x^2 = c_{11}\delta m_x + c_{12}\delta\sigma_x^2, \\ \delta q &= \frac{\partial q}{\partial m_x}\delta m_x + \frac{\partial q}{\partial \sigma_x^2}\delta\sigma_x^2 = c_{21}\delta m_x + c_{22}\delta\sigma_x^2. \end{aligned}\right\} \tag{5.146}$$

Let $\underset{\alpha}{m}$ denote the operator of averaging over α, in this case

$$\left.\begin{aligned}\delta m_x &= m_y, \\ \delta\sigma_x^2 &= \underset{\alpha}{m}\left\{(\widetilde{x}^0+y^0)^2\right\} - \underset{\alpha}{m}\left\{(\widetilde{x}^0)^2\right\} = 2\underset{\alpha}{m}\left\{\widetilde{x}^0 y^0\right\}\end{aligned}\right\} \tag{5.147}$$

provided that y^0 is small and only terms up to the first order of smallness are kept in the expressions for the variables.

Inserting eq. (5.146) into eqs. (5.144) and (5.145) and considering (5.147) yields

$$[Q(p) + c_{11}R(p)]\, m_y + c_{12}R(p)\delta\sigma_x^2 = 0, \tag{5.148}$$

$$[Q(p) + qR(p)]\, y^0 + R(p)\left[c_{21}m_y + c_{22}\delta\sigma_x^2\right]\widetilde{x}^0 = 0. \tag{5.149}$$

From the latter equation it follows that

$$y^0 = -\int_0^t g(t-\xi)\left\{c_{21}m_y\widetilde{x}^0(\xi+\alpha) + c_{22}\delta\sigma_x^2\widetilde{x}^0(\xi+\alpha)\right\}d\xi, \tag{5.150}$$

where $g\,(t)$ is the weight function which is the inverse Laplace transform of the operator

$$D(p) = \frac{R(p)}{Q(p) + qR(p)}. \tag{5.151}$$

We multiply eq. (5.150) by $\widetilde{x}^0(t+\alpha)x$ and average over α. Because the following function

$$\underset{\alpha}{m}\left\{\widetilde{x}^0(t+\alpha)\widetilde{x}^0(\xi+\alpha)\right\} = K_x(t-\xi) \tag{5.152}$$

is the correlation function for the solution we obtain

$$\underset{\alpha}{m}\left\{y^0\widetilde{x}^0\right\} = -\int_0^t g(t-\xi)K_x(t-\xi)\left[c_{21}m_y(\xi) + c_{22}\delta\sigma_x^2(\xi)\right]d\xi$$

or

$$\frac{1}{2}\delta\sigma_x^2(t) = -\int_0^t g(t-\xi)K_x(t-\xi)\left[c_{21}m_y(\xi) + c_{22}\delta\sigma_x^2(\xi)\right]d\xi. \tag{5.153}$$

Thus, functions m_y and $\delta\sigma_x^2$ describing the perturbed motion satisfy the system of equations which consists of the linear differential equation

(5.148) and the Volterra integral equation (5.153). Applying the Laplace transformation it is easy to obtain the characteristic equation of this system in the form

$$\begin{vmatrix} Q(p)+c_{11}R(p) & c_{12}R(p) \\ c_{21}L_0(p) & c_{22}L_0(p)+\dfrac{1}{2} \end{vmatrix} = 0. \tag{5.154}$$

Here $L_0(p)$ denotes the Laplace transform of the product of functions $g\,(t)$ and $K_x\,(t)$, i.e.

$$L_0(p) = L\left\{g\,(t)\,K_x\,(t)\right\} = \int_0^\infty e^{-pt} g\,(t)\,K_x\,(t)\,dt. \tag{5.155}$$

The condition for the asymptotic stability of the solution implies negativity of the real parts of all roots of eq. (5.154). As mentioned above, the fact that the mean values of $y\,(t,\alpha)$ and $\tilde{x}\,(t+\alpha)\,y\,(t,\alpha)$ tend to zero must follow from the stability of the solution $\tilde{x}\,(t+\alpha)$. For this reason, the conditions obtained can be treated as the necessary ones.

Let us consider some examples and establish the stability condition for the solution of the following equation

$$(p^2+2np+k^2)x + k_1 f(x) = F_0 \cos\omega t, \tag{5.156}$$

provided that the solutions are obtained by the method of harmonic linearisation. In this case

$$x = a\cos(\omega t+\theta) = \frac{F_0}{\sqrt{\left(\lambda^2-\omega^2\right)^2+4n^2\omega^2}}\cos(\omega t+\theta), \tag{5.157}$$

where $\lambda^2 = k_1 q$. Further,

$$D\,(p) = \frac{R\,(p)}{Q\,(p)+qR\,(p)} = \frac{k_1}{p^2+2np+\lambda^2},$$

hence,

$$g\,(t) = k_1 e^{-nt}\sin\lambda_1 t,\ \lambda_1^2 = \lambda^2 - n^2. \tag{5.158}$$

The correlation function for the solution is as follows

$$K_x\,(t) = \lim_{T\to\infty}\frac{1}{2T}\int_{-T}^{T} x\,(\tau+t)\,x\,(\tau)\,d\tau = \frac{a^2}{2}\cos\omega t. \tag{5.159}$$

Therefore

$$\begin{aligned} L_0(p) &= L\left\{k_1 e^{-nt}\sin\lambda_1 t\cdot\frac{a^2}{2}\cos\omega t\right\} \\ &= \frac{k_1 a^2}{2}\frac{p^2+2np+\lambda^2-\omega^2}{(p^2+2np+\lambda^2-\omega^2)^2+4\omega^2(p+n)^2}. \end{aligned} \tag{5.160}$$

Assume that function $f(x)$ is odd, then $f_0 = 0$ and $m_x = a_0 = 0$. Presuming that these equations hold true also for the perturbed motions, i.e. restricting ourselves to analysing the stability of the harmonic solution, we obtain

$$c_{11} = \frac{\partial f_0}{\partial a_0} = 0,\ c_{12} = \frac{\partial f_0}{\partial \sigma_x^2} = 0,\ c_{21} = \frac{\partial q}{\partial a_0} = 0.$$

Since

$$\sigma_x^2 = \frac{a^2}{2},$$

we have

$$c_{22} = \frac{\partial q}{\partial \sigma_x^2} = \frac{1}{a}\frac{dq}{da}. \tag{5.161}$$

The characteristic equation (5.154) is split into two equations

$$Q(p) = 0,\ \frac{1}{a}\frac{dq}{da}L_0(p) + \frac{1}{2} = 0. \tag{5.162}$$

The roots of the first equation have negative real parts if

$$n > 0,\ k^2 + k_1 q > 0.$$

The second equation can be rearranged into the following form

$$1 + k_1 a\frac{dq}{da}\frac{p^2+2np+\lambda^2-\omega^2}{(p^2+2np+\lambda^2-\omega^2)^2+4\omega^2(p+n)^2} = 0. \tag{5.163}$$

One of the stability conditions is positiveness of the absolute term of the characteristic equation. Setting $p = 0$ in eq. (5.163) we obtain

$$1 + k_1 a\frac{dq}{da}\frac{\lambda^2-\omega^2}{(\lambda^2-\omega^2)^2+4n^2\omega^2} > 0. \tag{5.164}$$

This condition has a clear physical meaning. The total derivative due to eq. (5.114) is

$$\frac{da}{dF_0} = \frac{1}{\sqrt{\left(\lambda^2-\omega^2\right)^2+4n^2\omega^2}} - \frac{\left(\lambda^2-\omega^2\right)F_0 k_1}{\left[\left(\lambda^2-\omega^2\right)^2+4n^2\omega^2\right]^{3/2}}\frac{dq}{da}\frac{da}{dF_0}.$$

Hence,

$$\frac{da}{dF_0} = \frac{1}{\sqrt{(\lambda^2 - \omega^2)^2 + 4n^2\omega^2}} \left\{ 1 + \frac{(\lambda^2 - \omega^2)k_1 a}{(\lambda^2 - \omega^2)^2 + 4n^2\omega^2} \frac{dq}{da} \right\}^{-1} . \tag{5.165}$$

Therefore, eq. (5.164) is equivalent to the following condition

$$\frac{da}{dF_0} > 0, \tag{5.166}$$

i.e. only that harmonic solution of the nonlinear equation can be stable for which an increase in the excitation amplitude leads to an increase in the vibration amplitude. Notice at this stage that stability condition (5.164) can be derived by other methods, for instance this condition is known in theory of nonlinear vibration, too, see [51].

Analogously, we can find the stability condition for the polyharmonic solutions obtained by linearising using the distribution function. Let us put

$$F(t) = \sum_{s=1}^{N} A_s \cos(\omega_s t + \varphi_s) \tag{5.167}$$

in eq. (5.112). Then,

$$x = a_0 + \sum_{s=1}^{N} a_s \cos(\omega_s t + \varphi_s + \theta_s), \tag{5.168}$$

$$\begin{aligned} \sigma_x^2 &= \frac{1}{2}\sum_{s=1}^{N} a_s^2 = \frac{1}{2}\sum_{s=1}^{N} \frac{A_s^2}{(c + k_1 q - m\omega_s^2)^2 + b^2\omega_s^2} \\ &= \frac{1}{2m^2}\sum_{s=1}^{N} \frac{A_s^2}{(\lambda^2 - \omega_s^2)^2 + 4n^2\omega_s^2}, \end{aligned} \tag{5.169}$$

where $q = q\left(a_0, \sigma_x^2\right)$ and $\lambda^2 = (c + k_1 q)/m$. The correlation function for the solution is

$$K_x(t) = \frac{1}{2}\sum_{s=1}^{N} a_s^2 \cos\omega_s t, \tag{5.170}$$

whereas the impulse response $g(t)$ is due to eq. (5.158). Thus,

$$L_0(p) = L\left\{ k_1 e^{-nt} \sin\lambda_1 t \sum_{s=1}^{N} \frac{a_s^2}{2} \cos\omega_s t \right\}$$

$$= \frac{k_1}{2}\sum_{s=1}^{N} a_s^2 \frac{p^2+2np+\lambda^2-\omega_s^2}{(p^2+2np+\lambda^2-\omega_s^2)^2+4\omega_s^2(p+n)^2}. \tag{5.171}$$

Introducing the expressions obtained into (5.154) and expanding the result yields

$$\left(p^2+2np+k^2+k_1\frac{\partial f_0}{\partial a_0}\right)\left[\frac{1}{2}+\frac{\partial q}{\partial \sigma_x^2}L_0(p)\right]-k\frac{\partial f_0}{\partial \sigma_x^2}\frac{\partial q}{\partial a_0}L_0(p)=0. \tag{5.172}$$

Since the constant term is absent in (5.167) we get

$$k^2 a_0 + k_1 f_0\left(a_0,\sigma_x^2\right)=0.$$

This equations allows $a_0\left(\sigma_x^2\right)$ to be determined which means that

$$k_1\frac{\partial f_0}{\partial \sigma_x^2}+\left(k^2+k_1\frac{\partial f_0}{\partial a_0}\right)\frac{da_0}{d\sigma_x^2}=0$$

and therefore

$$k_1 c_{12}+\left(k_1 c_{11}+k^2\right)\frac{da_0}{d\sigma_x^2}=0.$$

From this equation one obtains

$$\left(k^2+k_1 c_{11}\right)c_{22}-k_1 c_{12}c_{21}=\left(k^2+k_1 c_{11}\right)\frac{dq}{d\sigma_x^2}.$$

Considering this, eq. (5.172) can be rewritten in the following form

$$\left(p^2+np\right)\left[1+2\frac{\partial q}{\partial \sigma_x^2}L_0(p)\right]+\left(k^2+k_1\frac{\partial f_0}{\partial a_0}\right)\left[1+2\frac{dq}{d\sigma_x^2}L_0(p)\right]=0. \tag{5.173}$$

One of the necessary conditions for stability is obtained by setting $p=0$

$$\left(k^2+k_1\frac{\partial f_0}{\partial a_0}\right)\left[1+2\frac{dq}{d\sigma_x^2}L_0(0)\right]>0. \tag{5.174}$$

It is easy to prove that the first factor in (5.174) is positive for any nonlinear elastic characteristic and any polyharmonic solution. With this in view, the stability condition reads

$$1+2\frac{dq}{d\sigma_x^2}L_0(0)=1+k_1\frac{dq}{d\sigma_x^2}\sum_{s=1}^{N}a_s^2\frac{\lambda^2-\omega_s^2}{(\lambda^2-\omega_s^2)^2+4\omega_s^2 n^2}>0. \tag{5.175}$$

The total increment in σ_x^2 stipulated by changing the amplitudes of the harmonic components of the external excitation can be found by means of eq. (5.169). The result is

$$\Delta\sigma_x^2 = \frac{1}{2}\sum_{s=1}^{N} \frac{\Delta A_s^2}{(\lambda^2-\omega_s^2)^2+4n^2\omega_s^2} - k_1\frac{dq}{d\sigma_x^2}\sum_{s=1}^{N}\frac{A_s^2(\lambda^2-\omega_s^2)\Delta\sigma_x^2}{\left[(\lambda^2-\omega_s^2)^2+4n^2\omega_s^2\right]^2}.$$

Hence,

$$\Delta\sigma_x^2 = \frac{1}{2}\sum_{s=1}^{N} \frac{\Delta A_s^2}{(\lambda^2-\omega_s^2)^2+4n^2\omega_s^2}\left[1+2\frac{dq}{d\sigma_x^2}L_0\left(0\right)\right]^{-1}.$$

Provided that condition (5.174) holds true, the derivatives $\partial\sigma_x^2/\partial A_s^2$ are positive. An increase in the amplitude of any harmonic of the vibration excitation leads to an increase in the mean square value of the stable stationary solution.

Assume now that the driving force in eq. (5.156), $F\left(t\right)$, is a stationary normal centered random process with spectral density $S_0 = const.$ Let us establish the stability conditions for the stochastic solutions obtained by the method of stochastic linearisation. By virtue of the equations developed in Chapter 4 we get

$$f_0 = \frac{1}{\sqrt{2\pi}\sigma_x}\int_{-\infty}^{\infty} f(u)\exp\left[-\frac{\left(u-m_x\right)^2}{2\sigma_x^2}\right]du = 0, \tag{5.176}$$

$$S_x\left(\omega\right) = \frac{S_0}{\left|\lambda^2-\omega^2+2ni\omega\right|^2}, \tag{5.177}$$

$$\sigma_x^2 = \frac{1}{2\pi}\int_{-\infty}^{\infty}\frac{S_0\,d\omega}{(\lambda^2-\omega^2)^2+4n^2\omega^2}, \tag{5.178}$$

$$q = \frac{1}{\sqrt{2\pi\sigma_x^3}}\int_{-\infty}^{\infty} f(u)\left(u-m_x\right)\exp\left[-\frac{\left(u-m_x\right)^2}{2\sigma_x^2}\right]du. \tag{5.179}$$

Let us form the characteristic equation in the form of (5.154). The Fourier transform of eq. (5.177) is given by

$$K_x\left(t\right) = \frac{S_0\lambda_1}{4\lambda^2}e^{-nt}\left(\sin\lambda_1 t + \frac{\lambda_1}{n}\cos\lambda_1 t\right). \tag{5.180}$$

Defining $L_0(p)$ as the Laplace transform for the function

$$g\left(t\right)K_x\left(t\right) = \frac{S_0\lambda_1}{4\lambda^2}e^{-2nt}\left(\sin^2\lambda_1 t + \frac{\lambda_1}{n}\cos\lambda_1 t\sin\lambda_1 t\right),$$

we obtain

$$L_0(p) = \frac{S_0\lambda_1}{4\lambda^2}\left(\frac{1}{p+2n} - \frac{p+2n}{(p+2n)^2+4\lambda_1^2} + \frac{\lambda_1}{n}\frac{2\lambda_1}{(p+2n)^2+4\lambda_1^2}\right). \tag{5.181}$$

Expressions (5.176) and (5.179) yield

$$\begin{aligned} c_{11} &= \frac{\partial f_0}{\partial m_x} = \frac{1}{\sqrt{2\pi}\sigma_x}\int_{-\infty}^{\infty} f(u)\exp\left[-\frac{(u-m_x)^2}{2\sigma_x^2}\right]\frac{(u-m_x)}{\sigma_x^2}du = q, \\ c_{12} &= \frac{\partial f_0}{\partial \sigma_x^2}\frac{1}{\sqrt{2\pi}\sigma_x}\int_{-\infty}^{\infty} f(u)\frac{(u-m_x)^2}{2\sigma_x^4}\exp\left[-\frac{(u-m_x)^2}{2\sigma_x^2}\right]du \\ &= \frac{\partial q}{\partial m_x} = c_{21}. \end{aligned}$$

Introducing these expressions into (5.154) we have

$$\begin{vmatrix} p^2+2np+\lambda^2 & k_1 c_{12} \\ c_{12}L_0(p) & \dfrac{1}{2}+c_{22}L_0(p) \end{vmatrix} = 0. \tag{5.182}$$

The absolute term of this equation is obtained by setting $p = 0$ in the latter equation. The stability condition requires the absolute term to be positive, i.e.

$$\lambda^2\left[\frac{1}{2}+c_{22}L_0(0)\right] - c_{12}^2 k_1 L_0(0) > 0. \tag{5.183}$$

6

Optimal systems of vibration protection

6.1 Statement of the optimisation problem

Various claims imposed on vibration protection systems can be expressed in terms of certain constraints. In what follows we list a few of the most important classes of constraints which need to be accounted for in the synthesis of vibration protection systems.

1. *Absolute acceleration constraints.* If a vibration protection system is intended to protect an object from kinematic excitations, its main task is to reduce the absolute acceleration at certain points of the object. The corresponding constraints can be written down in the following general form

$$K_w \{w_i(t)\} \le K_{wi} \quad (i = 1, ..., s). \tag{6.1}$$

Here $w_i(t)$ denotes the components of the absolute acceleration of the body to be protected along certain given directions, e.g. the directions of the coordinate axes, and K_w stands for a functional whose form usually depends upon the character of the dynamic excitations. In the case of harmonic, polyharmonic or stationary random excitation the constraints are conventionally imposed on some mean values of acceleration, more often on the mean square

$$K_w \{w_i(t)\} = m\left\{w_i^2(t)\right\}, \tag{6.2}$$

where $m\{\}$ means time- and realisation-averaging for deterministic and random process, respectively. The maximum value of the absolute acceleration

$$K_w\{w_i(t)\} = \max_t |w_i(t)| \tag{6.3}$$

is natural to choose in the case of non-stationary or shock excitation. This value can be bounded by some value K_t.

2. *Force constraints.* As pointed out in Chapter 1, the main criterion of performance for the isolator which protects the base from the vibrating object is the value of the forces transmitted to the base

$$K_P\{P_r(t)\} \leq K_{P_r} \quad (r = 1, ..., N), \tag{6.4}$$

where P_r is the force in the $r-th$ isolator. The form of the functional K_P is prescribed in the same manner as was above.

Force constraints can be introduced in the case of kinematic excitations as well. For example, they may be related to the requirements of strength or reliability of some structural components.

3. *Relative displacement constraints.* It is known that in principle any strict requirements of the vibration protection system might be met by means of sufficiently soft isolators. The dynamic excitations, however, will cause large displacements relative to the base. In order to avoid impact with the stops, the latter must be placed sufficiently far from the object to be protected. In other words, vibration protection systems will have large dimensions and may not be realised in practice.

With this in mind, while designing a vibration protection system the relative displacement requirements are needed. The corresponding constraints may be represented as follows

$$K_u\{u_i(t)\} \leq K_{u_i} \quad (i = 1, ..., k), \tag{6.5}$$

with $u_i(t)$ being the relative displacements of particular points of the object. The form of the functional K_u depends upon the character of the external actions, e.g.

$$K_u\{u_i(t)\} = m\left\{u_i^2(t)\right\}, \tag{6.6}$$

when the excitation is stationary and

$$K_u\{u_i(t)\} = \max_t |u_i(t)|, \tag{6.7}$$

when the excitation is nonstationary.

4. *Control constraints.* When active facilities are incorporated, it is reasonable to impose constraints on the control forces. Either the control force

$$K_U = \max_t |U_i(t)| \leq K_{U_i} \tag{6.8}$$

or its mean square

$$K_U = m\left\{U_i^2(t)\right\} \le K_{U_i} \tag{6.9}$$

can be constrained. The value of the mean power of the actuator may also be constrained.

5. *Design constraints.* The necessity to impose constraints during the design process may occur for many reasons. For example, the use of, say, linear passive isolators or hydraulic actuators of a prescribed type, may be required. The necessity to bound the value of the system parameters may be actual. For instance, the stiffness and damping factors of the passive elements must lie within the certain bounds for design reasons. Ultimately, these requirements can all be expressed in terms of certain inequalities, the parameters of the vibration protection system satisfying them.

Assume that there exists a certain non-empty set of vibration protection systems satisfying all these given constraints for prescribed external dynamic excitations. Then, we can state the problem of choosing among them such a system which extremizes some quality criterion. This means that this particular system will be optimal only in a certain sense.

Any of the above functionals or their linear combination may be deemed as a quality criterion.

From the perspective of practice, the problem of synthesis of a vibration protection system ensuring some prescribed reduction of the level of dynamic actions under minimum dimensions is of crucial importance. In this case, constraints of the type (6.1)-(6.4) are imposed on the system and a functional governing the relative displacement must be minimised. There exist several approaches to the problem.

The first approach is based on the fact that the role of any vibration protection system reduces to forming some force actions $U_k(t)$ which may be viewed as control forces. In connection with this, the problem of synthesis of the optimal vibration protection system may be reduced to the problem of determining the optimal control. The problem of optimisation of vibration protection systems was first stated in this form for a single-degree-of-freedom system in [36]. Vibration of this system undergoing kinematic excitation $\xi(t)$ is governed by the following equation

$$m\ddot{x} = U, \tag{6.10}$$

where x is the absolute displacement, U is the force transmitted to the object by a vibration protection facility which is mounted between the object and the base. No matter what the facility is, its action on the object may be considered as a control force which is a time-dependent function $U(t)$. In this case, the requirements to the quality of vibration protection system reduce to the constraint imposed on the absolute acceleration

$$K\{\ddot{x}(t)\} \le K_0. \tag{6.11}$$

By virtue of (6.10) the latter equation is equivalent to the following constraint imposed on the control force

$$K\left\{\frac{1}{m}U(t)\right\} \leq K_0. \tag{6.12}$$

On the other hand, eq. (6.10) can be rewritten in the form

$$\ddot{u} = \frac{U}{m} - \ddot{\xi}(t), \tag{6.13}$$

where u denotes the relative displacement. Minimisation of the dimensions of the vibration protection system is equivalent to minimising a certain functional of $u(t)$

$$J\{u(t)\} = \min. \tag{6.14}$$

Under these circumstances the problem of synthesis of the optimal system of vibration protection can be stated as follows: among functions $U(t)$ belonging to a certain class D and satisfying condition (6.12) find such a function which minimises function (6.14) in which $u(t)$ is a solution of eq. (6.10) and belongs to a class E. In the following, this problem is referred to as *the simplest problem of the optimal synthesis of vibration protection system.* In [36] it was assumed that

$$K\{U(t)\} = \max_t \left|\frac{1}{m}U(t)\right|, \tag{6.15}$$

$$J\{u(t)\} = \max_t |u(t)|, \tag{6.16}$$

i.e. the Chebyshev norms of function $U(t)$ and $u(t)$ were chosen as the functionals J and K.

The class D includes piecewise-continuous functions within the interval $(0, \infty)$ whereas the class E contains functions satisfying the following conditions

$$u(0) = 0, \ \dot{u}(0) = 0.$$

This approach is natural for the synthesis of vibration isolation systems aimed at protection from shock excitations caused by a deterministic perturbation $\xi(t)$. The value

$$\Delta = \min_U \max_t |u(t)| \tag{6.17}$$

is called the *minimax displacement.*

Another statements of the problem of optimal synthesis involving other forms of functionals (6.15) and (6.16) are also possible. Some of these statements are considered below.

The problem of the practical synthesis is not solved even after the optimal control force $U(t)$ minimizing functional $J\{u(t)\}$ has been determined. It is necessary to design a system which will realize this control law, and this is not always feasible. Nevertheless, determining the optimal control and the corresponding minimum value of J is undoubtedly useful because this allows one to judge the limiting capabilities of the vibration protection. Clearly, that no design solution can render a value of J less than that of the optimal control. By comparing the value of J for a particular system with the minimum value of J one can judge how close this system is to the optimal system and whether it is expedient to complicate the system in order to lessen the value of the functional.

Another approach to the problem of optimisation of vibration protection systems suggested first in [78] is concerned with the search for optimal transfer functions for linear systems of vibration protection. Let us consider this approach for the example of a single-degree-of-freedom system. Assume that the control force $U(t)$ is related to the absolute $x(t)$ or relative $u(t)$ displacement as follows

$$U(t) = -w(p)x(t) \tag{6.18}$$

or

$$U(t) = -w_1(p)u(t)\,. \tag{6.19}$$

One seeks the transfer function $w(p)$ (or $w_1(p)$) which ensures that the solution of eq. (6.13) belongs to a certain class E and minimizes the functional $J\{u(t)\}$. Also, the solution of eq. (6.10) must satisfy condition (6.11).

This statement of the problem renders a minimum of the functional which is larger than that of the optimal control. However, the transfer function of the linear optimal system is obtained, i.e. the system can be realised in practice.

The third approach consists of determination of the optimal parameters of the system having the prescribed structure. Let, for example, the parameters of the simplest linear isolator, namely stiffness c and damping factor b, be sought. An isolator is sought for protection of the object from stationary kinematic excitation. The governing equation of motion is given by

$$(mp^2 + bp + c)u = -m\ddot{\xi}(t)\,. \tag{6.20}$$

Introducing $\ddot{x} = \ddot{u} + \ddot{\xi}(t)$ we obtain

$$\ddot{x} = \frac{bp + c}{mp^2 + bp + c}\ddot{\xi}(t)\,. \tag{6.21}$$

Assume that the time-averaged mean square of the deterministic stationary process $\ddot{x}(t)$ is constrained

$$\underset{t}{m} \left\{ \ddot{x}^2 (t) \right\} < \sigma_0^2. \tag{6.22}$$

Then, one can determine the values of b and c minimizing for example the following functional

$$J = \underset{t}{m} \left\{ u^2 (t) \right\}. \tag{6.23}$$

A statement of the problem of optimal vibration protection of an object subjected to random excitations was suggested in [12]. The problem of minimising the probability of violation of the constraints imposed on certain functionals describing the quality of vibration protection was stated there. This optimisation method is adequately covered in [12], [136] and [107] and for this reason is not considered in the present book.

6.2 Determination of the optimal control

We consider some methods of determinating the optimal control for the "simplest" problem. As mentioned above, this problem was originally solved for non-stationary (shock) excitations. Experience indicates that it is shock in vibration protection systems which causes large relative displacements, thus, minimisation of dimensions of the object is the most important task.

Let us formulate the problem of optimising shock protection in the form of the so called minimax statement. The equation of motion of a single-degree-of-freedom object is written in the form of eq. (6.13), i.e.

$$\ddot{u} = \ddot{x} - \ddot{\xi} = \frac{1}{m} U - \ddot{\xi} = v - \sigma (t). \tag{6.24}$$

Here $\xi (t)$ and $\sigma (t)$ denote the displacement and prescribed acceleration of the base, and v is the object acceleration. We look for a piecewise-continuous function $v (t)$ subject to the constraint

$$|v (t)| \leq v_0 \ \ (0 \leq t < \infty) \tag{6.25}$$

which minimizes the following functional

$$J = \max_t |u (t)| \ \ (0 \leq t < \infty), \tag{6.26}$$

where $u (t)$ is a solution of eq. (6.24) under the trivial initial conditions

$$u (0) = 0, \ \ \dot{u} (0) = 0. \tag{6.27}$$

The solution of the stated problem is elementary for some types of dynamic excitation. For instance, if at any time

$$|\sigma (t)| \leq v_0,$$

then there exists a solution

$$v(t) \equiv \sigma(t), \tag{6.28}$$

for which

$$\Delta = \min_v \max_t |u(t)| = 0.$$

This result implies that in those cases when the shock excitation is not needed to be reduced, one can design the "isolation system" with zero free running by means of rigid attachment of the object to the base.

One can easily find the solution for the case of a rectangular form of acceleration

$$\sigma(t) = \sigma_0 [\eta(t) - \eta(t - \tau)], \tag{6.29}$$

where $\sigma_0 > 0$, τ is the impact duration and $\eta(t)$ denotes the Heaviside step function. This case was studied in detail in [65] where the optimal control was shown to have the following form

$$v(t) \equiv v_0, \ 0 \le t \le t^*, \tag{6.30}$$

where

$$t^* = \frac{\sigma_0}{v_0}\tau = \frac{\tau}{\mu} > 0 \ \left(\mu = \frac{v_0}{\sigma_0}\right).$$

The absolute value of the relative displacement achieves its maximum value at $t = t^*$ and is given by

$$u = \Delta = \frac{\sigma_0 \tau^2 (1 - \mu)}{2\mu}. \tag{6.31}$$

This is the minimum possible value of the first maximum of the relative displacement. For $t > t^*$ the function $v(t)$ can be chosen so that all absolute values of $u(t)$ do not exceed $u(t^*)$. This can be done in many ways. For example, one can adopt

$$v(t) \equiv \sigma(t) \ \text{ for } t > t^*.$$

Then, for $t > t^*$

$$\ddot{u} = v(t) - \sigma(t) \equiv 0,$$

and therefore

$$\dot{u}(t) = \dot{u}(t^*) = 0,$$

since t^* extremizes $u(t)$. Hence,

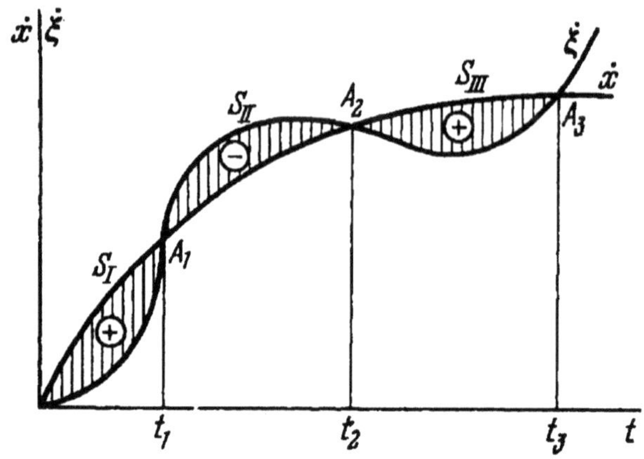

FIGURE 6.1.

$$u(t) = u(t^*) \text{ for } t > t^*.$$

Thus, under excitation (6.29) the free running of the isolator cannot be less than that given by eq. (6.31).

The search for optimal control becomes essentially more complicated for external actions of more complex form. The main complications are related to the properties of the functional (6.26). Let us seek the control which minimizes (6.26) over the interval (t_0, T_0). We split the interval into several subintervals and find the control minimizing the value of max $|u(t)|$ within each subinterval. However, it turns out that the set of these "optimal" controls does not produce the control which is optimal within the entire interval. The functionals possessing this feature are termed *non-additive functionals.* Non-additivity of functional (6.26) hinders the use of such modern optimal control methods as the maximum principle or dynamic programming. The graphic-analytical methods suggested in [36], [37] and [40] prove to be the most convenient for this purpose.

Let us integrate over t the left and right hand sides of eq. (6.24). The result is

$$\dot{u} = \int_0^t v(t)\,dt - \int_0^t \sigma(t)\,dt = \dot{x}(t) - \dot{\xi}(t). \tag{6.32}$$

Let $v(t)$ be some control (not necessarily the optimal one). Figure 6.1 shows the graph of $\dot{x}(t)$ for this control and $\dot{\xi}(t)$. The value of the relative displacement $u(t')$ is given by the area between these curves and the line $t = t'$. The area must be considered as positive on those parts where $\dot{x} > \dot{\xi}$ and negative where $\dot{x} < \dot{\xi}$. The points of intersection of these curves correspond

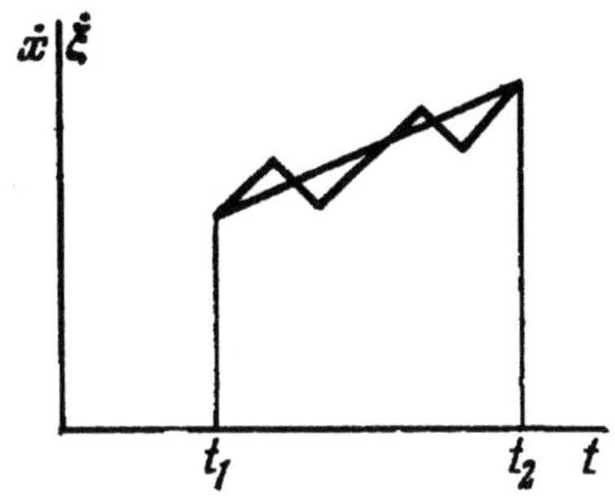

FIGURE 6.2.

to the extrema of $u(t)$. The value of these extrema are respectively given by

$$u(t_1) = s_I,\ u(t_2) = s_I - s_{II},\ u(t_3) = s_I - s_{II} + s_{III}, \ldots, \qquad (6.33)$$

where s_I, s_{II} and s_{III} are the absolute values of the areas shown in Fig. 6.1. We find the value of $\max_t |u(t)|$ after the maximum absolute value of the extrema has been determined.

In order to define the optimal control it is necessary to build a continuous curve $\dot{x}(t)$ which satisfies the condition

$$\ddot{x}\,|(t)| = |v(t)| \leq v_0$$

and minimizes the maximum absolute value of the sum of the areas in the series (6.33) for the given curve $\dot{\xi}(t)$. An algorithm for constructing the optimising curve was first suggested in [40] and [39]. The following basic assumption are made there:

1. It is adopted that action $\sigma(t)$ satisfies the following condition

$$|\sigma(t)| < v_0 \text{ for } t > T,$$

with T being a sufficiently large value. In other words, the absolute value of the action is assumed to achieve v_0 and remain smaller than v_0 for sufficiently large time t. This assumption allows us to limit the construction of the optimal control within some finite interval $(0, T^*)$ and adopt $v(t) = \sigma(t)$ for $t > T$.

2. It is proven that the optimal control, if it exists, can always be composed of the parts on which

$$v(t) = \pm v_0.$$

This means that the optimising curve $\dot{x}(t)$ can be a broken line whose segments have the slopes $\pm \arctan v_0$. Notice that the possibility of constructing the optimal control of another forms is not excluded because the

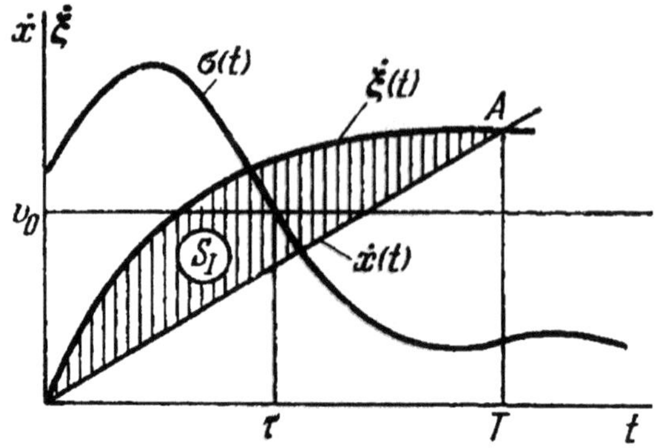

FIGURE 6.3.

optimal control for the stated problem is not unique in general. For example, let us assume that there exists such an optimal control that the curves $\dot{x}(t)$ and $\dot{\xi}(t)$ coincide within the interval (t_1, t_2), see Fig. 6.2. Clearly, $u(t) = const = u(t_1)$ on this interval. Assume that $u(t_1)$ is not equal to the minimax value. Then, on this interval curve $\dot{x}(t)$ can be replaced by a broken line whose segments have the slopes $\pm \arctan v_0$. In this case, $u(t)$ is no longer constant, but its value varies within some bounds, the latter can always be chosen so narrow that the value $|u(t)|$ remains smaller than the minimax value.

3. It is found that the algorithm for constructing the optimal control is defined by the numbers of intersections of curve $\sigma(t)$ with lines $v(t) = \pm v_0$, i.e. the number of roots of the equation

$$\sigma(t) = \pm v_0. \tag{6.34}$$

The problem is especially simple in the following case

$$\left.\begin{aligned} \sigma(t) > v_0 \text{ if } 0 \le t < \tau, \\ \sigma(t) \le v_0 \text{ if } t \ge \tau, \end{aligned}\right\} \tag{6.35}$$

i.e. there is only one root of eq. (6.34) which is equal to τ, see Fig. 6.3. In this case, the optimal control is given by

$$\left.\begin{aligned} v(t) &= v_0, & 0 \le t < T, \\ v(t) &= \sigma(t), & t > T, \end{aligned}\right\} \tag{6.36}$$

where $T > \tau$ denotes the time instant corresponding to the intersection point of the curve

$$\dot{x}(t) = v_0 t$$

with curve $\dot{\xi}(t)$. Indeed, this control satisfies constraint (6.25) by virtue of condition (6.35). The maximum absolute value of $u(t)$

$$\Delta = \max_t |u(t)| = \int_0^T \sigma(t)\,dt - \frac{v_0 T^2}{2}, \tag{6.37}$$

which is equal to area S_I in Fig. 6.3, cannot be decreased.

Table 6.1 shows the formulae for T and Δ for some particular forms of shock excitation satisfying condition (6.35).

It is more difficult to find the optimal control in the case in which the line $\sigma(t)$ intersects the level $\pm v_0$ twice, i.e. if there exists a part (τ_1, τ_2) where $|\sigma(t)| > v_0$. In this case, function $\dot{\xi}(t)$ has the slope v_0 at two points (A and B in Fig. 6.4a), i.e. the line $\dot{x} = v_0 t$ intersects $\dot{\xi}(t)$ either twice (except $t = 0$) producing areas s_1^0 and s_2^0 or never. Provided that in the first case

$$s_{II}^0 > 2s_I^0, \tag{6.38}$$

the built line determines the following optimal control

$$\left.\begin{aligned} v(t) &\equiv v_0, & t &\le T, \\ v(t) &\equiv \sigma(t), & t &> T. \end{aligned}\right\}$$

It is easy to prove that this control satisfies constraint (6.25). On the other hand, the maximum absolute value of the displacement is

$$\Delta = s_{II}^0 - s_I^0 > s_I^0.$$

In order to decrease this displacement, it is necessary either to increase area s_I^0 or to decrease s_{II}^0. Neither is feasible since line $\dot{x}(t)$ lies entirely under the constructed line for any admissible control. Assume now that

$$s_I^0 < s_{II}^0 < 2s_I^0. \tag{6.39}$$

We plot the following lines

$$\dot{x} = v_0 t - a \tag{6.40}$$

and

$$\dot{x} = -v_0 t, \tag{6.41}$$

see Fig. 6.4b. The broken line OAB defines certain admissible control

$$\left.\begin{aligned} v(t) &= -v_0, & 0 &\le t \le t_0, \\ v(t) &= v_0, & t_0 &< t \le T, \\ v(t) &= \sigma(t), & t &> T. \end{aligned}\right\} \tag{6.42}$$

$\sigma(t)$	T	Δ	Note
$\sigma_0 [\eta(t) - \eta(t-T)]$	$\frac{\tau}{\mu}$	$\frac{\sigma_0 \tau^2}{2\mu}(1-\mu)$	$\mu = \frac{v_0}{\sigma_0}$
$\sigma_0 e^{-kt}$	$\mu kT = 1 - e^{-kt}$	$\frac{\sigma_0 T}{k}\left[1 - \mu\left(1 + \frac{kT}{2}\right)\right]$	$\mu = \frac{v_0}{\sigma_0}$
$\sigma_0 \cos \frac{\pi t}{2\tau}$, $t \le \tau$	$T = \frac{2\tau}{\pi\mu}$, $\mu < \frac{2}{\pi}$	$\frac{4\sigma_0 \tau^2}{\pi^2}\left(1 + \frac{1}{2\mu} - \frac{\pi}{2}\right)$, $\mu < \frac{2}{\pi}$	$\mu = \frac{v_0}{\sigma_0}$
$0,\ t > \tau$	$\sin\alpha = \alpha$, $\mu > \frac{2}{\pi}$	$\frac{4\sigma_0 \tau^2}{\pi^2}\left(1 - \cos\alpha - \frac{1}{2}\mu\alpha^2\right)$, $\mu > \frac{2}{\pi}$	$\alpha = \frac{\pi T}{2\tau}$
$\sigma_0\left(1 - \frac{t}{\tau}\right)$, $t \le \tau$	$T = \frac{\tau}{2\mu}$, $\mu < \frac{1}{2}$	$\frac{\sigma_0 \tau^2}{2}\left(\frac{1}{4\mu} - \frac{1}{3}\right)$, $\mu < \frac{1}{2}$	$\mu = \frac{v_0}{\sigma_0}$
$0,\ t > \tau$	$T = 2\tau(1-\mu)$, $\mu > \frac{1}{2}$	$\frac{2}{3}\sigma_0 \tau^2 (1-\mu)^3$, $\mu > \frac{1}{2}$	

TABLE 6.1.

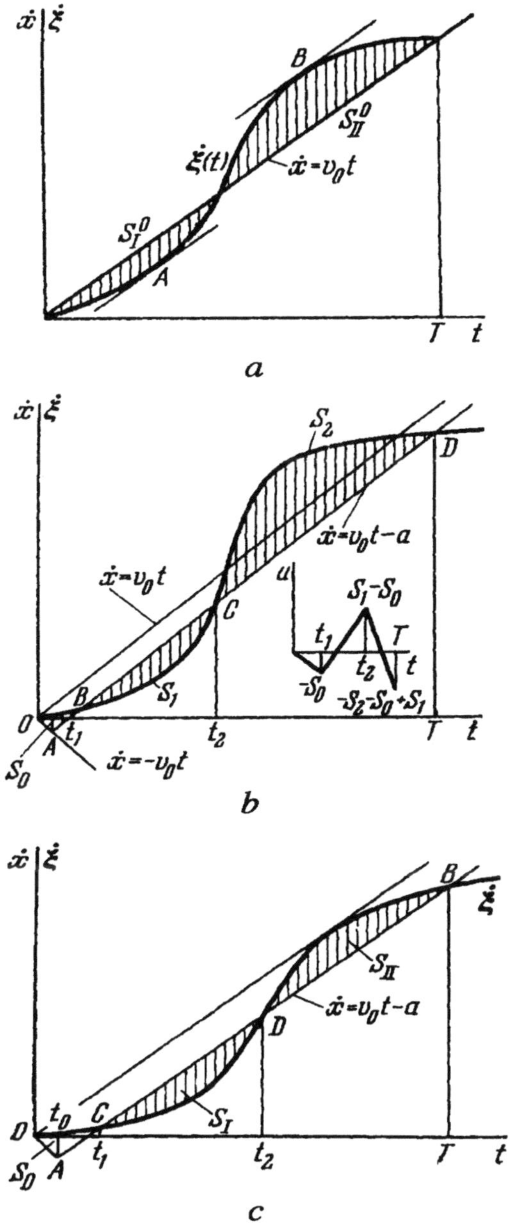

$\dot{x}$ | $\dot{\xi}$
B
S_{II}^0
$\dot{\xi}(t)$
$\dot{x} = v_0 t$
S_I^0
A
T
t
a
$\dot{x}$ | $\dot{\xi}$
S_2
D
$\dot{x} = v_0 t - a$
$\dot{x} = v_0 t$
u
$S_1 - S_0$
C
t_1
T
t_2
t
$-S_0$
$-S_2 - S_0 + S_1$
B
S_1
0
t_1
t_2
T
t
A
$\dot{x} = -v_0 t$
S_0
b
$\dot{x}$ | $\dot{\xi}$
B
$\dot{\xi}$
S_{II}
$\dot{x} = v_0 t - a$
D
t_0
C
S_I
0
S_0
A
t_1
t_2
T
t
c

FIGURE 6.4.

This broken line intersects curve $\dot{\xi}(t)$ at points C, D and B forming three areas s_0, s_I and s_{II}. The relative displacement $u(t)$ is additionally displayed in Fig. 6.4b.

Let us find that value of a which ensures that $s_{II} = 2(s_I - s_0)$. By virtue of condition (6.39) this value always exists. In this case

$$s_0 < \frac{s_I}{2},$$

since $s_{II} > s_I^0$ and hence

$$s_{II} > s_I.$$

The control (6.42) is then optimal. Indeed, it is easy to prove that in this case

$$\begin{aligned} |u(t_2)| &= s_I - s_0 = s_{II} + s_0 - s_I = |u(T)|, \\ |u(t_1)| &= s_0 < |u(t_2)|. \end{aligned}$$

The maximum displacement is thus

$$\Delta = s_I - s_0 = s_{II} + s_0 - s_I. \tag{6.43}$$

The curve $\dot{x}$ may lie within figure OAC, it is thus clear that any change in s_0 (for invariable s_I and s_{II}) leads to an increase in one of the extreme displacements, either $|u(t_2)|$ or $|u(T)|$. On the other hand, a decrease in the value of a causes an increase in $|u(t_2)|$, while a decrease in the value of $|u(t_2)|$ results in the growth of $|u(T)|$.

Let

$$s_{II}^0 < s_I^0. \tag{6.44}$$

In particular, this inequality holds if the line $\dot{x}(t) = v_0 t$ does not intersect curve $\dot{\xi}(t)$ at all, see Fig 6.4c. By analogy with the previous analysis it can be shown that the optimal control under condition (6.44) can be constructed as follows. We choose a so that $s_{II} = s_I$ and plot the lines (6.41) and (6.40). Let us assume that

$$2s_0 > s_I.$$

Then, we can find such an admissible control $v^*(t)$ on the part $0 \le t \le t_1$ which ensures that the area between the curve $x^*(t)$ corresponding to this control and the curve $\dot{\xi}(t)$ is equal to $s_I/2$. (Note that line $x^*(t)$ lies within figure OAC). Optimal control in this case is determined as follows

$$\left.\begin{aligned} &v(t) = v^*(t), && 0 \le t \le t_1, \\ &v(t) = v_0, && t_1 \le t \le T, \\ &v(t) = \sigma(t), && t > T, \end{aligned}\right\} \tag{6.45}$$

while the minimax displacement is given by

$$\Delta = |u(t_1)| = |u(t_2)| = |u(T)| = \frac{s_I}{2}.$$

Provided that $2s_0 < s_I$ under condition $s_{II} = s_I$ the value a must be increased so that relationship (6.43) holds. In this case the optimal control is given by eq. (6.42).

Consider the following action

$$\left.\begin{array}{ll} \sigma(t) = \sigma_0 \sin \frac{\pi}{\tau} t, & 0 \le t \le \tau, \\ \sigma(t) = 0, & t > \tau, \end{array}\right\} \tag{6.46}$$

which belongs to the above case. Here

$$\left.\begin{array}{ll} \dot{\xi}(t) = \frac{\sigma_0 \tau}{\pi} \left(1 - \cos \frac{\pi}{\tau} t\right), & 0 \le t \le \tau, \\ \dot{\xi}(t) = \frac{2\sigma_0 \tau}{\pi}, & t > \tau, \end{array}\right\} \tag{6.47}$$

Calculations show that for

$$\mu = \frac{v_0}{\sigma_0} < 0,6028$$

condition (6.38) holds, and hence, the optimal control is given by eq. (6.39) in which

$$T = \frac{2\tau}{\pi\mu}.$$

If

$$0,6028 < \mu < \frac{2}{\pi},$$

eq. (6.39) holds and the control of type (6.42) is optimal. Finally, if $\mu > 2/\pi$ the optimal control is defined by eq. (6.45).

If the number of excesses of function $\sigma(t)$ over the level v_0 turns out to be large, construction of the optimal control is carried out using the method of successive approximations described in detail in [40]. Numerical methods of solution are applied in complicated cases.

We enumerate now some other ways of solving the problem of optimal vibration protection in the minimax statement, i.e. for the quality criterion in form of eq. (6.26).

An approximate method of constructing the optimal control based on linear programming is suggested in [98]. The time interval $(0, T)$, on which the optimal control is sought, is split into N equal subintervals. The control is sought within the class of piecewise-constant functions, i.e. it is assumed that on the $i-th$ interval

$$v(t) = v_i = const, \tag{6.48}$$

where v_i are numerical values satisfying the condition $|v_i| < v_0$. The values of $u(t)$ are calculated at points t_i which are the boundaries of the subintervals

$$\begin{aligned} u(t_i) &= \int_0^{t_i} (t_i - t)\, v(t)\, dt - \xi(t_i) = \\ &= \frac{t_i T}{N} \sum_{k=1}^{i} v_k - \sum_{k=1}^{i} v_k \frac{t_k^2 - t_{k-1}^2}{2} - \xi(t_i). \end{aligned}$$

As $t_k = kT/N$ we obtain

$$\begin{aligned} u(t_i) &= i\frac{T^2}{N^2} \sum_{k=1}^{i} v_k - \frac{T^2}{N^2} \sum_{k=1}^{i} v_k \left(k - \frac{1}{2}\right) \cdot \xi(t_i) = \\ &= \frac{T^2}{N^2} \sum_{k=1}^{i} v_k \left(i - k + \frac{1}{2}\right) - \xi(t_i). \end{aligned} \tag{6.49}$$

Then we seek those values of v_k which minimise the functional

$$J = \max_i |u(t_i)|. \tag{6.50}$$

The problem stated is equivalent to a certain problem of linear programming. As shown in [98], the solution of this discrete problem may be viewed as an approximate solution of the continuous minimax problem formulated above. Moreover, it is proven that

$$\min_U \cdot \max_i |u(t_i)| \geq \min_U \cdot \max_t |u(t)|,$$

thus, one can estimate the upper minimax displacement by means of the minimum of functional (6.50).

The "simplest" problem is solved in [130] by methods of the variation calculus in the following statement. Consider the following equation

$$\ddot{u} + v(u, \dot{u}, t) = \sigma(t), \tag{6.51}$$

which is equivalent to eq. (6.24) under the assumption that the control force $v(u, \dot{u}, t)$ is constrained

$$|v(u,\dot{u},t)| \leq v_0. \tag{6.52}$$

Among the functions $u(t)$ and $v(u,\dot{u},t)$ satisfying equation of motion (6.51) under the following initial conditions

$$u(t_0) = u_0,\ \ \dot{u}(t_0) = \dot{u}_0$$

and condition (6.52) on the interval (t_0, T), we seek those functions which minimize the functional

$$J = u^2(t_i)\,, \tag{6.53}$$

where

$$t_i \in [t,T]\,,\ \ \dot{u}(t_i) = 0.$$

This statement is not quite equivalent to the minimax statement. The equivalency is guaranteed if the following functional

$$J_1 = \max_i u^2(t_i) \tag{6.54}$$

is minimised instead of (6.53). For this reason, the method of [130] turns out to be effective only in those cases in which it can be assumed that the first extremum of $u(t)$ has the maximum absolute value. On the other hand, this method allows us, in principle, not only to construct the optimal control but also determine the characteristic which realises the appropriate control force.

The construction of such a characteristic causes no difficulty in many simple cases. If, for example, the action satisfies condition (6.35), the optimal control (6.36) can be realised by means of the characteristic of the "Coulomb friction" type

$$v = v_0 \operatorname{sign}\dot{u}\ \ (|\dot{u}| > 0)\,,\ \ v(t) \leq v_0\ \ (\dot{u} = 0). \tag{6.55}$$

Since $\dot{u}(T) = 0$ and $\sigma(t) < v_0$ for $t > T$, the system is "locked" by the Coulomb friction for $t > T$, and the relative displacement remains invariable.

Paper [135] is concerned with computational methods of defining the optimal control. Adopting the constraint imposed on control in the form of eq. (6.25) the authors took the following functional to be minimised

$$J = \max_t u^{2p}(t)\ \ t \in (0,T)\,.$$

It is clear that the larger integer p, the greater the difference between the maximal maximum of function $u^{2p}(t)$ and the other maxima. This circumstance improves convergence of the iterative method of calculation of the optimal control proposed in the above paper. The method reduces to the following procedure:

1. One chooses an arbitrary control force $v(t)$, determines the value $t = t_m^{(0)}$ which maximises function $u^{2p}(t)$ (and $|u(t)|$) and calculates

$$\varphi_0 = u^{2p}\left(t_m^{(0)}\right). \tag{6.56}$$

2. Designating the variation of control as $\delta^{(0)}v(t)$ one calculates the corresponding increment in functional (6.56). As eq. (6.24) yields

$$u\left(t_m^{(0)}\right) = \int\limits_0^{t_m^{(0)}} (t_m^{(0)} - t)v(t)\,dt - \xi\left(t_m^{(0)}\right), \tag{6.57}$$

one obtains

$$\Delta\varphi_0 = 2pu^{2p-1}\left(t_m^{(0)}\right)\delta u = 2pu^{2p-1}\left(t_m^{(0)}\right)\int\limits_0^{t_m^{(0)}} (t_m^{(0)} - t)\delta^{(0)}v(t)\,dt. \tag{6.58}$$

Paper [135] discusses several variants of choice of $\delta^{(0)}v(t)$ ensuring a decrease in functional (6.56), in particular

$$\delta^{(0)}v(t) = \begin{cases} -\varepsilon v_0 \operatorname{sign} u\left(t_m^{(0)}\right), & \text{for } t \in \left(0, t_m^{(0)}\right), \\ 0, & \text{for } t \in \left(t_m^{(0)}, T\right). \end{cases} \tag{6.59}$$

3. For the control $v(t) + \delta^{(0)}v(t)$ one determines $t_m^{(1)}$ which maximises $u^{2p}(t)$, and the solution improvement process continues.

A "cycling" may occur, that is the value of t_m periodically repeats. In this case, the value of ε in eq. (6.59) is halved and the process is repeated as long as the value reduces up to a desired level.

The papers [115] and [116] address the substantiation of this method.

There exists a series of papers which states the problem of optimal control for periodic vibration actions. The minimax statement of the problem is as follows, [41]. Consider eq. (6.24) in which $\sigma(t)$ is a periodic function of time with period T. Among the T-periodic control forces which satisfy the following condition

$$v(t) \leq v_0 \quad t \in [0, T'] \tag{6.60}$$

we seek that solution which minimises the functional

$$J = \max_t |u(t)| \quad t \in [0, T],$$

where $u(t)$ is the T-periodic solution of eq. (6.24). In this case, the initial conditions are replaced by the condition of periodicity

$$u(T) = u(0), \quad \dot{u}(T) = \dot{u}(0). \tag{6.61}$$

It is worth mentioning that the following assumption is usually adopted in practical problems

$$\int_0^T \sigma(t)\,dt = 0, \tag{6.62}$$

which implies that $\sigma(t)$ has no constant component. This assumption is related to the fact that the constant component of the vibrational action presents no danger to the object as a rule. Thus, apart from constraint (6.60), the following condition is imposed

$$\int_0^T v(t)\,dt = 0. \tag{6.63}$$

In principle, the above approaches all are suitable to defining the optimal control. The graphic-analytic way of solving the problem is developed in [41]. To apply this approach, it is necessary to plot the graph of function $\xi(t)$ which is also periodic due to eq. (6.62), see Fig. 6.5. The next step is to plot $\dot{x}(t)$ in the form of a broken line, the segment slopes being given by eq. (6.25). The broken line must be chosen so that the periodicity condition holds, i.e.

$$\dot{x}(0) = \dot{x}(T)$$

and the maximum of the following sums

$$s_0, \; s_0 - s_I, \; s_0 - s_I + s_{II} + s_{III}, \ldots$$

(which are the areas of the figures between the curves, see Fig. 6.5) achieves the minimum.

The periodic optimal control is sought by methods of the classic variational calculus in [113]. Functional (6.9) is minimised, although this, as mentioned above, is not equivalent to the minimax problem. For this reason, the following additional conditions

$$u^2(t_i) - u^2(t_s) - a_s^2 = 0, \quad \dot{u}(t_s) = 0 \quad (s \neq i) \tag{6.64}$$

are imposed, where t_s denotes the time instant at which function $u^2(t)$ achieves its maximum. It is the authors' opinion that conditions (6.64) ensures the equivalency of the stated problem and the minimax statement.

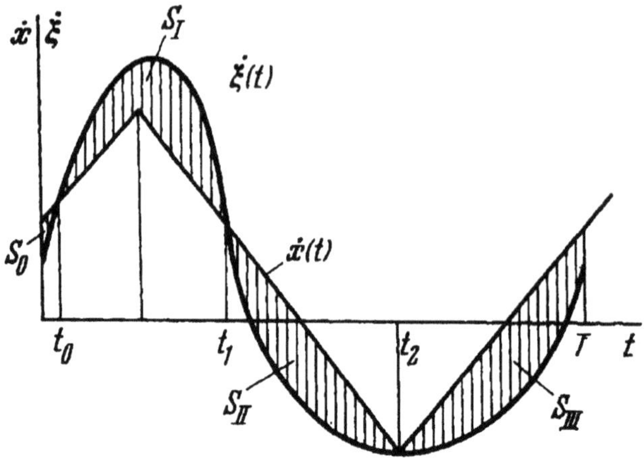

FIGURE 6.5.

Regretfully, only those examples when $u^2(t)$ has a single maximum within the period are considered in paper [113].

Up until now, we have dealt with non-additive functionals. If the functional is additive many other methods of determining the optimal control in closed form become applicable, e.g. the maximum principle of Pontryagin [100], the method of dynamic programming due to Bellman [6] and the methods of the classical calculus of variations.

In particular, these methods can be utilised in those cases in which the functional to be minimised is given in integral form. For example, if instead of functional (6.26), one minimises the following functional

$$J = \int_0^T u^2 dt, \tag{6.65}$$

subject to constraint (6.25), the solution can be obtained by means of the maximum principle. Under certain circumstances it is expedient to carry out the optimisation of shock protection systems by means of the quality criterion in the form of (6.65). Furthermore, it can be shown that minimising (6.65) means constraining the value of the functional (6.26).

In order to apply classic variational methods, it is common practice to go over to an open domain of control. With this aim, constraint (6.25) is replaced by the following condition

$$v^2 + w^2 - v_0^2 = 0, \tag{6.66}$$

which introduces a new control force w but allows one to go over to an open domain for v and w, see [113]. Another approach is to use the "penalty" system for violating condition (6.25) by introducing the control force into the functional to be minimised in the following way, see [4]

$$J = \int_0^T \left\{ u^2(t) + \lambda \left[|v(t)| - v_0 \right]^2 \right\} dt. \tag{6.67}$$

Clearly, the value of the weighting factor λ is of crucial importance.

It is worthwhile mentioning that considerable difficulties arise when one uses the above methods in practice. They are associated with non-autonomy of the problem of concern (time appears explicitly in eq. (6.24)), dependence of the functional on the phase coordinates, presence of the so-called singular control forces etc. In addition to this, these methods yield, as a rule, only necessary conditions for an optimum, i.e. they allow one to obtain only the extremum trajectories which must be additionally proven whether they are optimal. For some cases there arises the danger of obtaining the control providing a local minimum rather than the global one.

Before we proceed to some examples, let us recall the formulation and the basic statements of the maximum principle, [100]. Let the system of equations be given in the form

$$\dot{\mathbf{x}} = \mathbf{f}(\mathbf{x}, \mathbf{u}, t), \tag{6.68}$$

where $\mathbf{x}$ is the $n-$dimensional vector of the phase coordinate of the system and $\mathbf{u}$ is the $s-$dimensional vector of the control forces. One seeks the vector $\mathbf{u}$ which takes the system (6.68) from its initial state

$$\mathbf{x}(0) = \mathbf{x}_0 \tag{6.69}$$

to the final state

$$\mathbf{x}(T) = \mathbf{x}_T, \tag{6.70}$$

and minimises the following functional

$$J = \int_0^T F(\mathbf{x}, \mathbf{u}, t)\, dt. \tag{6.71}$$

We construct the scalar function

$$H = \psi_0 F(\mathbf{x}, \mathbf{u}, t) + \boldsymbol{\psi} \mathbf{f}(\mathbf{x}, \mathbf{u}, t) + \psi_{n+1}, \tag{6.72}$$

where ψ_0, $\boldsymbol{\psi}(\psi_1, ..., \psi_n)$ and ψ_{n+1} are auxiliary unknown functions of time. Using (6.72) as the Hamiltonian we write down the corresponding canonic equation

$$\left. \begin{aligned} \dot{x}_k &= \frac{\partial H}{\partial \psi_k}, \quad \dot{\psi}_k = -\frac{\partial H}{\partial x_k} \qquad (k = 0, ..., n), \\ \dot{x}_{n+1} &= 1 \qquad \dot{\psi}_{n+1} = -\frac{\partial H}{\partial t}. \end{aligned} \right\} \tag{6.73}$$

The first group of equations for $k = 1,..,n$ coincides with the system of equations in (6.68) and yields the following additional equation

$$\dot{x}_0 = F(\mathbf{x}, \mathbf{u}, t), \; x_0(0) = 0, \; x_0(T) = J. \tag{6.74}$$

The second group produces the following system of equations which are linear in ψ_k

$$\dot{\psi}_0 = 0, \; \dot{\psi}_k = -\psi_0 \frac{\partial F}{\partial x_k} - \boldsymbol{\psi} \frac{\partial \mathbf{f}}{\partial x_k}. \tag{6.75}$$

The third group of equations in (6.73) yields

$$x_{n+1} = t, \; \dot{\psi}_{n+1} = -\psi_0 \frac{\partial F}{\partial t} - \boldsymbol{\psi} \frac{\partial \mathbf{f}}{\partial t}. \tag{6.76}$$

With $\mathbf{u}_*(t)$ denoting the optimal control force, we obtain the corresponding solution $\mathbf{x}_*(t)$ of the system (6.68) under the initial condition (6.69). This solution satisfies eq. (6.70). Inserting $\mathbf{u}_*(t)$ and $\mathbf{x}_*(t)$ into eqs. (6.75) and (6.76) we obtain $\psi_0(t) = const$ and $\psi_k(t)$. The maximum principle states that (i) at any time instant $t \in [0, T]$ function H, considered as an explicit function of the control force $\mathbf{u}(t)$, achieves a maximum which is identically equal to zero, and (ii) $\psi_0 < 0$. Taking into account that functions $\psi_k(t)$ are determined up to an arbitrary constant factor, one can let

$$\psi_0(t) \equiv -1. \tag{6.77}$$

Consider an example. Let the motion of the system be governed by eq. (6.24) which can be cast in the form

$$\left.\begin{aligned} \dot{x}_1 &= x_2, \\ \dot{x}_2 &= v(t) - \sigma(t), \end{aligned}\right\} \tag{6.78}$$

where $x_1 = u$ and $x_2 = \dot{u}$. We will find the control force $\sigma(t)$ which satisfies condition (6.25), ensures that the system subjected to the trivial initial conditions

$$x_1(0) = 0, \; x_2(0) = 0 \tag{6.79}$$

returns, after the impact, to the original position

$$x_1(T) = 0, \; x_2(T) = 0, \tag{6.80}$$

and minimises the following functional

$$J = \int_0^T u^2(t)\, dt = \int_0^T x_1^2(t)\, dt. \tag{6.81}$$

In eqs. (6.79) and (6.81) the time instant T is an unknown value, which must be determined. The scalar function (6.72) is taken to have the form

$$H = \psi_0 x_1^2 + \psi_1 x_2 + \psi_2 \left[v - \sigma(t)\right] + \psi_3, \tag{6.82}$$

while the equation for the unknown variables ψ_0, ψ_1, ψ_2 and ψ_3, due to eqs. (6.75) and (6.76), are as follows

$$\dot{\psi}_0 = 0,\ \dot{\psi}_1 = -2\psi_0 x_1,\ \dot{\psi}_2 = -\psi_1,\ \dot{\psi}_3 = -\psi_2 \dot{\sigma}(t). \tag{6.83}$$

As the maximum principle yields

$$\psi_0 = const = -1,$$

we obtain from eq. (6.83)

$$\dot{\psi}_1 = 2x_1,\ \ddot{\psi}_2 = -2x_1,\ \psi_2^{(IV)} = -2\ddot{x}_1 = -2\dot{x}_2 = -2\left[v - \sigma(t)\right]. \tag{6.84}$$

On the other hand, function (6.82) viewed as an explicit function of v, (i.e. x_1, x_2, ψ_0, ψ_1, ψ_2 and ψ_3 are assumed to be constant) is maximal at any time instant if

$$v = v_0 \mathrm{sign}\psi_2. \tag{6.85}$$

Thus, the optimal control must in general consist of parts with $v(t) = \pm v_0$. "Switching" control, i.e. change in sign of $v(t)$, occurs at those time instants when function $\psi_2(t)$ changes its sign. For this reason, $\psi_2(t)$ can be termed the *switching function.* Substituting eq. (6.85) into the last equation in (6.84) yields the following differential equation for the switching function

$$\psi_2^{(IV)} = -2\left[v_0 \mathrm{sign}\psi_2 - \sigma(t)\right]. \tag{6.86}$$

As

$$\ddot{\psi}_2 = -2x_1,\ \dddot{\psi}_2 = -2\dot{x}_1,$$

function ψ_2 must satisfy the periodicity conditions which follow from eqs. (6.79) and (6.80)

$$\left.\begin{aligned} \ddot{\psi}_2(0) = 0, \quad \ddot{\psi}_2(T) = 0, \\ \dddot{\psi}_2(0) = 0, \quad \dddot{\psi}_2(T) = 0 \end{aligned}\right\} \tag{6.87}$$

and the additional condition

$$H(T) = 0, \tag{6.88}$$

following from the maximum principle and allowing T to be determined. Solving eq. (6.86) under conditions (6.87) and (6.88) presents a rather challenging problem which can hardly be solved even for the simplest types of function $\sigma(t)$.

Note that the maximum principle in the case under consideration does not allow the optimal control to be determined if there are control forces $v_*(t)$ for which the equation

$$\psi_2^{(IV)} = -2[v_*(t) - \sigma(t)] \tag{6.89}$$

possesses a trivial solution

$$\psi_2(t) \equiv 0 \tag{6.90}$$

on some of the above parts. These control forces are referred to as *singular control forces.* In the example under consideration the singular control force is as follows

$$v_*(t) \equiv \sigma(t), \tag{6.91}$$

provided that this control force is admissible. Indeed, let, for example, action $\sigma(t)$ satisfy the condition

$$|\sigma(t)| \le v_0, \ t \in (0, t^*).$$

Then, the control force due to eq. (6.91) is optimal since the solution of eq. (6.78) under the trivial initial condition is given by

$$x_1 \equiv 0, \ x_2 \equiv 0.$$

For this reason,

$$J(t^*) = 0,$$

ensuring optimality of the chosen control due to additivity and non-negativeness of functional (6.81). Therefore, while solving eq. (6.86) it must be guaranteed that $|\sigma(t)| > v_0$ at those time instants when x_1 and x_2 become zero. Otherwise, singular controls are needed to be "switched on" at these time instants.

We show now the way to determine the optimal control in the case when the action is given in the form (6.29) and furthermore

$$\sigma_0 > v_0.$$

Clearly, for the relative displacement to be reduced on the first part, we must require $v(t) > 0$ and, hence, $\psi_2 > 0$. On this part function $\psi_2(t)$ is governed by the following equation

$$\psi_2^{(IV)} = -2\left(v_0 - \sigma_0\right) = 2\left(\sigma_0 - v_0\right).$$

Accounting for the conditions $\ddot{\psi}_2(0) = \dddot{\psi}_2(0) = 0$, we obtain

$$\psi_2(t) = \frac{(\sigma_0 - v_0)\, t^4}{12} + C_1 t + C_2, \tag{6.92}$$

with C_1 and C_2 being the integration constants.

Equation (6.92) remains valid until $t = \tau$, after which

$$\psi_2^{(IV)} = -2v_0,$$

as $\sigma(t) \equiv 0$ for $t > \tau$. Hence,

$$\begin{aligned} \psi_2(t) &= -\frac{v_0\,(t-\tau)^4}{12} + D_1\,(t-\tau)^3 + D_2\,(t-\tau)^2 \\ &\quad + D_3\,(t-\tau) + D_4. \end{aligned} \tag{6.93}$$

Equating (6.92) and (6.93) at $t = \tau$ we obtain D_1, D_2, D_3 and D_4 and the following expression for $\psi_2(t)$

$$\begin{aligned} \psi_2(t) &= -\frac{v_0\,(t-\tau)^4}{12} + \frac{1}{3}\left(\sigma_0 - v_0\right)\tau\,(t-\tau)^3 \\ &\quad + \frac{1}{2}\left(\sigma_0 - v_0\right)\tau^2\,(t-\tau)^2 + \frac{1}{3}[(\sigma_0 - v_0)\tau^3 + C_1]\,(t-\tau) \\ &\quad + \frac{1}{12}\left(\sigma_0 - v_0\right)\tau^4 + C_1\tau + C_2. \end{aligned} \tag{6.94}$$

Next, it is necessary to define the time instant of switching $t^* = \tau + t_1$ from the condition $\psi_2(t^*) = 0$. The problem is thus reduced to solving the forth order algebraic equation

$$\begin{aligned} & -\frac{v_0 t_1^4}{12} + \frac{1}{3}\left(\sigma_0 - v_0\right)\tau t_1^3 + \frac{1}{2}\left(\sigma_0 - v_0\right)\tau^2 t_1^2 \\ & + \frac{1}{3}[(\sigma_0 - v_0)\tau^3 + C_1]t_1 + \frac{1}{12}\left(\sigma_0 - v_0\right)\tau^4 + C_1\tau + C_2 = 0 \end{aligned} \tag{6.95}$$

with two unknown parameters C_1 and C_2. Assume that we have obtained $t_1 = t_1(C_1, C_2)$, then on the part $t > t_1$

$$\psi_2^{(IV)}(t) = 2v_0$$

and, thus,

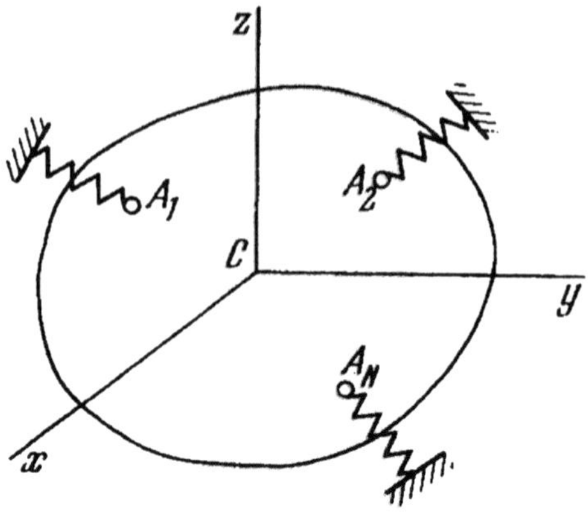

FIGURE 6.6.

$$\psi_2(t) = \frac{v_0 (t - t_1)^4}{12} + E_1 (t - t_1)^3 + E_2 (t - t_1)^2 + E_3 (t - t_1) + E_4,$$

where the integration constants E_1, E_2, E_3 and E_4 are determined from the continuity conditions for $\psi_2(t)$ and its derivative at $t = t_1$. Finally, we have

$$\begin{aligned}\psi_2(t) &= \frac{v_0 (t - t_1)^4}{12} - \left[\frac{v_0 t_1^2}{2} - \sigma_0 \tau t_1 + \frac{1}{2}(\sigma_0 - v_0)\tau^2\right](t - t_1)^2 \\ &+ \frac{1}{3}[(\sigma_0 - v_0) t_1^3 - \sigma_0 (t_1 - \tau)^3 + C_1](t - t_1) - \frac{v_0 (t_1 - \tau)^4}{12} \\ &+ \frac{1}{3}(\sigma_0 - v_0)\tau (t_1 - \tau)^3 + \frac{1}{2}(\sigma_0 - v_0)\tau^2 (t_1 - \tau)^2 \\ &+ \frac{1}{3}[(\sigma_0 - v_0)\tau^3 + C_1](t_1 - \tau) + \frac{1}{12}(\sigma_0 - v_0)\tau^4 + C_1 \tau + C_2.\end{aligned}$$

6.3 Optimal vibration isolation of a rigid body

The way to state the problem of the optimal isolation of a rigid body with 6 degrees of freedom which is mounted on N isolators is considered in this Section. Assume that the vibration of the system depicted in Fig. 6.6 is small and the axes of the coordinate system $Cxyz$ coincide with the principle central axes of inertia of the body. Let us derive the equation of motion for the system subjected to the kinematic excitation caused by motion of the base, the latter being considered as a rigid body. Let $\boldsymbol{\xi}$ designate the vector of displacement of that point of the base which coincides with the center of inertia of the object in the equilibrium position. The other notations are: $\boldsymbol{\eta}$ is the vector of small angles of rotation of the base, $\mathbf{s}$ is the vector of the relative displacement of point C, m is the mass of the body, and ρ_x, ρ_y and ρ_z are the radii of inertia of the body. The equations of motion can then be written in the form

$$\left.\begin{aligned}
m\left(\ddot{s}_x+\ddot{\xi}_x\right) &= \sum_{i=1}^{N} U_i \frac{\partial v_i}{\partial s_x}, \quad m\rho_x^2\left(\ddot{\theta}_x+\ddot{\eta}_x\right) = \sum_{i=1}^{N} U_i \frac{\partial v_i}{\partial \theta_x},\\
m\left(\ddot{s}_y+\ddot{\xi}_y\right) &= \sum_{i=1}^{N} U_i \frac{\partial v_i}{\partial s_y}, \quad m\rho_y^2\left(\ddot{\theta}_y+\ddot{\eta}_y\right) = \sum_{i=1}^{N} U_i \frac{\partial v_i}{\partial \theta_y},\\
m\left(\ddot{s}_z+\ddot{\xi}_z\right) &= \sum_{i=1}^{N} U_i \frac{\partial v_i}{\partial s_z}, \quad m\rho_z^2\left(\ddot{\theta}_z+\ddot{\eta}_z\right) = \sum_{i=1}^{N} U_i \frac{\partial v_i}{\partial \theta_z}.
\end{aligned}\right\} \tag{6.96}$$

Here U_i stands for the dynamic reaction of the $i-th$ isolator while v_i denotes the displacement of the point of attachment of the $i-th$ isolator. Considering expression (1.78) for v_i, we obtain

$$\left.\begin{aligned}
m\left(\ddot{s}_x+\ddot{\xi}_x\right) &= \sum_{i=1}^{N} U_i\alpha_i,\\
m\rho_x^2\left(\ddot{\theta}_x+\ddot{\eta}_x\right) &= \sum_{i=1}^{N} U_i\left(\gamma_i y_i-\beta_i z_i\right),\\
m\left(\ddot{s}_y+\ddot{\xi}_y\right) &= \sum_{i=1}^{N} U_i\beta_i,\\
m\rho_y^2\left(\ddot{\theta}_y+\ddot{\eta}_y\right) &= \sum_{i=1}^{N} U_i\left(\alpha_i z_i-\gamma_i x_i\right),\\
m\left(\ddot{s}_z+\ddot{\xi}_z\right) &= \sum_{i=1}^{N} U_i\gamma_i,\\
m\rho_z^2\left(\ddot{\theta}_z+\ddot{\eta}_z\right) &= \sum_{i=1}^{N} U_i\left(\beta_i x_i-\alpha_i y_i\right).
\end{aligned}\right\} \tag{6.97}$$

Since the reactions of the isolators U_i are forces acting on the body to be isolated and the goal of the isolation system is to reduce these reactions, any requirement to the system quality can be represented in the form of constraints imposed on these forces. Let us consider some variants of such constraints.

Constrained acceleration of some points of the isolated body is the most wide-spread requirement. It is a common requirement that the components of the absolute accelerations along the coordinate axes do not exceed some admissible values. Let $\mathbf{l}_j$ denote the radius vector of the $j-th$ point ($j=1,...,s$). The absolute accelerations are then as follows

$$\mathbf{w}_j = \ddot{\mathbf{s}} + \ddot{\boldsymbol{\xi}} + \left(\ddot{\boldsymbol{\theta}} + \ddot{\boldsymbol{\eta}}\right) \times \mathbf{l}_j \ (j = 1, ..., s), \tag{6.98}$$

which is equivalent to the following scalar equations

$$\left.\begin{aligned} w_{j1} &= \ddot{s}_x + \ddot{\xi}_x + \left(\ddot{\theta}_y + \ddot{\eta}_y\right) l_{j3} - \left(\ddot{\theta}_z + \ddot{\eta}_z\right) l_{j2}, \\ w_{j2} &= \ddot{s}_y + \ddot{\xi}_y + \left(\ddot{\theta}_z + \ddot{\eta}_z\right) l_{j1} - \left(\ddot{\theta}_x + \ddot{\eta}_x\right) l_{j3}, \\ w_{j3} &= \ddot{s}_z + \ddot{\xi}_z + \left(\ddot{\theta}_x + \ddot{\eta}_x\right) l_{j2} - \left(\ddot{\theta}_y + \ddot{\eta}_y\right) l_{j1}, \end{aligned}\right\} \tag{6.99}$$

where l_{j1}, l_{j2}, l_{j3} are the components of vector $\mathbf{l}_j$. On the other hand, it follows from eq. (6.97) that

$$\left.\begin{aligned} \ddot{s}_x + \ddot{\xi}_x &= V_1, \quad \ddot{\theta}_x + \ddot{\eta}_x = V_4, \\ \ddot{s}_y + \ddot{\xi}_y &= V_2, \quad \ddot{\theta}_y + \ddot{\eta}_y = V_5, \\ \ddot{s}_z + \ddot{\xi}_z &= V_3, \quad \ddot{\theta}_z + \ddot{\eta}_z = V_6, \end{aligned}\right\} \tag{6.100}$$

where

$$\left.\begin{aligned} V_1 &= \frac{1}{m}\sum_{i=1}^{N} U_i \alpha_i, & V_2 &= \frac{1}{m}\sum_{i=1}^{N} U_i \beta_i, \\ V_3 &= \frac{1}{m}\sum_{i=1}^{N} U_i \gamma_i, & V_4 &= \frac{1}{m\rho_x^2}\sum_{i=1}^{N} U_i \left(\gamma_i y_i - \beta_i z_i\right), \\ V_5 &= \frac{1}{m\rho_y^2}\sum_{i=1}^{N} U_i \left(\alpha_i z_i - \gamma_i x_i\right), & V_6 &= \frac{1}{m\rho_z^2}\sum_{i=1}^{N} U_i \left(\beta_i x_i - \alpha_i y_i\right). \end{aligned}\right\} \tag{6.101}$$

Inserting eq. (6.101) into eq. (6.99) yields

$$\left.\begin{aligned} w_{j1} &= V_1 + l_{j3}V_5 - l_{j2}V_6, \\ w_{j2} &= V_2 + l_{j1}V_6 - l_{j3}V_4, \\ w_{j3} &= V_3 + l_{j2}V_4 - l_{j1}V_5. \end{aligned}\right\} \tag{6.102}$$

The above constraints takes the form

$$\left.\begin{aligned} &|V_1 + l_{j3}V_5 - l_{j2}V_6| \le W_{j1}, \\ &|V_2 + l_{j1}V_6 - l_{j3}V_4| \le W_{j2}, \\ &|V_3 + l_{j2}V_4 - l_{j1}V_5| \le W_{j3}, \\ &\qquad (j = 1, ..., s). \end{aligned}\right\} \tag{6.103}$$

with W_{j1}, W_{j2} and W_{j3} being admissible values of the components of the absolute acceleration $\mathbf{w}_j$ along the coordinate axes. In some cases, the constraints can be imposed on the forces

$$|U_i| \leq R_{i0} \; (i = 1, ..., N). \tag{6.104}$$

For example, such constraints are encountered when it is necessary to ensure the strength of attachment of the isolators to the object.

Conditions (6.103) and (6.104) have two features: first, they relate accelerations and forces at the same instant of time, and second, they are linear in functions V_k and U_i. In the more general case, the constraints possessing such properties can be written in the form

$$\left| \sum_{i=1}^{N} \mu_i^{(j)} U_i \right| \leq R_{j0} \; (j = 1, ..., r). \tag{6.105}$$

Condition (6.103) is a particular case of condition (6.105) as V_k are all linearly dependent on U_i.

In some cases, the constraints can be imposed not on the instantaneous values of accelerations and reactions but on their values averaged over time. Such constraints are usually nonlinear. The more frequent form of such constraints is a restraint of the mean square value of acceleration within a time interval T

$$\frac{1}{T} \int_0^T w_{jk}^2 (t) \, dt \leq \sigma_{jk} \; (j = 1, ..., s, \; k = 1, 2, 3). \tag{6.106}$$

Consider now the criteria for optimising the isolation system. Choice of the form of the functional to be minimised for the system design depends essentially on the system purpose. Assume that the limitation in the dimensions of the isolation system plays a crucial role. This implies that the criterion for the system quality is a functional depending on the relative displacements at some points in the object

$$J = J \left\{ u^{(1)} (t), ..., u^{(m)} (t) \right\}. \tag{6.107}$$

Here $u^{(s)} (t)$ $(s = 1, ..., m)$ are components of displacement of the corresponding points along some chosen directions. In the simplest case, the functional to be minimised is simply the maximum value of the component of the relative displacement vector $\mathbf{u}$ along a prescribed direction $\mathbf{n}$.

Let us express the displacement to be minimised in terms of the generalised coordinates. The components of vector $\mathbf{u}$ along the direction $\mathbf{n}$ is given by

$$\begin{aligned} u_n &= (\mathbf{s} + \boldsymbol{\theta} \times \mathbf{p}) \cdot \mathbf{n} = (s_x + \theta_y p_3 - \theta_z p_2) n_1 + \\ &+ (s_y + \theta_z p_1 - \theta_x p_3) n_2 + (s_z + \theta_x p_2 - \theta_y p_1) n_3, \end{aligned} \tag{6.108}$$

where p_1, p_2, p_3 and n_1, n_2, n_3 denote the components of the radius vector $\mathbf{p}$ of the point under consideration and vector $\mathbf{n}$, respectively. Differentiating twice with respect to time and considering eq. (6.100) yields

$$\begin{aligned}\ddot{u}_n &= \left(\ddot{s}_x + \ddot{\theta}_y p_3 - \ddot{\theta}_z p_2\right) n_1 + \left(\ddot{s}_y + \ddot{\theta}_z p_1 - \ddot{\theta}_x p_3\right) n_2 \\ &+ \left(\ddot{s}_z + \ddot{\theta}_x p_2 - \ddot{\theta}_y p_1\right) n_3 = [V_1 - \ddot{\xi}_x + \left(V_5 - \ddot{\eta}_y\right) p_3 - \left(V_6 - \ddot{\eta}_z\right) p_2] n_1 \\ &+ [V_2 - \ddot{\xi}_y + \left(V_6 - \ddot{\eta}_z\right) p_1 - \left(V_4 - \ddot{\eta}_x\right) p_3] n_2 \\ &+ [V_3 - \ddot{\xi}_z + \left(V_4 - \ddot{\eta}_x\right) p_2 - \left(V_5 - \ddot{\eta}_y\right) p_1] n_3 \\ &= -\left[\ddot{\xi}_x n_1 + \ddot{\xi}_y n_2 + \ddot{\xi}_z n_3 + \ddot{\eta}_x (p_2 n_3 - p_3 n_2) + \ddot{\eta}_y (p_3 n_1 - p_1 n_3)\right. \\ &+ \ddot{\eta}_z (p_1 n_2 - p_2 n_1)] + [V_1 n_1 + V_2 n_2 + V_3 n_3 + V_4 (p_2 n_3 - p_3 n_2) \\ &+ V_5 (p_3 n_1 - p_1 n_3) + V_6 (p_1 n_2 - p_2 n_1)] .\end{aligned} \tag{6.109}$$

This equation can be recast as follows

$$\ddot{u}_n = -\ddot{\zeta}_n (t) + V (t) , \tag{6.110}$$

where

$$\begin{aligned}\ddot{\zeta}_n &= \ddot{\xi}_x n_1 + \ddot{\xi}_y n_2 + \ddot{\xi}_z n_3 \\ &+ \ddot{\eta}_x (p_2 n_3 - p_3 n_2) + \ddot{\eta}_y (p_3 n_1 - p_1 n_3) + \ddot{\eta}_z (p_1 n_2 - p_2 n_1),\end{aligned} \tag{6.111}$$

$$\begin{aligned}V (t) &= V_1 n_1 + V_2 n_2 + V_3 n_3 \\ &+ V_4 (p_2 n_3 - p_3 n_2) + V_5 (p_3 n_1 - p_1 n_3) + V_6 (p_1 n_2 - p_2 n_1),\end{aligned} \tag{6.112}$$

Here $\ddot{\zeta}_n (t)$ is a prescribed function of time characterising the dynamic excitations, namely, the first and second lines express the components of the acceleration of transportation and the absolute acceleration along the direction $\mathbf{n}$, respectively. $V (t)$ may be viewed as a control force which is a linear function of V_k, and, hence, of U_i. Comparing eqs. (6.110) and (6.24) it is easy to note that the problem of optimisation of the isolation system for the rigid body has reduced to the simplest problem. The only difference is that the constraints on function V_k are given in a more complex form than eq. (6.25). The requirements to the vibration protection system stated in the form of constraints define the set of admissible control forces $V (t)$, on which the functional

$$J = \max_t |u_n (t)| \tag{6.113}$$

must be minimised. Provided that the constraints have the form (6.105) the condition imposed on function $V(t)$ becomes

$$|V(t)| < V_*. \tag{6.114}$$

Indeed, the inequalities in eq. (6.105) form some closed domain in a N-dimensional space $\{U_i\}$. This domain is symmetric about the coordinate origin since each inequality defines a closed symmetric domain. In this domain function $V(t)$ which depends linearly on U_i may take any value satisfying inequality (6.114). It is easy to see that condition (6.105) imposes no constraints on function $V(t)$, this means that this control force can be any piecewise-continuous function satisfying condition (6.114).

Therefore, if the constraints have form (6.105) and the functional (6.113) is to be minimised, the problem reduces to the simplest problem investigated in Section 6.1., namely, among piecewise-continuous functions $V(t)$ satisfying condition (6.114) we seek a function which minimises functional (6.113) in which $u_n(t)$ is solution of eq. (6.110). Furthermore, it is necessary to define the class E to which function $u_n(t)$ must belong. For example, in the case of a shock protection system, this class may be given by the following conditions

$$u_n(0) = 0, \ \dot{u}_n(0) = 0, \tag{6.115}$$

whereas in the case of protection against stationary excitations it is reasonable to consider only the stationary functions $V(t)$ and $u_n(t)$.

It is essential that some problems of synthesis of optimal systems of vibration protection can be solved in the very process of reducing the original problem to the simplest one regardless of the character of the dynamic action. Indeed, for any function $\ddot{\zeta}_n(t)$ in (6.110) an increase in V_* broadens the domain of the admissible control forces and can lead only to a decrease in minimum value of J. Correspondingly, a decrease in V_* can result only in an increase in the minimum value of J. Thus, the value of V_* can serve as a quality criterion for the chosen system of vibration protection. The fact that the value of V_* is independent of the type of dynamic action makes this approach convenient for choosing the principal scheme of an isolating suspension, especially in the case of unpredictable excitations which are acting on the system under design.

As shown above, the problem of determining V_* reduces to the search for the minimum value of linear function (6.112) under linear constraints. The number of variables is six if all constraints have the form given by eq. (6.103). The number of variables increases up to N if V_k depends on U_i under constraints of the type (6.104). In both cases, determination of V_* presents a classical problem of linear programming, namely, defining the maximum value of a linear function subject to linear constraints. The methods of solving this problem are covered in detail in e.g. [33]. The maximum of the linear function is known to be achieved only on the boundary of the convex domain bounded by linear constraints.

The above-said remains valid also in the case in which the functional to be minimised is given by

$$J = \max_t \left| \sum_{\beta=1}^{m} \gamma_\beta u^{(\beta)}(t) \right|, \tag{6.116}$$

where γ_β are the constants which are actually the weighting coefficients, $u^{(\beta)}(t)$ are components of the relative displacements of some body points with coordinates $p_1^{(\beta)}$, $p_2^{(\beta)}$ and $p_3^{(\beta)}$ along the direction $\mathbf{n}^{(\beta)}$ (given by the direction cosines $n_1^{(\beta)}$, $n_2^{(\beta)}$ and $n_3^{(\beta)}$). Taking the second derivative of the expression

$$u(t) = \sum_{\beta=1}^{m} \gamma_\beta u^{(\beta)}(t) = \sum_{\beta=1}^{m} \gamma_\beta \left(\mathbf{s} + \boldsymbol{\theta} \times \mathbf{p}^{(\beta)} \right) \cdot \mathbf{n}^{(\beta)},$$

we arrive at the following equation

$$\ddot{u} = F(t) + V, \tag{6.117}$$

where

$$\begin{aligned} F(t) = -\sum_{\beta=1}^{m} \gamma_\beta \Big[& \ddot{\xi}_x n_1^{(\beta)} + \ddot{\xi}_y n_2^{(\beta)} + \ddot{\xi}_z n_3^{(\beta)} + \ddot{\eta}_x (p_2^{(\beta)} n_3^{(\beta)} - p_3^{(\beta)} n_2^{(\beta)}) \\ & + \ddot{\eta}_y (p_3^{(\beta)} n_1^{(\beta)} - p_1^{(\beta)} n_3^{(\beta)}) + \ddot{\eta}_z (p_1^{(\beta)} n_2^{(\beta)} - p_2^{(\beta)} n_1^{(\beta)}) \Big], \end{aligned} \tag{6.118}$$

$$\begin{aligned} V(t) \; = \; & V_1 \sum_{\beta=1}^{m} \gamma_\beta n_1^{(\beta)} + V_2 \sum_{\beta=1}^{m} \gamma_\beta n_2^{(\beta)} + V_3 \sum_{\beta=1}^{m} \gamma_\beta n_3^{(\beta)} \\ & + V_4 \sum_{\beta=1}^{m} \gamma_\beta \left[p_2^{(\beta)} n_3^{(\beta)} - p_3^{(\beta)} n_2^{(\beta)} \right] \\ & + V_5 \sum_{\beta=1}^{m} \gamma_\beta \left[p_3^{(\beta)} n_1^{(\beta)} - p_1^{(\beta)} n_3^{(\beta)} \right] \\ & + V_6 \sum_{\beta=1}^{m} \gamma_\beta \left[p_1^{(\beta)} n_2^{(\beta)} - p_2^{(\beta)} n_1^{(\beta)} \right]. \end{aligned} \tag{6.119}$$

If the restrictions are not linear but relate the values of V_k (or U_i) at the same time instant, the problem can also be reduced to the simplest problem by minimising functional (6.116). However the restrictions on $V(t)$ take another form. In the general case they reduce to prescribing one of several domains for admissible values of $V(t)$

$$V_{*s} \leq V(t) \leq V_{**s} \ (s = 1, ..., l). \tag{6.120}$$

If the original constraints have more complex form, e.g. (6.106), or if functional J differs from (6.116), it is not feasible to reduce the problem to the simplest one. This takes place in the case of statement of the problem on optimal isolation of the elastic body, see [91]. Here we limit our analysis to the problems reducible to the simplest one under constraints on $V(t)$ in the form (6.114).

Let us clarify which parameters the value of V_* is dependent on. This depends on the coefficients in eqs. (6.103)-(6.105), (6.112) and (6.119). If all constraints are given in the form of (6.103), the value of V_* then depends only on the coordinates $l_{j1}, l_{j2}, l_{j3}, p_1^{(\beta)}, p_2^{(\beta)}, p_3^{(\beta)}$ and direction cosines. In other words, in this case the value of V_* does not depend on the location of the isolator attachment points and the choice of the optimal suspension scheme by means of this criterion is not feasible.

Let $p_k^{(\beta)} = l_{\beta k}$ $(\beta = 1, ..., m, \ k = 1, 2, 3)$, i.e. the components of displacement of those points whose accelerations are constrained are chosen as functions $u^\beta(t)$. Expression (6.119) then can be reduced to the following form

$$\begin{aligned} V &= \sum_{\beta=1}^{m} \gamma_\beta \left[\left(V_1 + l_{\beta_3} V_5 - l_{\beta_2} V_6\right) n_1^{(\beta)} + \left(V_2 + l_{\beta_1} V_6 - l_{\beta_3} V_4\right) n_2^{(\beta)} \right. \\ & \quad \left. + \left(V_3 + l_{\beta_2} V_4 - l_{\beta_1} V_5\right) n_3^{(\beta)} \right]. \end{aligned} \tag{6.121}$$

Provided that the constraints are partly given in form (6.104), V_k in eqs. (6.112) and (6.119) must be expressed in terms of U_i by means of (6.112). In this case, the equation for $V(t)$ will contain the coordinates of the isolator attachment points and the direction cosines of the isolator's axes. The value of V_* will depend on these parameters, that is, there appears a possible optimal choice ensuring a maximum of V_k.

We proceed now to some examples which illustrate the method suggested. Consider the isolation system depicted in Fig. 6.7. Assume that the requirements of the system are stated as follows:

(i) the horizontal component of the absolute acceleration of any point of the body to be isolated cannot exceed W_0 and

(ii) the following constraints are imposed on the isolators' reactions

$$|U_i| < R \ (i = 1, 2, 3, 4). \tag{6.122}$$

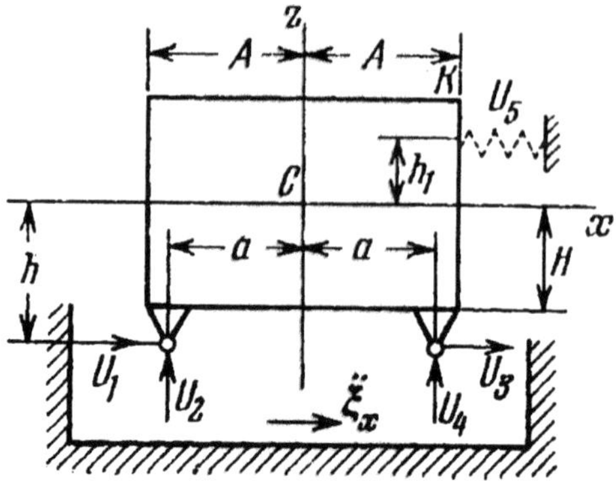

FIGURE 6.7.

The functional to be minimised is the horizontal displacement of point K with the coordinates $x = A$ and $z = H$. Let us determine the value of V_*. As in the case under consideration

$$p_1 = A, \ p_2 = 0, \ p_3 = H, \ n_1 = 1, \ n_2 = n_3 = 0,$$

eq. (6.39) yields

$$V(t) = V_1 + HV_5. \tag{6.123}$$

On the other hand, by virtue of eq. (6.103)

$$|V_1 + zV_5| \leq W_0. \tag{6.124}$$

Here z is the ordinate of the point whose acceleration must not exceed W_0. Clearly, z may take any value from the interval $-H \leq z \leq H$ as any point of the object to be isolated can play the role of this point. The maximum of $|V_1 + zV_5|$ can be achieved either at $z = H$ or at $z = -H$. Hence, inequality (6.124) will be satisfied for any $|z| < H$ if

$$|V_1 + HV_5| \leq W_0, \ |V_1 - HV_5| \leq W_0. \tag{6.125}$$

Comparing eq. (6.123) with eq. (6.125) we see that the value of V_* cannot exceed W_0. If there were only constraints on the acceleration, V_* would be equal to W_0. The constraints on the isolators' reactions may lead to a decrease in V_*. We express V, V_1 and V_5 in terms of the isolators' reactions and substitute them into eqs. (6.123) and (6.125). As

$$\begin{aligned} V_1 &= \frac{1}{m}(U_1 + U_3) = \frac{Q_1}{m}, \\ V_5 &= -\frac{1}{m\rho_y^2}\left[(U_1 + U_3)h + (U_4 - U_2)a\right] = -\frac{1}{m\rho_y^2}(hQ_1 + aQ_2), \end{aligned}$$

where $Q_1 = U_1 + U_3$ and $Q_2 = U_4 - U_2$ we obtain from eq. (6.123)

$$V = -\frac{1}{m}\left[\left(\frac{Hh}{\rho_y^2} - 1\right) Q_1 + \frac{ah}{\rho_y^2} Q_2\right] \tag{6.126}$$

and from eq. (6.125)

$$\left|\left(\frac{Hh}{\rho_y^2} \mp 1\right) Q_1 + \frac{Ha}{\rho_y^2} Q_2\right| \leq mW_0. \tag{6.127}$$

Constraints (6.122) may be cast in the form

$$|Q_1| \leq 2R,\ |Q_2| \leq 2R. \tag{6.128}$$

The other constraints can be ignored since V depends only on Q_1 and Q_2. In other words, if the control forces $Q_1(t)$ and $Q_2(t)$ satisfying eqs. (6.127) and (6.128) are found, we can always sort out a value of $U_i(t)$ which satisfies constraints (6.122).

Let us find the value of R which ensures that the constraints do not decrease V. In other words, this value must guarantee that the maximum of V is equal to W_0 under conditions (6.127) and (6.128). By introducing the notation

$$\begin{aligned} \frac{Hh}{\rho_y^2} - 1 &= \delta_1,\ \frac{Hh}{\rho_y^2} + 1 = \delta_2,\ \frac{Ha}{\rho_y^2} = \varepsilon,\ \frac{2R}{mW_0} = k, \\ \frac{Q_1}{mW_0} &= k_1,\ \frac{Q_2}{mW_0} = k_2,\ \frac{V}{W_0} = \nu \end{aligned}$$

eqs. (6.126)-(6.128) take the following form

$$\nu = -\left(\delta_1 k_1 + \varepsilon k_2\right), \tag{6.129}$$

$$|\delta_1 k_1 + \varepsilon k_2| \leq 1,\ |\delta_1 k_1 \pm \varepsilon k_2| \leq 1, \tag{6.130}$$

$$|k_1| \leq k,\ |k_2| \leq k. \tag{6.131}$$

Condition (6.130) sorts out the domain in the plane (k_1, k_2), one of the boundaries being the following segment

$$\delta_1 k_1 + \varepsilon k_2 = 1.$$

The coordinates of the segment ends are given by

$$k_1 = 0,\ k_2 = \frac{1}{\varepsilon};\ k_1 = -1,\ k_2 = \frac{1 + \delta_1}{\varepsilon} = \frac{h}{a}.$$

Within this segment $\nu = 1$ whereas in the rest of the domain $\nu < 1$. Thus, for $W_0 = V_*$, at least one point of the segment must satisfy condition (6.131). Evidently, the minimum value which meets this condition can be defined as follows

$$k_{\min} = \min \max \{|k_1|, |k_2|\},$$

i.e. the minimum value of the maximum of two numbers $|k_1|$ and $|k_2|$ at each point of the segment. Solving this problem yields

$$k_{\min} = \frac{1}{\varepsilon} \quad \text{if} \quad \frac{h}{a} > \frac{1}{\varepsilon} \qquad \left(h > \frac{\rho_y^2}{H}\right),$$

$$k_{\min} = \frac{h}{a} \quad \text{if} \quad 1 < \frac{h}{a} < \frac{1}{\varepsilon} \qquad \left(a < h < \frac{\rho_y^2}{H}\right),$$

$$k_{\min} = \frac{1}{\varepsilon - \delta_1} \quad \text{if} \quad \frac{h}{a} < 1, \ \frac{h}{a} < \frac{1}{\varepsilon} \qquad \left(a > h, \ h < \frac{\rho_y^2}{H}\right).$$

Let us consider now the system shown in Fig. 6.7 with an additional isolator 5 depicted by the dotted line. Assume that the constraint and the functional to be minimised are the same as before. Additionally, there must be

$$|U_5| < R. \tag{6.132}$$

Since in this case

$$V_1 = \frac{1}{m}(Q_1 - U_5), \ V_5 = -\frac{1}{m\rho_y^2}(hQ_1 + aQ_2 + h_1 U_5),$$

we obtain

$$V = \left(\frac{Hh_1}{\rho_y^2} + 1\right) U_5 - \left(\frac{Hh}{\rho_y^2} - 1\right) Q_1 - \frac{Ha}{\rho_y^2} Q_2. \tag{6.133}$$

The constraints are cast in the following form

$$\left| \left(\frac{Hh}{\rho_y^2} \pm 1\right) Q_1 + \frac{Ha}{\rho_y^2} Q_2 - \left(\frac{Hh}{\rho_y^2} \mp 1\right) U_5 \right| \le mW_0, \tag{6.134}$$

$$|Q_1| \le 2R, \ |Q_2| \le 2R, \ |U_5| \le R, \tag{6.135}$$

or in the nondimensional form

$$\nu = -\delta_1 k_1 - \varepsilon k_2 + \eta_1 k_3, \tag{6.136}$$

$$|\delta_1 k_1 + \varepsilon k_2 - \eta_1 k_3| \le 1, \ |\delta_1 k_1 + \varepsilon k_2 - \eta_2 k_3| \le 1, \tag{6.137}$$

$$|k_1| \le k, \ |k_2| \le k, \ |k_3| \le \frac{k}{2}, \tag{6.138}$$

where

$$k_3 = \frac{U_5}{mW_0}, \ \eta_1 = \frac{Hh_1}{\rho_y^2} + 1, \ \eta_2 = \frac{Hh_1}{\rho_y^2} - 1.$$

Conditions (6.137) select a domain in the space of variables k_1, k_2 and k_3. One of the boundaries of this domain is the following plane

$$\delta_1 k_1 + \varepsilon k_2 - \eta_1 k_3 = -1.$$

Comparing this equation with (6.136) indicates that ν achieves the maximum value equal to 1. The boundaries of this plane are the following lines

$$\begin{aligned} a)\ k_1 + k_3 &= 0,\ (\delta_1 + \eta_1) k_1 + \varepsilon k_2 = -1, \\ b)\ k_1 + k_3 &= 1,\ (\delta_1 + \eta_1) k_1 + \varepsilon k_2 = -1 + \eta_1. \end{aligned}$$

In order to ensure that conditions (6.138) do not decrease the value of V_*, the value of k must satisfy the following condition

$$k \ge \min \max \{|k_1|, |k_2|, 2|k_3|\},$$

the minimum being sought over all points of the plane. This problem is convenient to solve when the prescribed values are taken. Let, for instance

$$\delta_1 = -0,2,\ \delta_2 = -1,8,\ \varepsilon = 0,8,\ \eta_1 = 1,4,\ \eta_2 = 0,6.$$

Inequalities (6.137) are then written down in the form

$$\left.\begin{aligned} -0,2k_1 + 0,8k_2 - 1,4k_3 \le 1, \\ 0,2k_1 - 0,8k_2 + 1,4k_3 \le 1, \\ 1,8k_1 + 0,8k_2 + 0,6k_3 \le 1, \\ -1,8k_1 - 0,8k_2 - 0,6k_3 \le 1. \end{aligned}\right\} \tag{6.139}$$

Condition $-0,2k_1 + 0,8k_2 - 1,4k_3 = 1$ will not be satisfied if $k_{\min} < 0,59$, as $(0,2 + 0,8 + 0,5 \cdot 1,4) \cdot 0,59 = 1$. On the other hand, if

$$k_1 = -0,59, \quad k_2 = 0,59, \quad k_3 = -0,295,$$

the conditions (6.139) are all met. Therefore, in the case considered $k_{\min} = 0,59$. When isolator 5 is absent, we have $k_{\min} = 1$, as in this case

$$\frac{1}{\varepsilon} = 1,25,\ \frac{h}{a} = \frac{1 + \delta_1}{\varepsilon} = 1,$$

and therefore,

$$\frac{h}{a} < \frac{1}{\varepsilon}.$$

Thus, introducing an additional isolator allows, in principle, the maximum reactions of all isolators to be reduced by 41%, the same value of V being retained. Clearly, the reduction of the maximum reaction can be achieved only under the optimal control, the control force $V(t)$ being determined by solving the corresponding simplest problem.

6.4 Approximate synthesis of the optimal isolator characteristics

As already mentioned, the optimal controls constructed by the above methods, may turn out to be non-realisable in practice. The real synthesis of the optimal system of vibration protection must consist of determination of the optimal characteristics of passive and active elements ensuring minimisation of a functional under prescribed constraints. In this Section we consider some methods of synthesis of passive elastic-dissipative characteristics based upon the search of the optimal characteristics among the functions belonging to a certain class which is given in the form of a family depending on several parameters.

Typically, this statement of the problem yields the characteristics which are optimal for a particular external excitation, i.e. in general another optimal characteristic is obtained for another excitation. The essential feature of the method developed below is that it allows one to obtain the characteristics which are optimal on average for a certain class of external excitations which is quite satisfactory from the perspective of practical synthesis.

Consider a linear passive system of vibration protection whose motion is described by the following equation

$$\ddot{u} + r\dot{u} + qu + s = F(t), \tag{6.140}$$

where u is the isolator's deflection, q, r and s are constants, and $F(t)$ is a stationary dynamic excitation. Assume that the requirement of the isolation system is reduced to constraining the value of dispersion of the absolute acceleration of the body to be isolated

$$\mu_w^2 = \lim_{T\to\infty} \frac{1}{2T} \int_{-T}^{T} (r\dot{u} + qu + s)^2\, dt = m\left\{(r\dot{u} + qu + s)^2\right\} \le w_0^2. \tag{6.141}$$

Let us determine the values of parameters q, r and s which minimise the following functional

$$J = \lim_{T\to\infty} \frac{1}{2T} \int_{-T}^{T} u^2 dt = m\left\{u^2(t)\right\} = \sigma_u^2 + m_u^2. \tag{6.142}$$

We impose the additional constraints on q and r

$$q \geq 0, \ r \geq 0, \tag{6.143}$$

which are necessary to guarantee that the isolator characteristic is realisable in practice and the stationary solution of eq. (6.140) is stable.

Assume that the mean value m_F and the spectral density of the process $F(t)$ are given. Then under the assumption that $u = m_u + u^0(t)$, we obtain

$$m_u q + s = m_F, \tag{6.144}$$

$$\sigma_u^2 = \frac{1}{2\pi} \int_{-\infty}^{\infty} \frac{S_F(\omega)\, d\omega}{\left|q - \omega^2 + ri\omega\right|^2} = \frac{1}{2\pi} \int_{-\infty}^{\infty} \frac{S_F(\omega)\, d\omega}{\left(q - \omega^2\right)^2 + r^2\omega^2}, \tag{6.145}$$

$$\mu_w^2 = \frac{1}{2\pi} \int_{-\infty}^{\infty} \frac{q^2 + r^2\omega^2}{\left(q - \omega^2\right)^2 + r^2\omega^2} S_F(\omega)\, d\omega + (qm_u + s)^2. \tag{6.146}$$

Since σ_u^2 does not depend upon s, we can choose s so that $m_u = 0$. Equation (6.144) then yields

$$s = m_F.$$

The problem thus reduces to the choice of q and r which minimises σ_u^2 under the following constraint

$$\sigma_w^2 = \frac{1}{2\pi} \int_{-\infty}^{\infty} \frac{q^2 + r^2\omega^2}{\left(q - \omega^2\right)^2 + r^2\omega^2} S_F(\omega)\, d\omega \leq w_0^2 - m_F^2 = \sigma_0^2. \tag{6.147}$$

We solve this problem for two particular cases.

a) Let $F(t) = A\cos\omega_0 t$, then $S_F(\omega) = 0,5A^2\delta(\omega - \omega_0)$. Substituting this value into eqs. (6.145) and (6.147) we have

$$\sigma_u^2 = \frac{A^2}{2\left[\left(q - \omega_0^2\right)^2 + r^2\omega_0^2\right]}, \tag{6.148}$$

$$\sigma_w^2 = \frac{A^2\left(q^2 + r^2\omega_0^2\right)}{2\left[\left(q - \omega_0^2\right)^2 + r^2\omega_0^2\right]}. \tag{6.149}$$

Assume now that the minimum of σ_u^2 is achieved on the boundary of the domain given by condition (6.147), i.e. under the following condition

$$\frac{A^2\left(q^2+r^2\omega_0^2\right)}{2\left[\left(q-\omega_0^2\right)^2+r^2\omega_0^2\right]}=\sigma_0^2. \tag{6.150}$$

From this equation we express r^2 in terms of q

$$r^2=\frac{\alpha^2\left(q-\omega_0^2\right)^2-q^2}{\omega_0^2\left(1-\alpha^2\right)},\ \alpha^2=\frac{2\sigma_0^2}{A^2}. \tag{6.151}$$

Factor α is less than unity, otherwise no protection from dynamic excitation is needed, and the rigid attachment of the object $\left(\sigma_u^2=0\right)$ is the optimal one. Inserting eq. (6.151) into eq. (6.148) we obtain

$$\sigma_u^2=\frac{A^2\left(1-\alpha^2\right)}{2\left[\left(q-\omega_0^2\right)^2-q^2\right]}=\frac{A^2\left(1-\alpha^2\right)}{2\omega_0^2\left(\omega_0^2-2q\right)}. \tag{6.152}$$

It is evident that σ_u^2 is positive only if $q<\omega_0^2/2$, otherwise $\alpha<1$ is not feasible. The minimum value of σ_u^2 is obtained at $q=0$. In this case

$$\left(\sigma_u^2\right)_{\min}=\frac{A^2\left(1-\alpha^2\right)}{2\omega_0^4}. \tag{6.153}$$

The optimal value of r is obtained by substituting $q=0$ into eq. (6.151)

$$r_{opt}^2=\frac{\alpha^2\omega^2}{1-\alpha^2}. \tag{6.154}$$

What remains is to prove whether the minimum of functional (6.148) is achieved when condition (6.150) is satisfied, i.e. in the case in which the acceleration dispersion is equal to the limiting value. To prove this, we take derivative $\partial\left(\sigma_u^2\right)/\partial\left(\alpha^2\right)$ due to eq. (6.153)

$$\frac{\partial\left(\sigma_u^2\right)}{\partial\left(\alpha^2\right)}=-\frac{A^2}{2\omega_0^4}<0.$$

Therefore, a decrease in σ_0^2 which is equivalent to a decrease in α^2 leads to an increase in $\left(\sigma_u^2\right)_{\min}$, and for this reason the solution obtained is optimal indeed. It is worth noting that the solution obtained ensures minimisation of the root-mean-square of the relative displacement rather then minimisation of the system dimensions. It is not allowed to use the minimum amplitude $u(t)$ obtained for choosing the value of free running since buffers make the system nonlinear and lead to the occurrence of the resonance phenomena. However, the minimisation of the relative displacements is in many cases important regardless if this results in minimisation of the free running. So, for example, a decrease in the relative oscillation of the device scale is an actual problem since this oscillation impedes utilising the device.

b) Let $F(t)$ be white noise having spectral density S_0. Evaluating the integrals with help from [105] yields

$$\sigma_u^2 = \frac{1}{2\pi} \int_{-\infty}^{\infty} \frac{S_0 d\omega}{\left(q-\omega^2\right)^2 + r^2\omega^2} = \frac{S_0}{2qr}, \tag{6.155}$$

$$\sigma_w^2 = \frac{1}{2\pi} \int_{-\infty}^{\infty} \frac{\left(q^2 + r^2\omega_0^2\right) S_0 d\omega}{\left(q-\omega^2\right)^2 + r^2\omega^2} = \frac{S_0\left(q+r^2\right)}{2r}. \tag{6.156}$$

Equating the latter expression to σ_0^2 we obtain

$$q = \frac{2r\sigma_0^2}{S_0} - r^2. \tag{6.157}$$

Inserting this expression into eq. (6.155) yields

$$\sigma_u^2 = \frac{S_0}{4r^2\sigma_0^2 - 2S_0 r^3}. \tag{6.158}$$

This allows us to find the value of r which extremises the latter expression

$$r_{opt} = \frac{4\sigma_0^2}{3S_0}. \tag{6.159}$$

It is easy to prove that this value of r corresponds to the minimum of σ_u^2. By virtue of eq. (6.157) we obtain

$$q_{opt} = \frac{8\sigma_0^2}{9S_0^2}. \tag{6.160}$$

Finally, eq. (6.155) yields

$$\left(\sigma_u^2\right)_{\min} = \frac{27S_0^4}{64\sigma_0^6}. \tag{6.161}$$

This expression shows that a decrease in σ_0^2 for prescribed S_0 can lead only to an increase in $\left(\sigma_u^2\right)_{\min}$, thus, the minimum value of σ_u^2 is indeed achieved for the maximum admissible acceleration dispersion.

Up until now, we have looked for the optimal parameters of linear systems. The results obtained are the manifestation of the dependence of the optimal values of the factors q, r and s on the excitation parameters. Assume next, that q, r and s are the linearisation factors for a nonlinear characteristic $f(u, \dot{u})$ which provide a minimum of the mean square of the error due to the approximation

$$\varphi(u, \dot{u}) = f(u, \dot{u}) - qu - r\dot{u} - s.$$

Let us ascertain whether it is possible to choose a nonlinear function $f(u, \dot{u})$ so that for any excitation belonging to a certain class the values of the linearisation factors coincide with those of the optimal linear system under the same excitation. This nonlinear characteristic may be considered as optimal for the given class of excitations.

As shown in Chapter 3, the factors of the equivalent linearisation are functions of the moments of the processes $u(t)$ and $\dot{u}(t)$. For this reason, synthesis of the nonlinear optimal system requires the relationship between the moments and the optimal factors q, r and s. Having this relationship one can try to find a nonlinear function which realises this relationship.

Let us explain the above-said on the following examples. The characteristic which is optimal for excitations of the type $F(t) = A \cos \omega_0 t$ is sought. As shown above, for fixed values of A and ω_0 we have

$$q_{opt} = 0, \; r_{opt}^2 = \frac{\alpha^2 \omega_0^2}{1 - \alpha^2}, \; s_{opt} = 0.$$

On the other hand, comparing eqs. (6.145) and (6.147) yields

$$\sigma_w^2 = q^2 \sigma_u^2 + r^2 \frac{1}{2\pi} \int_{-\infty}^{\infty} \frac{\omega^2 S_F(\omega) \, d\omega}{\left(q - \omega^2\right)^2 + r^2 \omega^2} = q^2 \sigma_u^2 + r^2 \sigma_{\dot{u}}^2. \tag{6.162}$$

As in the optimal case

$$\sigma_w^2 = \sigma_0^2, \tag{6.163}$$

and $q_{opt} = 0$, eq. (6.162) yields

$$r_{opt}^2 = \frac{\sigma_0^2}{\sigma_{\dot{u}}^2} = \frac{2\sigma_0^2}{a^2 \omega_0^2}, \tag{6.164}$$

where a designates the amplitude of the relative displacement. Thus, the optimal isolator must not contain any elastic element while its damper must have the characteristic with the harmonic linearisation factor due to eq. (6.164). In accordance with (3.23) we find that dry friction possesses this property provided that the friction force is given by

$$h = \frac{\pi \sigma_0}{\sqrt{2}}. \tag{6.165}$$

In the case of white noise excitation, eqs. (6.160) and (6.161) yield

$$q_{opt} = \frac{\sigma_0}{\sqrt{3} \sigma_u}. \tag{6.166}$$

Since

$$\left(\sigma_{\dot{u}}^2\right)_{opt} = \frac{1}{2\pi} \int_{-\infty}^{\infty} \frac{S_0 \omega^2 d\omega}{\left(q_{opt} - \omega^2\right)^2 + r_{opt}^2 \omega^2} = \frac{S_0}{2 r_{opt}} = \frac{3 S_0^2}{8 \sigma_0^2}, \tag{6.167}$$

we obtain from eq. (6.159)

$$r_{opt} = \sqrt{\frac{2}{3}} \frac{\sigma_0}{\sigma_{\dot{u}}}. \tag{6.168}$$

Dependence (6.166) corresponds to the elastic characteristic of the type

$$f(u) = F_0 \text{sign} u, \tag{6.169}$$

which is referred to as a *system with initial compressive force.* The dependence between the value of the initial compression F_0 and σ_0 is determined by the distribution function for the stationary process $\sigma(t)$. Assuming the process to be close to being normal, we obtain, due to eq. (3.42), in which $m_u = 0$, that

$$q_{opt} = \frac{F_0}{\sqrt{2\pi\sigma_u^3}} \int_{-\infty}^{\infty} u \,\text{sign} u \exp\left(-\frac{u^2}{2\sigma_u^2}\right) du = \frac{\sqrt{2} F_0}{\sqrt{\pi} \sigma_u}. \tag{6.170}$$

Comparing eqs. (6.166) and (6.170) we have

$$F_0 = \sqrt{\frac{\pi}{6}} \sigma_0. \tag{6.171}$$

The optimal damper is, as before, the dry friction damper. Since for the Gaussian distribution

$$r = \frac{\sqrt{2} h}{\sqrt{\pi} \sigma_{\dot{u}}},$$

it follows from (6.168) that

$$h = \sqrt{\frac{\pi}{3}} \sigma_0. \tag{6.172}$$

Hence, the optimal characteristic of the passive isolator is given by

$$f(u, \dot{u}) = \sqrt{\frac{\pi}{6}} \sigma_0 \text{sign} u + \sqrt{\frac{\pi}{3}} \sigma_0 \text{sign} \dot{u}. \tag{6.173}$$

The optimal characteristics in another particular case can be found by analogy. The method developed can be applied also to systems having several elastic isolators.

We consider now the problem of the optimal isolation of an elastic object, the analysis being limited to the case of the simplest isolator of rigidity c and damping factor r. For the sake of simplicity, we assume that kinematic excitation $\xi(t)$ does not contain a constant component. The relative displacement

$$u = z_A - \xi$$

due to eq. (1.149) takes the form

$$u(t) = -\left[1 + e_A(p)(q + rp)\right]^{-1} \xi(t). \tag{6.174}$$

The isolator's reaction

$$U(t) = d_A(p) z_A, \tag{6.175}$$

is then given by

$$U(t) = \frac{(q + rp)}{1 + e_A(p)(q + rp)} \xi(t). \tag{6.176}$$

Following the method in question we impose the constraint of dispersion of $U(t)$

$$\sigma_U^2 = \frac{1}{2\pi} \int_{-\infty}^{\infty} \left| \frac{q + ri\omega}{1 + e_A(i\omega)(q + ri\omega)} \right|^2 S_\xi(\omega)\, d\omega \leq G_0^2, \tag{6.177}$$

where $S_\xi(\omega)$ is dispersion of the kinematic action $\xi(t)$. Thus, we seek the values of q and r which minimise the functional

$$\sigma_u^2 = \frac{1}{2\pi} \int_{-\infty}^{\infty} \left| 1 + e_A(i\omega)(q + ri\omega) \right|^{-2} S_\xi(\omega)\, d\omega \tag{6.178}$$

under condition (6.177).

Consider the case of harmonic action

$$S_\xi(\omega) = \frac{\xi_0^2}{2} \delta(\omega - \omega_0).$$

Equations (6.177) and (6.178) yield

$$\sigma_U^2 = \frac{\left(q^2 + r^2\omega_0^2\right) k^2(\omega_0)}{\left[q + k(\omega_0)\right]^2 + r^2\omega_0^2} \cdot \frac{\xi_0^2}{2} \leq G_0^2, \tag{6.179}$$

$$\sigma_u^2 = \frac{k^2(\omega_0)}{\left[q + k(\omega_0)\right]^2 + r^2\omega_0^2} \cdot \frac{\xi_0^2}{2} = \min. \tag{6.180}$$

Here $k(\omega_0) = d(i\omega)$ is a real-valued number since no damping in the object is considered. Presuming that the optimal values of q and r lie on the boundary of domain (6.179) we express r^2 in terms of q, to get

$$r^2 = \frac{g_0^2 \left[q + k(\omega_0)\right]^2 - q^2 k^2(\omega_0)}{\omega_0^2 \left[k^2(\omega_0) - g_0^2\right]} \quad \left(g_0^2 = \frac{2G_0^2}{\xi_0^2}\right). \tag{6.181}$$

If $k^2(\omega_0) < g_0^2$ then the rigid attachment is admissible for which $u(t) \equiv 0$. Indeed, in this case we obtain from eq. (6.175) that

$$U(t) = -d(p)\,\xi(t),$$

and, therefore,

$$\sigma_U^2 = \frac{k^2(\omega_0)\,\xi_0^2}{2} < G_0^2. \tag{6.182}$$

If $k^2(\omega) > g_0^2$ then $q_{opt} = 0$ for $k(\omega_0) < 0$. In this case, eq. (6.181) yields

$$r_{opt}^2 = \frac{g_0^2 k^2(\omega_0)}{\omega_0^2 \left[k^2(\omega_0) - g_0^2\right]}. \tag{6.183}$$

We obtain from eq. (6.182) that

$$\left(\sigma_u^2\right)_{\min} = \frac{k^2(\omega_0) - g_0^2}{k^2(\omega_0)} \cdot \frac{\xi_0^2}{2}. \tag{6.184}$$

Comparing eqs. (6.183) and (6.184) and taking into account that $\sigma_{\dot{u}}^2 = \omega_0^2 \sigma_u^2$ ($u(t)$ is harmonic process) yields

$$r_{opt}^2 = \frac{g_0^2}{\sigma_{\dot{u}}^2} \cdot \frac{\xi_0^2}{2} = \frac{G_0^2}{\sigma_{\dot{u}}^2}. \tag{6.185}$$

This means that the optimal characteristic is again Coulomb friction

$$f(u, \dot{u}) = \frac{\pi}{4} G \operatorname{sign} \dot{u}. \tag{6.186}$$

If $k(\omega_0) < 0$ the value of σ_u^2 decreases as q increases. However, q grows until condition (6.181) gives positive values of r^2. The maximal and thus optimal value of q can be obtained from condition $r = 0$. Due to eq. (6.181) this condition is given by

$$g_0^2 \left[q_{opt} + k(\omega_0)\right]^2 - q_{opt}^2 k^2(\omega_0) = 0. \tag{6.187}$$

This yields

$$q_{opt} = \frac{k(\omega_0)\, g_0}{k(\omega_0) - g_0} \tag{6.188}$$

and $r_{opt} = 0$. The second root of eq. (6.187) is negative for $k(\omega_0) > g_0$.

Let us calculate the increase in dispersion σ_u^2 provided that the dry friction damper (6.186) is used instead of the optimal characteristic and $k(\omega_0) > 0$. The result is

$$\frac{\sigma_u^2}{(\sigma_u^2)_{\min}} = \frac{\sigma_u^2(0)}{\sigma_u^2(q_{opt})} = \frac{2q_{opt}k(\omega_0) + k^2(\omega_0)}{k^2(\omega_0)} = 1 + \frac{g_0}{k(\omega_0) - g_0}. \quad (6.189)$$

Therefore, utilising a dry friction damper instead of the optimal characteristic increases the dispersion of the isolator deflection insignificantly at those values of ω_0 for which $k(\omega_0) \gg g_0$.

Assume, for example, that the body to be isolated consists of two masses M and m connected by a spring of rigidity c. Then,

$$k(\omega_0) = \frac{Mm\omega_0^4 - c(m+M)\omega_0^2}{c - m\omega_0^2} = M\omega_0^2 \frac{\omega_0^2 - \lambda_0^2}{\lambda_0^2 - \omega_0^2}, \quad (6.190)$$

where

$$\lambda_0^2 = \frac{c(m+M)}{mM}, \quad \lambda^2 = \frac{c}{m}.$$

In this case we have $k(\omega_0) > 0$ only if

$$1 < \frac{\omega_0^2}{\lambda^2} < 1 + \frac{m}{M}.$$

From this range one must extract the domain in which $k(\omega_0) > g_0$. Only in the latter domain dry friction is not optimal.

Bibliography

[1] Ageev, M.D., Optimal structure of systems of vibration isolation under random excitation. Second Congress on Theoretical and Applied Mechanics of the USSR, abstracts of presentations (in Russian). Nauka, Moscow, 1964.

[2] Aizerman, M.A., Theory of automatic control (in Russian). Nauka, Moscow, 1966.

[3] Babakov, I.M., Theory of vibration (in Russian). Nauka, Moscow, 1965.

[4] Babitsky, V.I. and Izrailovich, M.Ya., On one problem of optimal vibration isolation (in Russian). Inzh. Zhurnal Mekhanika Trerdogo Tela, No. 5, 1968.

[5] Babitsky, V.I. and Kobrinsky, A.E., Electrodynamic damper (in Russian). Izv. AN SSSR OTN Mekhanika i Mashinostroenie, No. 3, 1962.

[6] Bellman, R., Dynamic programming (in Russian). Inostrannaya Literatura, Moscow, 1960.

[7] Bessekersky, V.A., Dynamic synthesis of systems of automatic control (in Russian). Nauka, Moscow, 1965.

[8] Bessekersky, V.A. and Popov, E.P., Theory of systems of automatic control (in Russian). Nauka, Moscow, 1972.

[9] Bies, D.A., Feasibility study of a hybrid vibration isolation system. Express information Test rigs and stands, No. 13, 1969.

[10] Blekhman, I.I., Synchronisation of dynamic systems (in Russian). Nauka, Moscow, 1971.

[11] Bogomolov, A.I. and Stepanov, P.A., Determination of the optimal transfer function for vibration isolation systems (in Russian). Izv. Vuzov, Mashinostroenie, No. 7, 1970.

[12] Bolotin, V.V., Application of the methods of probability theory and reliability theory to structural design (in Russian), Izdatelstvo Literatury po Stroitelstvu, Moscow, 1971.

[13] Bolychevtsev, E.M., Choice of the optimal law of isolation in the case of shock excitation (in Russian). Izv. AN SSSR Mekhanika Trerdogo Tela, No. 5, 1971.

[14] Bolychevtsev, E.M., Synthesis of the optimal control forces minimizing the maximum deviation subject to external excitation (in Russian). Transactions of Institute of Mechanics of the Moscow State University, 1975.

[15] Bryson, A.E. and Ho, Yu-Chi, Applied optimal control. Blaisdel Publ. Company, London, 1969.

[16] Bukhtina, T.M., Potemkin, B.A., Sinev, A.V. and Frolov, K.V., Synthesis of optimal system of vibration protection for the human operator with dynamic properties of the human body being taken into account (in Russian). "Problems of reliability and protection of devices from vibration", collection of papers, Irkutsk, 1972.

[17] Bykhovsky, M.L., On one problem of the dynamics of machines with electric drive (in Russian). Proceedings of Seminar on Theory of Machines and Mechanisms, vol. 18, No. 71, Institute of Machine Design, 1959.

[18] Calcaterra, P.C. and Schubert, D.W., Isolation of helicopter rotor-induced vibrations using active elements. Shock and Vibration Bull. Vol. 37, No. 6, 1968.

[19] Cavanaugh, R.D., Air suspension and servo-controlled isolation systems. Shock and Vibration Handbook, Vol. 2, Chapter 33, New York, 1961.

[20] Chen, C.T., Control system design. Pond Wood Press, New York, 1987.

[21] Crandall, S., Random vibrations of nonlinear systems with prestressed springs (in Russian). Mekhanika, collection of translations, No. 3, 1962.

[22] Crandall, S., ed, Random vibrations. The MIT Press, Cambridge, Mass, 1963.

[23] Crandall, S. and Mark, W.D., Random vibration in mechanical systems, Academic Press, New York, 1963.

[24] Demidovich, B.P., Lectures on the mathematical theory of stability (in Russian), Nauka, Moscow, 1967.

[25] Eliseev, S.V. and Olkov, V.V., Ways of changing the dynamic properties and principles of constructing active systems of vibration protection (in Russian). "Problems of reliability and protection of devices from vibration", collection of papers, Irkutsk, 1972.

[26] Eliseev, S.V. and Lontsikh, P.A., Experimential investigation of an active pneumatic system of vibration protection (in Russian). "Problems of reliability and protection of devices from vibration", collection of papers, Irkutsk, 1972.

[27] Feldbaum, A.A., Foundations of the theory of optimal control systems (in Russian), Nauka, Moscow, 1966.

[28] Feldbaum, A.A. and Butkovsky, A.G., Methods of the theory of automatic control (in Russian), Nauka, Moscow, 1971.

[29] Ferrers, M., Extension of Lagrange's equations. Quart. Journ. Vol. 12, No. 4, 1872.

[30] Furunzhiev, R.I., Design of optimal systems of vibration protection (in Russian), Vysheishaya Shkola, Minsk, 1961.

[31] Furman, F.A., Active hydraulic systems of vibration protection (in Russian). Vestnik Mashinostroeniya No. 5, 1972.

[32] Gantmakher, F.R., Lectures on analytical mechanics (in Russian). Nauka, 1966.

[33] Gass, S., Linear programming (methods and applications), McGraw-Hill, Toronto, London, 1958.

[34] Gelman, A.S., Frolov, K.V., Furman, F.A. and Voronchikhin, F.G., A tracking hydraulic absorber of vibration in mechanical systems (in Russian). In: Vibration and stability of devices, machines and elements of control systems, collection of papers. Nauka, Moscow, 1968.

[35] Gnoensky, L.S., Kamensky, G.A. and Elsgolts, L.E., Mathematical foundations of the system of automatic control (in Russian), Nauka, Moscow, 1969.

[36] Guretsky, V.V., On one problem of optimal control (in Russian). Izv. AN SSSR Mekhanika, No. 1, 1965.

[37] Guretsky, V.V., Limiting capabilities of protection of equipment from shock action (in Russian). Izv. AN SSSR Mekhanika, No. 2, 1965.

[38] Guretsky, V.V., On a choice of the optimal characteristic parameters of the shock-proofed isolators (in Russian). Inzh. Zhurnal Mekhanika Trerdogo Tela, No. 1, 1966.

[39] Guretsky, V.V., On the number of switches of the optimal control in the problem of minimisation of maximum deviation (in Russian). Inzh. Zhurnal Mekhanika Trerdogo Tela, No. 1, 1968.

[40] Guretsky, V.V., On the problem of minimising the maximum deviation (in Russian). Transactions of the Leningrad Polytechnical Institute , No. 307, Mashinostroenie, Leningrad, 1969.

[41] Guretsky, V.V., On the limiting capabilities of isolation under vibrational loadings (in Russian). Izv. AN SSSR Mekhanika Trerdogo Tela, No. 1, 1969.

[42] Guretsky, V.V., On optimisation of parameters of isolation system under stationary random excitation (in Russian). Mashinovedenie, No. 5, 1971.

[43] Guretsky, V.V. and Mazin, L.S., On one problem of optimal stabilisation (in Russian). Izv. AN SSSR Mekhanika Trerdogo Tela, No. 1, 1970.

[44] Guretsky, V.V. and Mazin, L.S., On optimal isolation of elastic bodies (in Russian). Mashinovedenie, No. 3, 1970.

[45] Guretsky, V.V. and Mazin, L.S., On the limiting capabilities of vibration protection accounting for the inertial properties of isolators (in Russian). Mashinovedenie, No. 1, 1973.

[46] Hayashi, G., Forced oscillations in nonlinear systems. Nippon, Osaka, 1953.

[47] Iorish, Yu.I., Protection of aircraft equipment from vibration (in Russian). Oboronizdat, Moscow, 1949.

[48] Iorish, Yu.I., Measurements of vibration (in Russian). Mashgiz, Moscow, 1963.

[49] Kalman, R.E., On the general theory of control systems (in Russian).Proceedings of 1st Int. Congress for Automatic Control, Vol. 2, Publishers of the USSR Academy of Science, Moscow, 1960.

[50] Katkovnik, V.Ya. and Poluektov, R.A., Multidimensional discrete systems of control (in Russian), Nauka, Moscow, 1966.

[51] Kats, A.M., On forced nonlinear vibrations (in Russian). Proceedings of the Leningrad Industrial Institute, GONTI, Leningrad, 1939.

[52] Kats, A.M., Biharmonic oscillations of dissipative nonlinear system caused or supported by harmonic excitation forces (in Russian). Prikladnaya Matematika i Mekhanika, No. 4, 1954.

[53] Kats, A.M., Forced vibration in nonlinear nearly-conservative single-degree-of-freedom-systems (in Russian). Prikladnaya Matematika i Mekhanika, No. 1, 1955.

[54] Kaufman, S., Lapinsky, W.L. and McCaa, R.C., Response of a single-degree-of-freedom isolator to a random disturbance. Journal of the Acoust. Society of America, Vol. 33, No. 8, 1961.

[55] Kazakov, I.E. and Dostupov, B.G., Statistical dynamics of nonlinear control systems (in Russian), Fizmatgiz, Moscow, 1962.

[56] Kemper, I.D. and Ayre, R.S., Optimum damping and stiffness in nonlinear single-degree-of-freedom systems. Journal of the Acoust. Society of America, Vol. 47, No. 3, 1970.

[57] Khodzhaev, K.Sh., Electromagnet-excited oscillations in linear mechanic systems (in Russian). Inzh. Zhurnal Mekhanika Trerdogo Tela, No. 5, 1968.

[58] Kobrinsky, A.E., Mechanisms with elastic constraints (in Russian), Nauka, Moscow, 1964.

[59] Kolesnikov, K.S., Liquid-propellant rocket as the control object (in Russian). Mashinostroenie, Moscow, 1969.

[60] Kolovsky, M.Z., On some variants of vibroisolators (in Russian). Scientific-Technical Bulletin of the Leningrad Polytechnical Institute, No. 8, Publishers of the Leningrad Polytechnical Institute, Leningrad, 1959.

[61] Kolovsky, M.Z., Forced vibrations in nonlinear isolators in the presence of a resistance force proportional to relative displacement (in Russian). Transactions of the Leningrad Polytechnical Institute , No. 210, Mashgiz, Leningrad, 1960.

[62] Kolovsky, M.Z., On the application of the method of small parameters to determining discontinuous periodic solutions (in Russian). Proceedings of the International Symposium on nonlinear oscillations, vol. 1, Publishers of the Academy of Science of the Ukraine, Kiev, 1963.

[63] Kolovsky, M.Z., On the influence of high frequency perturbations on the resonant vibrations of nonlinear system (in Russian). Transactions of the Leningrad Polytechnical Institute, No. 226, Mashinostroenie, Leningrad, 1963.

[64] Kolovsky, M.Z., On the choice of distribution law for linearising nonlinear differential equations (in Russian), Transactions of the Leningrad Polytechnical Institute, No. 235, Mashinostroenie, 1964.

[65] Kolovsky, M.Z., Nonlinear theory of vibration protection systems (in Russian). Nauka, Moscow, 1966.

[66] Kolovsky, M.Z., On the application of linearisation methods for investigation of forced vibrations in nonlinear oscillatory systems (in Russian). Inzh. Zhurnal Mekhanika Trerdogo Tela, No. 1, 1966.

[67] Kolovsky, M.Z., On excitation of high frequency vibrations in nonlinear systems of vibration protection (in Russian). Inzh. Zhurnal Mekhanika Trerdogo Tela, No. 3, 1966.

[68] Kolovsky, M.Z., On optimal vibration isolation (in Russian). Mashinovedenie, No. 5, 1966.

[69] Kolovsky, M.Z., On estimation of the accuracy of solutions obtained by the method of statistical linearisation (in Russian). Avtomatika i Telemekhanika, No. 10, 1966

[70] Kolovsky, M.Z., On vibrations of elastic body mounted on nonlinear isolators (in Russian). Mashinovedenie, No. 1, 1969.

[71] Kolovsky, M.Z., On the estimation of the accuracy of approximate stochastic solutions of nonlinear differential equations (in Russian). Nonlinear and Optimal Systems, collection of papers, 1971.

[72] Kolovsky, M.Z., Theory of vibration protection systems (in Russian). Mashinovedenie, No. 4, 1971.

[73] Kolovsky, M.Z., On the vibration of rigid bodies mounted on linear active isolators (in Russian). "Problems of reliability and protection of devices from vibration", collection of papers, Irkutsk, 1972.

[74] Kolovsky, M.Z., Pervozvansky, A.A., On linearisation by means of the distribution function in problems of the theory of nonlinear vibration (in Russian). Izv. AN SSSR OTN Mekhanika i Mashinostroenie, No. 5, 1962.

[75] Kolovsky, M.Z., Pervozvansky, A.A., Application of the method of distribution functions for determining polyharmonic solutions of nonlinear equations (in Russian). Izv. Vuzov "Mashinostroenie" No. 4, 1963.

[76] Koltunov, M.A. and Maiboroda, V.A., Condition for the design of isolating and vibration protecting systems (in Russian). Mekhanika Polimerov, No. 6, 1972.

[77] Korablev, S.S., Theory of electromechanical vibration absorbers (in Russian). Prikladnaya Mekhanika, Vol. 4, No. 3, 1968.

[78] Larin, V.B., On the isolation of devices mounted on moving objects (in Russian). Inzh. Zhurnal Mekhanika Trerdogo Tela, No. 2, 1966.

[79] Larin, V.B., Choice of parameters of system of vibroisolation of devices (in Russian). Prikladnaya Mekhanika, Vol. 4, No. 3, 1968.

[80] Larin, V.B., Statistical problems of vibration protection (in Russian), Naukova Dumka, Kiev, 1974.

[81] Loitsansky, L.G. and Lurie, A.I., A course of theoretical mechanics (in Russian), Vol. 2, Gostekhizdat, Moscow, 1938.

[82] Lurie, A.I., Operational calculus (in Russian), Fizmatgiz, Moscow, 1961.

[83] Lyon, R., Equivalent linearisation of the hard-spring oscillator. Journal of the Acoust. Society of America, Vol. 32, No. 9 1960.

[84] Lyon, R., On the vibration statistics of a randomly excited hard-spring oscillator. Journal of the Acoust. Society of America, Vol. 32, No. 6, 1960; Vol. 33, No. 11, 1961.

[85] Malkin, I.G., Some problems of the theory of nonlinear oscillations (in Russian), Gostekhizdat, Moscow, 1961.

[86] Mandelstam, L.I. and Papaleksi, N.D., On phenomena of resonance of order n (in Russian). In: Mandelstam, L.I. Complete collection of papers, Vol. 2, Publishers of the USSR Academy of Science, Moscow, 1947.

[87] Maximovich, Yu.P., Control of vibration in pneumatic vibroisolators (in Russian), Dinamika Mashin, collection of papers, Mashinostroenie, 1969.

[88] Maximovich, Yu.P., On the achievable quality of vibration protection from periodic excitation (in Russian). Mashinovedenie, No. 4, 1970.

[89] Maximovich, Yu.P., On optimal vibration protection (in Russian). Izv. AN SSSR Mekhanika Trerdogo Tela, No. 5, 1970.

[90] Mazin, L.S., Influence of errors in the determination of inertial characteristics of rigid body on system of its isolation (in Russian). Izv. AN SSSR Mekhanika Trerdogo Tela, No. 1, 1969.

[91] Mazin, L.S., On optimal isolation of elastic bodies (in Russian). Transactions of the Leningrad Polytechnical Institute, No. 307, Mashinostroenie, Leningrad, 1969.

[92] Morozov, B.I. and Raikhlin, R.I., Active vibration protection by means of systems of automatic control (in Russian). Vibrazionnaya Tekhnika, MDNTP No. 2, 1967.

[93] Neimark, Yu.I., Calculation of the optimal vibration protection (in Russian). Inzh. Zhurnal Mekhanika Trerdogo Tela, No. 2, 1966.

[94] Okumura, A., On a method for vibration problems of branchtype systems. Proceedings of 2nd Japan National Congress of Applied Mechanics, Tokyo, 1953.

[95] Palmov, V.A., Vibrations of elasto-plastic bodies. Springer-Verlag, Berlin Heidelberg, 1998.

[96] Panovko, Ya. G., Internal friction of vibrating elastic systems (in Russian). Fizmatgiz, 1960.

[97] Pervozvansky, A.A., Random processes in nonlinear control systems (in Russian). Fizmatgiz, 1962.

[98] Pervozvansky, A.A., On minimum of maximum deviation of controlled linear systems (in Russian). Izv. AN SSSR Mekhanika, No. 2, 1965.

[99] Pontryagin, L.S., Ordinary differential equations (in Russian), Fizmatgiz, Moscow, 1962.

[100] Pontryagin, L.S., Boltyansky, V.G., Gamkrelidze, R.V. and Mishchenko, E.F., Mathematical theory of optimal processes (in Russian), Nauka, Moscow, 1969.

[101] Popov, E.P., Dynamics of systems of automatic control (in Russian). Gostekhizdat, Moscow, 1954.

[102] Popov, E.P., Paltov, I.P., Approximate methods of investigation of nonlinear control systems (in Russian). Fizmatgiz, Moscow, 1960.

[103] Porter, B., Synthesis of optimal suspension systems. Engineer, Vol. 223, No. 5805, 1967.

[104] Porter, B. and Crossley, T.R., Modal control. Taylor and Francis Ltd., London, 1972.

[105] Pugachev, V.S., Theory of random functions and its application to problems of automatic control (in Russian). Fizmatgiz, Moscow, 1962.

[106] Pugachev, V.S., ed, Foundations of automatic control (in Russian). Fizmatgiz, Moscow, 1963.

[107] Radin, V.P., On optimisation of linear systems of vibration protection by means of its reliability (in Russian). Prikladnaya Mekhanika, Vol. 8, No. 9, 1972.

[108] Roberts, J.B. and Spanos, P.D., Random vibration and statistical linearisation. John Wiley and Sons, Chichester, 1990.

[109] Rozenvasser, E.N., Oscillations of nonlinear systems (in Russian), Nauka, Moscow, 1969.

[110] Ruzicka, J.E., 1969. Active vibration and shock isolation. Express Information Test Rigs and Stands, No. 10, 1969.

[111] Ruzicka, J.E., Passive shock isolation. Sound and Vibration, Vol. 1, No. 8, 9, 1970.

[112] Saranchuk, V.G., On the minimum deviation of one-degree-of-freedom oscillatory systems (in Russian). Transactions of the Leningrad Polytechnical Institute, No. 318, Mashinostroenie, Leningrad, 1971.

[113] Saranchuk, V.G. and Troitsky, V.A., Vibration protection facilities having minimum free running (in Russian). Transactions of the Leningrad Polytechnical Institute, No. 307, Mashinostroenie, Leningrad, 1969.

[114] Saranchuk, V.G. and Troitsky, V.A., Synthesis of optimal vibration isolators (in Russian). Transactions of the Leningrad Polytechnical Institute, No. 318, Mashinostroenie, Leningrad, 1971.

[115] Shatrovsky, L.N., A numerical method for solving problems of optimal control (in Russian). Zhurnal Vychislitelnoy Matematiki i Matematicheskoi Fiziki, Vol. 2, No. 3, 1962.

[116] Shatrovsky, L.N., Method of successive improvement of control function and parameters in problems of optimisation of control (in Russian). Prikladnye Zadachi Tekhnicheskoy Kibernetiki, collection of papers, Sovetskoe Radio, Moscow, 1966.

[117] Shmakov, V.G., Investigation of pneumatic active systems of vibration isolation for precise machines (in Russian). Thesis 1970.

[118] Shmakov, V.G., Levelling pneumatic vibration supports for precise machines (in Russian). Stanki i Instrumenty, No. 1, 1970.

[119] Schubert, D.W. and Ruzicka, J.E., Theoretical and experimental investigation of electric-hydraulic systems of vibration protection (in Russian). Translation of Transaction of ASME, Mir, No. 4, 1969.

[120] Sevin, E. and Pikley, W.D., Optimum shock and vibration isolation. New York, 1971.

[121] Sinev, A.B., Optimal spectral properties of random input actions for passive and active systems of vibration protection (in Russian). Mashinovedenie, No. 1, 1973.

[122] Sinev, A.B. and Furunzhiev, R.I., Optimisation of systems of vibration protection (in Russian). "Problems of reliability and protection of devices from vibration", collection of papers, Irkutsk, 1972.

[123] Sinev, A.B., Potemkin, B.A. and Safronov, Yu.G., On the capabilities of realising optimal transfer functions by means of electric-hydraulic systems of vibration protection (in Russian). Influence of vibrations on the human organism and problems of vibration protection, Nauka, Moscow, 1972.

[124] Solodovnikov, V.V., ed., Foundations of automatic control, Vol. 1 - Theory, Vol. 2, Part 1 - Sensitive amplifying and actuating elements (in Russian). Mashgiz, Moscow, 1959.

[125] Solodovnikov, V.V., Statistical dynamics of linear systems of automatic control (in Russian), Fizmatgiz, Moscow, 1960.

[126] Spasokukotsky, O.K. and Sud-Zlochevsky, A.I., Elements of electroautomatics (in Russian). Tehknika, Kiev, 1965.

[127] Stocker, J.J., Nonlinear vibrations in mechanical and electrical systems. John Wiley and Sons, New York, 1992.

[128] Sveshnikov, A.A., Applied methods of the theory of random functions (in Russian). Nauka, Moscow, 1968.

[129] Troitsky, V.A., Matrix methods of computation of vibrations in systems consisting of rods (in Russian). Transactions of the Leningrad Polytechnical Institute, No. 210, Mashgiz, Leningrad, 1960.

[130] Troitsky, V.A., On synthesis of optimal vibration isolators (in Russian). Prikladnaya Matematika i Mekhanika, No. 4, 1967.

[131] Troitsky, V.A., Some optimal problems of the theory of vibrations (in Russian). Transactions of the Leningrad Polytechnical Institute, No. 318, Mashinostroenie, Leningrad, 1971.

[132] Veiz, V.L., Dynamics of machinery aggregates (in Russian), Mashinostroenie, Moscow, 1969.

[133] Viktorov, E.D., and Kolovsky, M.Z., Approximate synthesis of optimal systems of vibration isolation. Second National Conference on Problems of reliability in civil engineering (in Russian). Vilnius, 1968.

[134] Viktorov, E.D., and Larin, R.M., Methods of successive improvements of the control functions in the problem of the optimal vibration isolation (in Russian). Transactions of the Leningrad Polytechnical Institute, No. 307, Mashinostroenie, Leningrad, 1969.

[135] Viktorov, E.D., and Larin, R.M., Some feature of the problem of the minimum of the maximum deviation (in Russian). Inzh. Zhurnal Mekhanika Trerdogo Tela, No. 4, 1967.

[136] Volokovsky, V.Yu. and Radin V.P., On the choice of optimal parameters on nonlinear systems of vibration protection subject to random excitation (in Russian). Izv. AN SSSR Mekhanika Trerdogo Tela, No. 2, 1972.

[137] Voronov, A.A., Foundations of the theory of automatic control (in Russian), Energiya, Moscow, Vol. 1, 1965; Vol. 2, 1966.

[138] Vulfson, I.I. and Kolovsky, M.Z., Nonlinear problems of dynamics of machines (in Russian), Mashinostroenie, Leningrad, 1968.

[139] Wicker, J. and Ziemba, S., On a certain approximate method for determining optimum dynamic characteristics of a linear system. Zagadn. gragan nieliniowych, Vol. 10, 1969.

[140] Wiener, N., The Fourier integral and its applications (in Russian). Fizmatgiz, 1963.

[141] Yablonsky, V.V., Genkin, M.D. and Elezov, V.G., Methods of active absorption of vibrations in mechanisms (in Russian), Nauka, Moscow, 1971.

[142] Zalmanzon, L.A., Theory of the elements of pneumonics (in Russian), Nauka, Moscow, 1969.

[143] Zasyadko, A.A., Kiselev, V.M. and Eliseev, S.V., An active electric-hydraulic system of vibration protection (in Russian). "Problems of reliability and protection of devices from vibration", collection of papers, Irkutsk, 1972.

Index

Foundations of Engineering Mechanics

Series Editors: Vladimir I. Babitsky, Loughborough University
Jens Wittenburg, Karlsruhe University

Babitsky	Theory of Vibro-Impact Systems and Applications (1998, ISBN 3-540-63723-0)
Guz	Fundamentals of the Three-Dimensional Theory of Stability of Deformable Bodies (1999, ISBN 3-540-63721-4)
Kolovsky	Nonlinear Dynamics of Active and Passive Systems of Vibration Protection (1999, ISBN 3-540-65661-8)
Kovaleva	Optimal Control of Mechanical Oscillations (1999, ISBN 3-540-65442-9)
Palmov	Vibrations of Elasto-Plastic Bodies (1998, ISBN 3-540-63724-9)
Skrzypek / Ganczarski	Modelling of Material Damage and Failure of Structures (1999, ISBN 3-540-63725-7)

Springer and the environment

At Springer we firmly believe that an international science publisher has a special obligation to the environment, and our corporate policies consistently reflect this conviction.

We also expect our business partners – paper mills, printers, packaging manufacturers, etc. – to commit themselves to using materials and production processes that do not harm the environment. The paper in this book is made from low- or no-chlorine pulp and is acid free, in conformance with international standards for paper permanency.

Springer

MIX
Papier aus verantwortungsvollen Quellen
Paper from responsible sources
FSC® C105338

If you have any concerns about our products,
you can contact us on
ProductSafety@springernature.com

In case Publisher is established outside the EU,
the EU authorized representative is:
Springer Nature Customer Service Center GmbH
Europaplatz 3, 69115 Heidelberg, Germany

Printed by Libri Plureos GmbH
in Hamburg, Germany